Studies in Computational Intelligence

Volume 1202

The series "Studies in Computational Intelligence" (SCI) publishes new developments and advances in the various areas of computational intelligence—quickly and with a high quality. The intent is to cover the theory, applications, and design methods of computational intelligence, as embedded in the fields of engineering, computer science, physics and life sciences, as well as the methodologies behind them. The series contains monographs, lecture notes and edited volumes in computational intelligence spanning the areas of neural networks, connectionist systems, genetic algorithms, evolutionary computation, artificial intelligence, cellular automata, self-organizing systems, soft computing, fuzzy systems, and hybrid intelligent systems. Of particular value to both the contributors and the readership are the short publication timeframe and the world-wide distribution, which enable both wide and rapid dissemination of research output.

Indexed by SCOPUS, DBLP, WTI AG (Switzerland), zbMATH, SCImago.

All books published in the series are submitted for consideration in Web of Science.

M. F. Mridha · Nilanjan Dey
Editors

Machine Vision in Plant Leaf Disease Detection for Sustainable Agriculture

Springer

Editors
M. F. Mridha
Department of Computer Science
American International University-Bangladesh
Dhaka, Bangladesh

Nilanjan Dey
Department of Computer Science
and Engineering
Techno International New Town
Kolkata, West Bengal, India

ISSN 1860-949X ISSN 1860-9503 (electronic)
Studies in Computational Intelligence
ISBN 978-981-96-4519-0 ISBN 978-981-96-4520-6 (eBook)
https://doi.org/10.1007/978-981-96-4520-6

This Springer imprint is published by the registered company Springer Nature Singapore Pte Ltd.
The registered company address is: 152 Beach Road, #21-01/04 Gateway East, Singapore 189721, Singapore

Preface

"Machine Vision in Plant Leaf Disease Detection for Sustainable Agriculture" offers a comprehensive exploration of the intersection between advanced technology and agricultural sustainability. With a focus on leveraging machine vision techniques for the early detection and management of plant diseases, this book serves as a vital resource for researchers, practitioners, and stakeholders in the agricultural sector.

The book begins by providing an overview of the challenges posed by plant diseases to global food security and agricultural sustainability. It highlights the limitations of traditional disease detection methods and underscores the need for innovative approaches that can offer timely and accurate diagnosis.

Through a systematic examination of machine vision principles and methodologies, the book delves into the various stages of disease detection, from image acquisition to feature extraction and classification. Key concepts such as image preprocessing, feature selection, and machine learning algorithms are discussed in detail, with emphasis on their practical implementation in real-world scenarios.

Drawing upon a wealth of case studies and examples, the book showcases the diverse applications of machine vision in plant disease detection across different crops and regions. From detecting common foliar diseases like powdery mildew and leaf rust to identifying emerging threats such as bacterial blight and viral infections, the capabilities of machine vision are demonstrated in a variety of contexts.

Moreover, the book explores the potential of machine vision to contribute to sustainable agriculture practices. By enabling early detection and targeted intervention, machine vision technologies can help minimize the use of agrochemicals, reduce crop losses, and optimize resource allocation, thereby promoting environmental stewardship and economic viability.

In conclusion, "Machine Vision in Plant Leaf Disease Detection for Sustainable Agriculture" offers a comprehensive overview of the state-of-the-art in this rapidly evolving field. It provides valuable insights and practical guidance for researchers, practitioners, and policymakers seeking to harness the power of machine vision for the advancement of agricultural sustainability and food security on a global scale.

This book reflects a rich diversity of contributions from experts across the globe, highlighting the collaborative efforts of researchers from countries including Bangladesh, the USA, Canada, Japan, Australia, and Saudi Arabia. The chapters explore a wide range of topics, as outlined below.

Chapter "Impact of Deep Learning and Computer Vision on Plant Leaf Disease Detection" by Islam et al. provides an in-depth analysis of the transformative role of deep learning and computer vision in identifying plant leaf diseases, highlighting their potential to revolutionize agricultural practices.

Chapter "Plant Leaf Disease Detection Using Deep Stacking: Integrating CNNs and Gradient Boosting for Enhanced Classification Accuracy" by Hossain et al. showcases a novel approach combining convolutional neural networks (CNNs) with gradient boosting techniques to achieve superior accuracy in plant disease classification.

Chapter "Revolutionizing Agriculture: Deep Learning Models for Crop Pest and Disease Analysis" by Eva et al. explores the application of deep learning models in managing crop health by identifying pests and diseases with remarkable precision.

Chapter "BananaLeaf-Net: A Proposed Deep Learning Model for Accurate Classification of Banana Leaf Diseases" by Sum et al. introduces a dedicated deep learning framework tailored for the classification of banana leaf diseases.

Chapter "Improving Agronomic Disease Detection and Classification: The Superiority of Hybrid Inception-Xception Ensemble Model for Rice Leaf Disease Classification" by Hafiz et al. presents a hybrid ensemble model that leverages the strengths of Inception and Xception architectures to enhance rice leaf disease classification.

Chapter "A Hybrid CNN Architecture for Efficient Detection of Maize Plant Diseases" by Asaduzzaman et al. proposes an innovative CNN architecture for the rapid and accurate identification of maize plant diseases.

Chapter "Detection of Corn Leaf Diseases Using Deep CNN with SHAP-Based Explainability for Enhanced Agriculture" by Afroj et al. combines deep CNN models with SHAP explainability techniques to ensure transparent and interpretable corn leaf disease detection.

Chapter "Harnessing Convolutional Neural Networks for Potato Leaf Disease Detection: A Proposed Model" by Ahmed et al. focuses on developing a CNN-based solution for the efficient detection of potato leaf diseases.

Chapter "Automated Plant Diseases Analysis Using Lightweight Deep Learning Models" by Opee et al. introduces lightweight deep learning models designed for scalable and resource-efficient plant disease analysis.

Chapter "Improving Bean Disease Detection with Explainable Vision Transformers and Machine Learning Classifiers" by Islam et al. highlights the integration of vision transformers with machine learning classifiers for explainable and accurate bean disease detection.

Chapter "Deploying CNN-ResNet50-BiLSTM for Paddy Leaf Disease Detection" by Bappi et al. combines CNN-ResNet50 with BiLSTM to enhance the detection and classification of paddy leaf diseases.

Chapter "Advancing Agriculture Through Irrigation Automation: Smart Management for Sustainable Farming" by Khaled et al. explores the integration of automation and smart technologies in irrigation systems to promote sustainable agricultural practices.

Chapter "Deep Learning Algorithms for Detecting Banana Leaf Spot Diseases" by Nabil et al. focuses on the application of deep learning algorithms for the precise detection and classification of banana leaf spot diseases.

The editors extend their gratitude to the exceptional authors and referees for their valuable contributions to this book. Their dedication and expertise have been instrumental in creating this comprehensive resource. Special thanks are also due to the Springer team for their unwavering support.

Dhaka, Bangladesh — M. F. Mridha
Kolkata, India — Nilanjan Dey

Contents

About the Editors

M. F. Mridha (Senior Member IEEE, Professional Member ACM) is currently working as Associate Professor in the Department of Computer Science, American International University-Bangladesh (AIUB). He also worked as Associate Professor and Chairman in the Department of Computer Science and Engineering, Bangladesh University of Business and Technology (BUBT) from 2019 to 2022 and as CSE Department Faculty Member at the University of Asia Pacific and as Graduate Head from 2012 to 2019. He is Founder and Director of Advanced Machine Intelligence Research Lab (AMIR Lab). He received his Ph.D. in the domain of AI from Jahangirnagar University in the year 2017. For more than 18 years, he has been with the master's and undergraduate students as Supervisor of their thesis work. He has authored/edited several books with Springer and published more than 280 journal and conference papers. He has served as Program Committee Member in several international conferences/workshops.

Nilanjan Dey received the B.Tech., M.Tech. in information technology from West Bengal Board of Technical University and Ph.D. degrees in electronics and telecommunication engineering from Jadavpur University, Kolkata, India, in 2005, 2011, and 2015, respectively. Currently, he is Associate Professor with the Techno International New Town, Kolkata, and Visiting Fellow of the University of Reading, UK. He is Editor-in-Chief of International Journal of Ambient Computing and Intelligence, Associate Editor of IEEE Transactions on Technology and Society, and Series Co-editor of Springer Tracts in Nature-Inspired Computing and Data-Intensive Research from Springer Nature and Advances in Ubiquitous Sensing Applications for Healthcare from Elsevier, etc. He is Fellow of IETE and Senior Member of IEEE.

Impact of Deep Learning and Computer Vision on Plant Leaf Disease Detection

Md. Raisul Islam, Md. Zakir Hossain Zamil, Md. Anisur Rahman, Md. Nahid Hasan, and Md. Mohsin Kabir

Abstract

Plant leaf disease detection is vital in ensuring agricultural manufacturing and food security. Conventional methods for identifying and diagnosing plant diseases are often time-consuming, and there is a possibility of human error. In recent years, deep learning (DL) and computer vision (CV) techniques have revolutionized the field of agriculture, offering automated, precise, and scalable solutions for disease detection. The comprehensive review explores these technologies' valuable impact on plant leaf disease detection. It examines the significant advancements made in the application of convolutional neural networks (CNNs), transfer learning, and other DL architectures, which have markedly improved the results and speed of disease identification. The review also addresses the challenges encountered in real-world implementations. Furthermore, the paper discusses the necessary trends and future directions, including the integration of Internet of Things (IoT) devices and edge computing, which promise to enhance the effectiveness of disease detection systems further providing a thorough analysis of the current state and future potential of DL and CV in agriculture.

Keywords

Crop pest disease categorization • Deep learning • Agriculture • Convolutional neural networks (CNNs) • Image classification

1 Introduction

Plant leaf disease detection is vital in current agriculture because it affects food security worldwide. Their prevalence can lead to a significant drop in crop yield [1]. The traditional way primarily relies on manual inspection by experts [2]. While this approach may work in some cases, however, this is fundamentally limited in terms of flexibility and accuracy. Relying on these traditional methods is not only time-consuming but also vulnerable to subjective errors, as the accuracy of diagnosis can greatly depend on the inspector's level of experience and knowledge [3]. On the other hand, more dynamic and reliable diagnostic solutions for disease detection are becoming increasingly important in the agricultural industry as farming scales up and agricultural systems become more complex [4].

Over time, deep learning (DL) and computer vision (CV) have shown to be transformative solutions to these problems. Recently, DL models, particularly using Convolutional Neural Networks (CNNs), have proven tools for complicated image classification problems, such as plant disease identification from leaf photos [5]. Since CNNs are designed to learn the hierarchical structure automatically from raw pixel data, they can capture subtle patterns that may be indicative of disease [6]. Besides providing better accuracy, such a transition from conventional methods to automated systems enables the monitoring of crops in real time and helps with interventions during that period. Several models, such as CNN [7], hybrid architectures, and Recurrent Neural Networks (RNN)

Md. R. Islam · Md. A. Rahman
School of Engineering and Technology, Western Illinois University, Macomb, Illinois, USA
e-mail: raisulislamjony@gmail.com

Md. A. Rahman
e-mail: eng.anisurrahman92@gmail.com

Md. Z. H. Zamil
School of Computer Sciences, Western Illinois University, Macomb, Illinois, USA
e-mail: mdzakirhossainzamil@gmail.com

Md. N. Hasan
University of Wisconsin-Milwaukee, University of Wisconsin System, Madison, Wisconsin, USA
e-mail: hasan26@uwm.edu

Md. M. Kabir (✉)
School of Innovation, Design and Engineering, Mälardalens University, Västerås, Sweden
e-mail: md.mohsin.kabir@mdu.se

M. F. Mridha and N. Dey (eds.), *Machine Vision in Plant Leaf Disease Detection for Sustainable Agriculture*, Studies in Computational Intelligence 1202, https://doi.org/10.1007/978-981-96-4520-6_1

[8], are used for deep-based plant disease detection [9]. Some transferred learning techniques in this domain use pre-trained models such as ResNet and InceptionNet, which achieve high accuracy even with a limited training set [10].

The studies have been conducted on various datasets to benchmark these models and they are traditional image processing techniques. Moreover, the use of computer vision methods such as segmentation and data augmentation integrated increased model performance through outperforming training image quality and diversity. It is hard to overstate the importance of available datasets, which all DL models are ultimately based on. Having large data of high quality and different natures is highly critical to achieving higher accuracy and generalization of the built models. The generation of massive-scale annotated datasets presents some unique issues, such as the acquisition of annotated examples and the inter-image variation of environmental conditions. However, the projection of more advanced DL models, along with the enhancement of annotated datasets, are on the right track, leading to plant disease detection techniques that are more accurate and efficient. For the survey, we reviewed research articles published between 2022 and 2024. The main goal of the survey is to determine how much DL and computer vision increase flexibility and reduce issues in the field of detecting plant disease. The main goals of this survey are

- Evaluate the current advancements and applications of DL and computer vision in detecting plant leaf diseases.
- Review and categorize the various DL architectures and computer vision algorithms and their results.
- Different types of frequently used dataset and their performance have been measured.
- Propose potential improvements, emerging technologies, and future research directions to advance the field of disease detection on plant leaves using DL and CV.

The structure of the paper is as follows: Section 2 describes the application of plant leaf diseases. Section 3 shows the popular algorithms used in this field. In Sect. 4, the benchmark dataset is described. The analysis of various research works is present in Sect. 5. The next challenges and future researcher direction are in Sect. 6, and the last one is Sect. 7 which is the conclusion of the paper.

2 Plant Leaf Disease: Application of Deep Learning and Computer Vision Base

2.1 Plant Abnormalities Identify and Infestations

Artificial Intelligence (AI) has recently been utilized in plant pathology to identify and manage plant diseases and infestations. These innovative technologies are poised to revolutionize how plant health issues are detected, diagnosed, and treated [11]. This debate looks into the many AI technologies used to detect plant abnormalities and infestations, exploring their benefits, limitations, and overall impact on the discipline of plant pathology. DL techniques are among the most commonly used AI methods in this industry. CNN, Yolo, U-Net, and other techniques [12] have been used to classify plant diseases using digital imagery. These algorithms can be trained to recognize distinct disease patterns and symptoms, making them useful for disease classification at the early stages. However, ML models require substantial, annotated datasets for training and may struggle with previously unknown diseases.

Haridasan et al. proposed an automated method that can effectively detect and classify rice plant illnesses using a CV-based technique that incorporates image processing. They employed picture preprocessing and segmentation to detect diseased regions, as well as a combination of SMV classifiers and CNNs to recognize and classify diseases. The suggested DL strategy, utilizing ReLU and activation functions, achieved a validation accuracy of 91.45%. They proposed a predictive cure to help farmers and agricultural groups efficiently combat these diseases [13]. Xu et al. proposed RFE-CNN is a DNN algorithm. This approach combines the elliptic metric learning, residual attention block, CNN, and feedback block. CNNs are used to feature extraction from good and diseased leaves, followed by RCAB optimization, FB refinement, and EML and CNN classification. The model outperformed and obtained an overall accuracy of 98.83% [14]. Further, a novel hybrid model was introduced by Bedi et al. that combines a Convolutional Autoencoder (CAE) network with a CNN for automatic plant disease detection. They tested detecting bacterial spot disease in peach plants using leaf images from the PlantVillage dataset. Their used system performed with 99.35% training accuracy and 98.38% testing accuracy with just 9,914 training parameters, making it more efficient in terms of training time and disease identification compared to existing methods [15].

Sunil et al. investigated two major diseases affecting cardamom—Colletotrichum Blight and phyllosticta leaf spot. For their work, they used U2-Net to eliminate unwanted backgrounds in input images by selecting multi-scale features. The EfficientNetV2 model was utilized to detect cardamom plant disease and they achieved a detection accuracy of 98.26% [16]. Tomatoes, a globally popular crop and essential kitchen ingredient, face quality and quantity issues due to various diseases. India ranks second in tomato production. A DL-based approach for disease detection and classification was proposed by Agarwal et al. using a CNN with three convolutional layers, three max-pooling layers, and two fully connected layers. The model outperforms pre-trained models. They achieved classification accuracy ranging from 76 to 100% across different classes, with an average accuracy of 91.2% [17]. Moupojou et al. introduced FieldPlant, a dataset of 5,170 images collected directly from plantations, annotated under the supervision of plant pathologists. They used DL for their work and achieved an accuracy of 82.54% for the inception V3 model [18].

2.2 Leaf and Disease Type Identification

Early crop illnesses can assist farmers in managing and mitigating difficulties in a timely manner by Gajjar et al. They used a system for identifying crop diseases from leaf images using DL. They used a single-shot detector (SSD) for leaf recognition and localization and a deep CNN (DCNN) for illness classification. For field use in real time, they installed on Nvidia Jetson TX1 hardware. Their suggested system beats current CNN architectures and obtained an approximate 96.88% disease classification accuracy [19].

Plant leaf diseases hinder the growth of plant species, and early, accurate diagnosis can minimize further damage. Identifying these diseases from leaf images requires time, expertise, and a sophisticated approach. DL offers significant potential for enhancing accuracy in this area. Abd Algani et al. focused a DL method called ant colony optimization with CNN (ACO-CNN). Their approach to ant colony optimization (ACO) can improve disease diagnosis by analyzing color, leaf arrangement from images, and texture. The proposed approach includes several steps: image acquisition, segmentation, noise removal, and classification [20]. Singh et al. focused on using DL, specifically CNN, to automatically detect leaf diseases. A CNN model was developed and trained using a database of potato leaf images, with 1,700 images for training and 600 for testing. Their model was also applied to identify citrus diseases using CNN, DL, base learning, and transfer learning techniques. Experimental results showed that the proposed architecture achieving a ResNet model accuracy of 99.62% [21].

To address this, Hassan et al. proposed using depthwise separable convolutions to reduce both parameter count and computation requirements. They trained their models on an open dataset of 14 plant species with 38 disease classes and healthy leaves. They achieved impressive accuracies of 98.42% with InceptionV3, 99.11% with InceptionResNetV2, 97.02% with MobileNetV2, and 99.56% with EfficientNetB0, surpassing traditional handcrafted-feature methods. Their results highlighted the potential of deep CNN models for effective disease identification and their applicability in real-time agricultural systems [22]. Arun et al. proposed Complete Concatenated DL (CCDL) for detecting diseases across multiple crops. Their model introduced a Complete Concatenated Block (CCB) as its core unit, where a point convolution layer is added before each convolution layer to reduce the number of parameters. The complete concatenation path enhances feature map utilization, improving classification accuracy. The model is trained on a reorganized Plant Village dataset, and a pruned version, PCCDL, is developed for size reduction. Among its variants, the PCCDL with Partial Standard Convolution Technique (PCCDL-PSCT) achieved the highest classification accuracy of 98.14% [23].

3 Algorithms

Algorithms are a vital part of training a dataset for better results and performance. In recent years, many DL algorithms that developed for advanced work, and researchers are using them to reduce costs and automatically detect. Various algorithms have been used for the detection of the issue of plant leaves and other crops, and a few of them are described here.

3.1 Convolutional Neural Networks (CNNs)

CNNs [7] are vital in DL for detecting plant diseases. Because of the automatic feature extraction provided by the convolutional layers, this architecture is often utilized for image processing. A CNN's base architecture is made up of several critical layers that are placed in a certain order to learn from data. Figure 1 presents the CNN architecture. Input layer that accepts raw data, like pixel values from an image. This input is subsequently fed through several convolutional layers that use various filters to extract features and detect patterns. An activation function, usually a ReLU (Rectified Linear Unit), is used to provide nonlinearity and allows to learn the network's more complicated patterns. Pooling layers, such as max-pooling, are used to minimize dimensionality in data while retaining the most important attributes. The output of these layers is flattened and fed into fully connected layers, which use the collected features to perform classification or regression tasks. Finally, the output layer predicts the result.

3.2 YOLO (You Only Look Once)

YOLO (You Only Look Once) [24] is an object identification system that works on real time. It divides an input image into a grid and predicts bounding boxes and class probabilities for each grid cell. Figure 2 shows the architecture of YOLO. The basic architecture is a single CNN that predicts multiple bounding boxes and their associated class scores from the full image in a single forward pass. This unified technique allows for fast and accurate identification by considering object detection as a regression problem rather than a classification problem using region recommendations. YOLO's architecture is made up of convolutional layers and fully connected layers, with a final output layer that predicts class

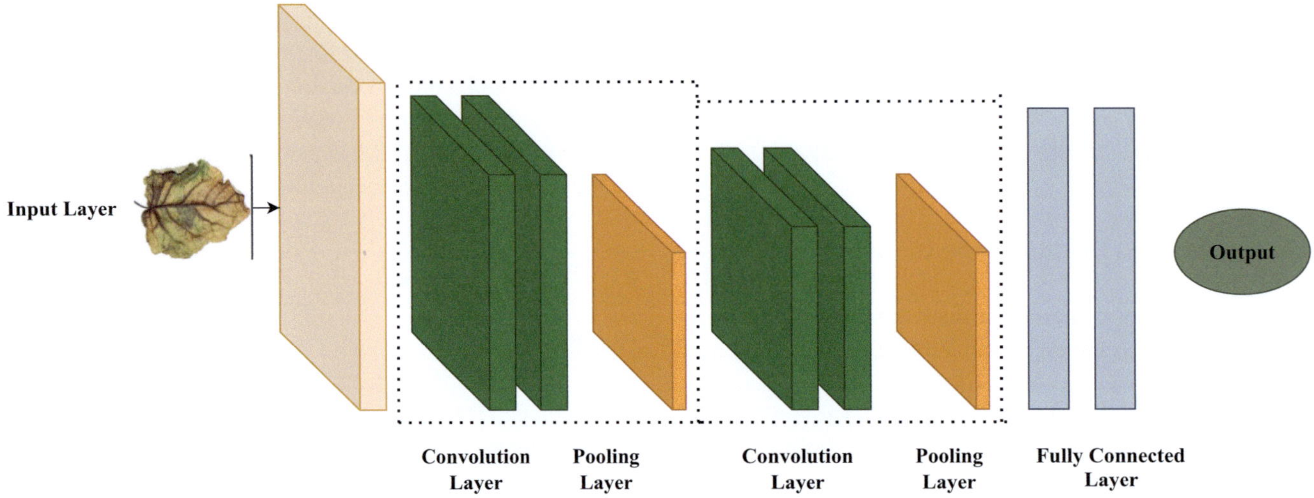

Fig. 1 Architecture of the convolutional neural network (CNN)

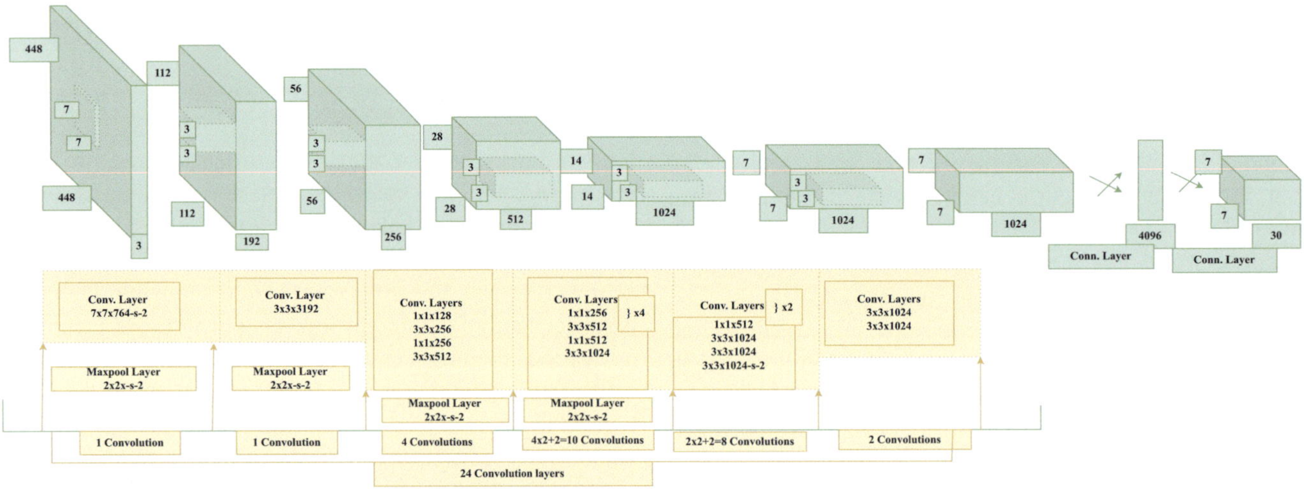

Fig. 2 Architecture of the you only look once (YOLO)

probabilities. The main novelty is its end-to-end training and detection capability, which dramatically improves speed and efficiency over existing techniques.

3.3 Single Shot Multibox Detector (SSD)

SSD architecture [25] is intended for object detection in pictures. It consists of a basic network, usually a modified VGG16 or ResNet, that extracts feature mappings from the input image. SSD predicts item classes and bounding box locations by applying a succession of convolutional layers on top of these feature maps. The model uses multi-scale feature maps by adding more convolutional layers to the underlying network, allowing it to detect objects of different sizes. The final predictions are refined with various aspect ratios and scales on bounding boxes.

3.4 Faster R-CNN

Faster R-CNN [26] is a two-stage object identification that includes a region proposal network (RPN) and a rapid R-CNN detector. The RPN creates candidate object proposals by sliding a tiny network over feature maps generated by a backbone convolutional network. These ideas are then passed into the rapid R-CNN detector, which refines and classifies the objects by extracting fixed-size feature maps through a RoI pooling layer. The network may be trained end to end, optimizing both the RPN and the rapid R-CNN detector simultaneously.

3.5 Mask R-CNN

Mask R-CNN [27] expands Faster R-CNN by adding a parallel branch for predicting segmentation masks, bounding box,

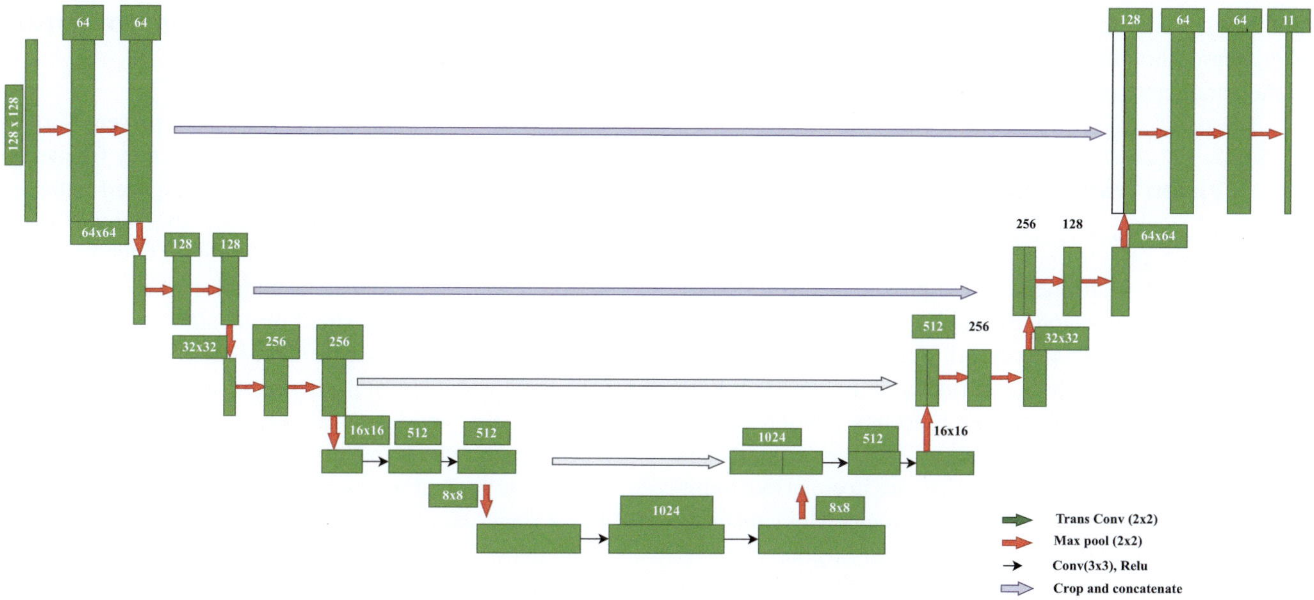

Fig. 3 Architecture of the U-Net

and class labels. The basic design includes a backbone network, usually a ResNet or a ResNeXt, that extracts feature maps from input images. These feature maps are then passed through a Region Proposal Network (RPN) to produce candidate object regions. A RoIAlign layer captures precise characteristics from each region, which are then input into a fully connected layer for classification and bounding box regression. Simultaneously, a mask branch predicts binary masks for each object instance in the suggested regions, enabling pixel-wise segmentation.

3.6 DANet

The Dynamic Attention Network (DANet) represents a significant advancement in plant disease detection, enhancing feature representation through dynamic attention mechanisms [28]. Its architecture includes decoupled detection heads that improve both localization and classification accuracy, making it effective for identifying plant diseases from leaf images. Studies have shown that DANet outperforms traditional models like RetinaNet, achieving superior average precision (AP) metrics in challenging scenarios. In plant pathology, DANet has facilitated more accurate and timely disease diagnosis, supporting better crop management by adaptively weighing the importance of different features in response to environmental factors.

3.7 MFF-Net

The Multi-Scale Feature Fusion Network (MFF-Net) is an advanced model for plant leaf disease detection, utilizing multi-scale feature extraction to enhance detection and classification across different scales and resolutions [29]. By integrating features from various network layers, MFF-Net captures both fine and coarse details of leaf images. Recent research shows that MFF-Net effectively detects diseases in crops like maize and tomato, achieving high accuracy in classifying multiple disease types. Its ability to fuse features from different scales makes it robust against variations in leaf appearance caused by factors such as lighting and orientation.

3.8 U-Net

The U-Net architecture has become highly effective for plant disease detection due to its robust image segmentation capabilities [30]. Figure 3 presents the architecture of U-Net. U-Net's encoder–decoder structure excels at capturing both contextual information and precise localization, making it well suited for identifying disease symptoms on plant leaves. Recent studies have demonstrated its effectiveness in segmenting diseased leaf images, distinguishing healthy from unhealthy regions with high accuracy. U-Net has been successfully applied to classify diseases in crops like tomatoes and maize, enhancing diagnostic accuracy and enabling automated monitoring systems. Its integration into plant disease detection workflows improves early detection and intervention, supporting more sustainable agricultural practices. The equation is

$$X = U(D(Y; \theta_D); \theta_U) \quad (1)$$

Y is input size, $D(Y;\ \Theta_D$ represents the downsampling (encoder) path, where θ_D are the parameters (weights and biases) of the encoder. X is the output segmentation map and $U(;\theta_U$ represents the upsampling (decoder) path, where θ_U are the parameters of the decoder.

3.9 Vision Transformers (ViTs)

By processing picture patches utilizing self-attention processes, Vision Transformers (ViTs) provide a revolutionary method to computer vision [31]. This enables ViTs to outperform conventional CNNs in capturing contextual information and long-range relationships. Recent research shows that ViTs enhance plant disease classification accuracy, often outperforming conventional CNN models. ViTs have demonstrated superior performance in detecting diseases in crops like tomatoes and durians thanks to their ability to focus. This capability makes ViTs particularly effective for identifying and classifying plant diseases across varied leaf regions. Their integration into plant disease detection systems promises more robust and scalable solutions, aiding farmers with advanced technologies for better crop management and disease control.

To conclude, the DL algorithmic landscape for plant disease detection is variegated, with CNNs and fast detectors like YOLO variants, on one hand, and highly accurate models, such as SSD or Faster R-CNN/Mask R-CNN, on the other. The choice of algorithm should be made with the particular needs of the agricultural application in mind, such as speed, accuracy, and complexity of image data processing (Fig. 4).

4 Dataset

For this chapter, we will explain a few popular datasets that are often used for plant leaf disease identification. There are various types of datasets available for plant diseases or the field of agriculture, and different sizes of data are available and suitable for DL and ML models. Different sources are used for collecting datasets such as Kaggle, Mendeley, or databases. Table 1 provides the few datasets and the features of the dataset and Fig. 5 presents the few data from the PlantVillage dataset.

5 Result Analysis of Plant Disease Detection

In this section, we are providing results on various works on leaf disease detection of plant from 2022 to 2024. Table 2 presents the existing work analysis for the plant leaf disease detection field.

6 Challenges and Opportunities for Future work

- **Limited Datasets**: Insufficient datasets present serious problems for plant disease detection since they frequently lead to models that perform poorly when exposed to unseen data because they are unable to generalize across many conditions [72]. The risk of overfitting, in which a model performs well on training data but fails in real-world applications, is increased by small or unbalanced datasets that do not fully cover the spectrum of illness symptoms, environmental variables, or plant species changes [73]. Developing reliable models is also hampered by the lack of labeled data, which makes it challenging to differentiate between related illnesses and adjust to novel ones.
  These issues can be resolved by using data augmentation techniques, which use rotation, scaling, and flipping to create differences in the current photos, so artificially expanding the dataset. Furthermore, the model's capacity to generalize can be improved by applying transfer learning from previously trained models to bigger, related datasets [74]. The limitations of small datasets can also be addressed by working with agricultural organizations to collect more extensive datasets and by using synthetic data produced by methods like Generative Adversarial Networks (GANs), which will result in plant disease detection systems that are more accurate and dependable.
- **High Intra-Class Variability**: High intra-class variability is the wide range of appearances that a single disease can exhibit, environmental, and the stage of infection. This variability makes it difficult for models to recognize and classify the disease consistently, as the symptoms might appear differently across various instances [75]. For example, a disease might cause spots on leaves that vary in color, size, or shape, making it challenging for the model to identify the disease accurately [76].
  Transfer learning and sophisticated data preprocessing enable models better generalize across various disease presentations in order to overcome this difficulty. Through the use of modifications like rearranging, scaling, or color alterations, data augmentation can enhance the diversity of data. However, transfer learning enables the model to use previously learned information from related tasks, improving its capacity to identify different symptoms [77]. Additionally, model accuracy and resistance against significant intra-class variability can be enhanced by putting into practice strong feature extraction techniques that concentrate on key disease traits rather than surface-level fluctuations.
- **Occlusion and Overlapping Leaves**: Occlusion and overlapping leaves present significant challenges in plant

Fig. 4 Few leaf images from the PlantVillage dataset

disease detection as they can obscure critical visual cues needed to identify diseases accurately. The detection model may miss or misinterpret key features, leading to false negatives or inaccurate classification [78]. This issue is particularly problematic in dense plant canopies, where multiple leaves and branches can block each other, making it difficult to isolate and analyze individual leaves [79].

To overcome this challenge, advanced techniques such as multi-view imaging, where plants are photographed from different angles, can help create a more complete representation of the plant. Additionally, incorporating depth information through stereo cameras or LiDAR can assist in distinguishing between overlapping leaves. DL models can also be trained to recognize partially occluded leaves by using augmented datasets that simulate these conditions, improving the robust models. Another approach is to employ segmentation algorithms that can separate overlapping leaves and isolate the relevant regions for disease analysis.

Table 1 Description about the popular dataset for plant leaf disease detection

References	Dataset	Description
[32]	PlantVillage	The PlantVillage collection includes approximately 50,000 images of both healthy and damaged plant leaves from diverse plants. It contains tagged photos of several plant illnesses, a valuable resource for creating ML models for plant disease identification
[33]	IP102	The dataset includes over 75,000 images across 102 categories. It contains 19,000 annotated images for detection. The dataset features taxonomy, grouping insect pests by their impact on specific agricultural products
[34]	LeafSnap	The LeafSnap dataset comprises over 185,000 images of leaves from more than 2,600 plant species. It features high-quality photos with annotations for plant species and leaf attributes, aiding research in plant identification and ML applications
[35]	UCI Leaf	The UCI Leaf dataset includes 3400 images of leaves from 40 different plant species. Every image is represented by a set of features taken from the form and texture of the leaves, providing data for classification and pattern recognition tasks in plant species identification
[36]	PlantNet	The PlantNet dataset contains over 1.3 million images of plants across 10,000 species. It is designed for plant identification and includes labeled images with detailed annotations
[37]	Tomato leaf disease	The dataset contains photos of tomato plant leaves classified into various disease groups. The dataset contains 14,531 pictures, each depicting a different illness sign, to help train models for the effective identification and classification of leaf diseases
[38]	Wheat disease	The wheat disease dataset contains images of wheat leaves affected by various diseases, including rust and mildew. It features multiple classes representing different disease types and healthy leaves and contains 208 stripe rust-affected leaves, 102 healthy leaves, and 97 septoria-affected leaf disease-infected images
[39]	Soybean leaf disease	The Soybean leaf disease dataset comprises six types of soybean disease leaves, including rust, downy mildew, and bacterial blight. It contains 1620 annotated disease images categorized by disease type and severity
[40]	Apple leaf disease	The dataset consists of four different classes of 2,000 images of apple leaves affected by various diseases. Each image is labeled according to the specific disease present, providing an invaluable resource for developing and assessing ML models for identifying and diagnosing apple leaf diseases
[41]	Cassava leaf disease	The Cassava leaf disease dataset includes nearly 21,000 photos of cassava leaves, divided into four disease kinds and one healthy category. This dataset is used to train and evaluate ML models that detect and categorize diseases affecting cassava crops, hence assisting agricultural diagnostics and management
[42]	Rice leaf disease	The collection contains 5932 photos of four different types of rice leaf diseases. The dataset contains tagged images for training and testing models that recognize and categorize various rice leaf diseases
[43]	Banana leaf disease	The dataset that contains 937 images of banana leaves affected by various diseases, including Sigatoka, Panama disease, and Black Sigatoka. It features labeled images with annotations for different disease types and severity levels, aiding in the development of automated disease detection systems and plant health monitoring

- **Real-Time Processing Requirements**: Because DL models require a lot of calculation, plant disease detection is quite difficult, which can hinder their deployment in resource-constrained environments, including farms. These models often require substantial processing power and memory, making it difficult to implement capabilities, such as mobile or edge devices [80].
  To overcome these challenges, optimizing models through techniques and the use of lightweight architectures such as MobileNet or GhostNet can significantly reduce computational requirements [81]. Additionally, deploying models on edge devices with specialized hardware accelerators, such as GPUs or TPUs, can enhance processing speeds. Taking cloud computing for heavy processing tasks while using edge devices for data capture and preliminary analysis can also help get good accuracy on real time enabling efficient and timely disease detection in agricultural settings.
- **High Computational Costs**: High computational costs in plant disease detection present significant challenges, as DL models require substantial processing power. This high demand can be a barrier to widespread adoption, particularly in resource-limited settings [82].
  Several tactics can be used to address this. First, model optimization methods like knowledge distillation, quantization, and pruning can minimize the size and complexity of neural networks without sacrificing performance [83]. Utilizing lightweight architectures such as MobileNet

Fig. 5 Few leaf images from the tomato leaf disease dataset

or EfficientNet can also lower computational demands. Moreover, cloud computing and edge devices for distributed processing can help manage resources more efficiently. Finally, adopting efficient data processing pipelines and hardware accelerators, such as GPUs or TPUs, can significantly speed up both training and inference times, making high-performance plant disease detection more accessible.

- **Edge Computing**: Edge computing can significantly enhance plant leaf disease detection by enabling real-time analysis directly at the source of data collection, such as on farms or in greenhouses. Smartphones, drones, or IoT sensors are edge devices. It is possible to process plant photos locally without sending massive volumes of data to centralized cloud servers. This reduces latency, allows for immediate identification of diseases, and facilitates rapid response to prevent the spread of infections. Additionally, edge computing can operate in environments with limited or unreliable Internet connectivity, making it ideal for remote or rural areas. By limiting the exposure of sensitive agricultural data, localized processing also improves data security and privacy. All things considered, edge computing provides a scalable and effective way to incorporate cutting-edge plant disease detection techniques into useful agricultural applications.

7 Conclusion

The revolutionary effects of DL and computer vision technology on the identification of plant leaf diseases are highlighted in this thorough review. Through intensive analysis of various kinds of DL models, including CNNs, YOLO, SSD, and Mask R-CNN, among others, it is evident these advanced approaches improve accuracy, efficiency, and scalability in detecting diseases in agricultural applications. The integration of these techniques allows for the early, precise identification and rapid intervention in plant diseases for sustainable agriculture. Addressing the challenges in plant leaf disease detection, such as data availability, model robustness, and real-time processing in diverse conditions, remains crucial. Issues like high intra-class variability, leaf occlusion, and computational costs require innovative solutions. Future research should focus on multi-modal information, lightweight models, and unsupervised learning to improve generalization in real-world farming. Developing large, diverse, well-annotated datasets that capture plant species variability is essential for enhancing model performance. The field is dynamic and evolving, necessitating collaboration among computer scientists, agricultural experts, and industry stakeholders to bridge gaps and develop robust, deployable solutions. Real-time, on-field dis-

Table 2 Analysis of the result of existing work of plant leaf disease detection between 2022 and 2024

References	Year	Model	Dataset	Preprocess	Result (%)
[44]	2024	DenseNet-121	4590 Leaf images	Data augmentation, resizing, rescaling	Accuracy: 99.9,
[44]	2024	MobileNetV2	Cherry dataset	Data augmentation, resizing, rescaling	Accuracy: 98.9
[45]	2022	YOLOv5	PlantVillage	Image vector normalization, data augmentation	Precision: 93.73, Recall: 92.94, F1-score: 92.97
[46]	2023	Hybrid DL	IARI-TomEBD	Data augmentation	Accuracy: 87.55–100
[47]	2022	ResTS	PlantVillage	–	F1-score: 99.1
[48]	2022	Region fully convolutional network (RFCN)	NZDLPlant Disease-v1	Data augmentation	Mean average precision: 93.80
[49]	2022	DenseNet169-MLP	Rice plant image	Feature extraction, image segmentation, normalization	Accuracy: 97.68, Sensitivity: 100, Precision: 90.47, FPR: 5.19
[50]	2022	Deep learning	Cotton image	Rescaling, image and contrast enhancement	Accuracy: 98.53
[51]	2022	MobileNet	Apple leaf diseases	Data augmentation	Accuracy: 73.50
[52]	2022	1D CNN	PlantVillage	–	Accuracy: 100, 97
[53]	2023	AX-RetinaNet (CAHA-AXRNet)	Rice plant	Data normalization	Rice plant Accuracy: 98.1
[54]	2022	Optimal mobile network-based CNN (OMNCNN)	Plant leaf disease	–	Recall: 98.5, Accuracy: 98.92, F1-score: 98.7
[55]	2023	Full convolutional dual-path networks (FC-SNDPN)	PlantVillage	Image cropping, data augmentation, edge segmentation	Accuracy: 97.59, F1-score:98.75
[56]	2024	Federated DL (FDL)	PlantVillage	Data augmentation	Precision:95.7, Recall: 95.4, F1-score: 95.3
[57]	2024	CoffeeNet	Coffee plant leaf	Data augmentation	mAP: 97
[58]	2022	Fine-tuning DenseNet-121	PlantVillage	Data augmentation	Accuracy: 99.81
[59]	2022	CNN	Tomato disordered samples	K-means clustering technique	Accuracy: 99.6
[60]	2022	Xception, Faster-R-CNN	Apple leaf disease	Data augmentation, Data annotation	Overall: MAP: 42, Accuracy: 88
[61]	2023	CNN	Cotton leaf diseases		Accuracy: 99.39
[62]	2022	CNN	3000 Real-time image		Accuracy: Binary Class: 94.33, Multi-class: 95.3
[63]	2022	Discrete wavelet transformation (DWT)	PlantVillage	Data augmentation	Accuracy: 98.63
[64]	2022	DCNN	240,000 plant leaves images	Image augmentation	Accuracy: 98.41, Recall: 100, Precision: 94, F1-score: 97
[65]	2022	YOLO v5	PlantVillage	–	MAP: 90
[66]	2022	PlantDiseaseNet-MV	Turkey-PlantDataset	–	Accuracy: 97.56
[67]	2022	Ensemble CNNs	PlantVillage	–	Accuracy: 98.1
[68]	2023	Deeper lightweight CNN (DLMC-Net)	Crop leaf original dataset	Image augmentation	Accuracy: Grapes: 99.50, Tomato: 96.56
[69]	2022	Convolutional RNN-region-based CNN (CRNN-R-CNN)	Banana dataset	Edge normalization	Precision: 97, Recall: 97.7, sensitivity: 98.69
[70]	2022	YOLOv5	2375 Images of rubber disease	Data enhancement precision: 86.5	
[71]	2024	ResNet50 and DenseNet201	Infected date palm leaves	Data augmentation	Accuracy: 98.67

ease monitoring and management may be made possible by integrating cutting-edge models into IoT systems and edge computing devices, transforming disease management and promoting global security through precision agriculture.

References

1. Kabir, M.M., Hafiz, M.S., Bandyopadhyaa, S., Jim, J.R., Mridha, M.: Tea leaf age quality: age-stratified tea leaf quality classification dataset. Data Brief **54**, 110462 (2024)
2. Munir, N., Gulzar, W., Abideen, Z., Hancock, J.T., El-Keblawy, A., Radicetti, E.: Nanotechnology improves disease resistance in plants for food security: applications and challenges. Biocatal. Agric. Biotechnol. **51**, 102781 (2023)
3. Meena, S.D., Susank, M., Guttula, T., Chandana, S.H., Sheela, J.: Crop yield improvement with weeds, pest and disease detection. Procedia Comput. Sci. **218**, 2369–2382 (2023)
4. Kabir, M.M., Ohi, A.Q., Mridha, M.F.: A multi-plant disease diagnosis method using convolutional neural network. In: Computer Vision and Machine Learning in Agriculture, pp. 99–111 (2021)
5. Shi, T., Liu, Y., Zheng, X., Hu, K., Huang, H., Liu, H., Huang, H.: Recent advances in plant disease severity assessment using convolutional neural networks. Sci. Rep. **13**(1), 2336 (2023)
6. Khanna, M., Singh, L.K., Thawkar, S., Goyal, M.: Planet: a robust deep convolutional neural network model for plant leaves disease recognition. Multimedia Tools Appl. **83**(2), 4465–4517 (2024)
7. LeCun, Y., Boser, B., Denker, J.S., Henderson, D., Howard, R.E., Hubbard, W., Jackel, L.D.: Backpropagation applied to handwritten zip code recognition. Neural Comput. **1**(4), 541–551 (1989)
8. Rumelhart, D.E., Hinton, G.E., Williams, R.J.: Learning representations by back-propagating errors. Nature **323**(6088), 533–536 (1986)
9. Kotwal, J., Kashyap, R., Pathan, S.: Agricultural plant diseases identification: from traditional approach to deep learning. Mater. Today: Proc. **80**, 344–356 (2023)
10. Rai, C.K., Pahuja, R.: Classification of diseased cotton leaves and plants using improved deep convolutional neural network. Multimedia Tools Appl. **82**(16), 25307–25325 (2023)
11. Demilie, W.B.: Plant disease detection and classification techniques: a comparative study of the performances. J. Big Data **11**(1), 5 (2024)
12. Kaur, N., Devendran, V.: A novel framework for semi-automated system for grape leaf disease detection. Multimedia Tools Appl. **83**(17), 50733–50755 (2024)
13. Haridasan, A., Thomas, J., Raj, E.D.: Deep learning system for paddy plant disease detection and classification. Environ. Monitor. Assess. **195**(1), 120 (2023)
14. Xu, L., Cao, B., Zhao, F., Ning, S., Xu, P., Zhang, W., Hou, X.: Wheat leaf disease identification based on deep learning algorithms. Physiol. Mol. Plant Pathol. **123**, 101940 (2023)
15. Bedi, P., Gole, P.: Plant disease detection using hybrid model based on convolutional autoencoder and convolutional neural network. Artif. Intell. Agric. **5**, 90–101 (2021)
16. Sunil, C., Jaidhar, C., Patil, N.: Cardamom plant disease detection approach using efficientnetv2. IEEE Access **10**, 789–804 (2021)
17. Agarwal, M., Singh, A., Arjaria, S., Sinha, A., Gupta, S.: Toled: tomato leaf disease detection using convolution neural network. Procedia Comput. Sci. **167**, 293–301 (2020)
18. Moupojou, E., Tagne, A., Retraint, F., Tadonkemwa, A., Wilfried, D., Tapamo, H., Nkenlifack, M.: Fieldplant: a dataset of field plant images for plant disease detection and classification with deep learning. IEEE Access **11**, 35398–35410 (2023)
19. Gajjar, R., Gajjar, N., Thakor, V.J., Patel, N.P., Ruparelia, S.: Real-time detection and identification of plant leaf diseases using convolutional neural networks on an embedded platform. In: The Visual Computer, pp. 1–16 (2022)
20. Abd Algani, Y.M., Caro, O.J.M., Bravo, L.M.R., Kaur, C., Al Ansari, M.S., Bala, B.K.: Leaf disease identification and classification using optimized deep learning. Meas.: Sens. **25**, 100643 (2023)
21. Singh, G., Yogi, K.K.: Performance evaluation of plant leaf disease detection using deep learning models. Arch. Phytopathol. Plant Protect. **56**(3), 209–233 (2023)
22. Hassan, S.M., Maji, A.K., Jasiński, M., Leonowicz, Z., Jasińska, E.: Identification of plant-leaf diseases using CNN and transfer-learning approach. Electronics **10**(12), 1388 (2021)
23. Arun, R.A., Umamaheswari, S.: Effective multi-crop disease detection using pruned complete concatenated deep learning model. Expert Syst. Appl. **213**, 118905 (2023)
24. Redmon, J., Divvala, S., Girshick, R., Farhadi, A.: You only look once: unified, real-time object detection. In: Proceedings of the IEEE Conference on Computer Vision and Pattern Recognition, pp. 779–788 (2016)
25. Liu, W., Anguelov, D., Erhan, D., Szegedy, C., Reed, S., Fu, C.-Y., Berg, A.C.: Ssd: Single shot multibox detector. In: Computer Vision—ECCV 2016: 14th European Conference, Amsterdam, The Netherlands, October 11–14, 2016, Proceedings, Part I 14, pp. 21–37. Springer (2016)
26. Ren, S., He, K., Girshick, R., Sun, J.: Faster R-CNN: towards real-time object detection with region proposal networks. IEEE Trans. Pattern Anal. Mach. Intell. **39**(6), 1137–1149 (2016)
27. He, K., Gkioxari, G., Dollár, P., Girshick, R.: Mask R-CNN. In: Proceedings of the IEEE International Conference on Computer Vision, pp. 2961–2969 (2017)
28. Dai, X., Chen, Y., Yang, J., Zhang, P., Yuan, L., Zhang, L.: Dynamic detr: end-to-end object detection with dynamic attention. In: Proceedings of the IEEE/CVF International Conference on Computer Vision, pp. 2988–2997 (2021)
29. Luo, Y., Sun, J., Shen, J., Wu, X., Wang, L., Zhu, W.: Apple leaf disease recognition and sub-class categorization based on improved multi-scale feature fusion network. IEEE Access **9**, 95517–95527 (2021)
30. Ronneberger, O., Fischer, P., Brox, T.: U-net: convolutional networks for biomedical image segmentation. In: Medical Image Computing and Computer-assisted intervention–MICCAI 2015: 18th International Conference, Munich, Germany, October 5-9, 2015, Proceedings, Part III 18, pp. 234–241. Springer (2015)
31. Dosovitskiy, A.: An image is worth 16 × 16 words: transformers for image recognition at scale (2020). arXiv:2010.11929
32. https://www.kaggle.com/datasets/emmarex/plantdisease/data (2018)
33. Wu, X., Zhan, C., Lai, Y.-K., Cheng, M.-M., Yang, J.: Ip102: A large-scale benchmark dataset for insect pest recognition. In: 2019 IEEE/CVF Conference on Computer Vision and Pattern Recognition (CVPR), pp. 8779–8788 (2019). https://doi.org/10.1109/CVPR.2019.00899
34. Kumar, N., Belhumeur, P.N., Biswas, A., Jacobs, D.W., Kress, W.J., Lopez, I.C., Soares, J.V.: Leafsnap: a computer vision system for automatic plant species identification. In: Computer Vision—ECCV 2012: 12th European Conference on Computer Vision, Florence, Italy, October 7-13, 2012, Proceedings, Part II 12, pp. 502–516. Springer (2012)
35. Silva, P., Maral, A.: Leaf. UCI Machine Learning Repository (2014). https://doi.org/10.24432/C53C78 (2014)
36. Garcin, C., Joly, A., Bonnet, P., Servajean, M., Salmon, J.: Pl@ntNet-300K image dataset. https://doi.org/10.5281/zenodo.4726653
37. Huang, M.-L., Chang, Y.-H.: Dataset of tomato leaves. Mendeley Data (2020). https://doi.org/10.17632/ngdgg79rzb.1
38. Getachew, H.: Wheat leaf dataset. Mendeley Data (2020). https://doi.org/10.17632/wgd66f8n6h.1
39. Hang, Y.: Soybean disease data set in natural scenes. https://doi.org/10.21227/tc50-b707

40. Feng, J., Chao, X.: Apple tree leaf disease segmentation dataset. Science Data Bank (2022). https://doi.org/10.11922/sciencedb.01627
41. Divyanth, L.G., Soni, P., Machavaram, R.: Cassava leaf disease dataset. Mendeley Data (2021). https://doi.org/10.17632/3832tx2cb2.1
42. sethy, P.K.: Rice leaf disease image samples. Mendeley Data (2020). https://doi.org/10.17632/fwcj7stb8r.1
43. Arman, S.E., Bhuiyan, M.A.B., Abdullah, H.M., Islam, S., Chowdhury, T.T., Hossain, M.A.: Bananalsd: a banana leaf images dataset for classification of banana leaf diseases using machine learning. Data Brief **50**, 109608 (2023)
44. Srivastava, M., Meena, J.: Plant leaf disease detection and classification using modified transfer learning models. Multimedia Tools Appl. **83**(13), 38411–38441 (2024)
45. Wang, H., Shang, S., Wang, D., He, X., Feng, K., Zhu, H.: Plant disease detection and classification method based on the optimized lightweight yolov5 model. Agriculture **12**(7), 931 (2022)
46. Chug, A., Bhatia, A., Singh, A.P., Singh, D.: A novel framework for image-based plant disease detection using hybrid deep learning approach. Soft Comput. **27**(18), 13613–13638 (2023)
47. Shah, D., Trivedi, V., Sheth, V., Shah, A., Chauhan, U.: Rests: residual deep interpretable architecture for plant disease detection. Inf. Process. Agric. **9**(2), 212–223 (2022)
48. Saleem, M.H., Potgieter, J., Arif, K.M.: A performance-optimized deep learning-based plant disease detection approach for horticultural crops of New Zealand. IEEE Access **10**, 89798–89822 (2022)
49. Narmadha, R., Sengottaiyan, N., Kavitha, R.: Deep transfer learning based rice plant disease detection model. Intell. Autom. Soft Comput. **31**(2) (2022)
50. Memon, M.S., Kumar, P., Iqbal, R.: Meta deep learn leaf disease identification model for cotton crop. Computers **11**(7), 102 (2022)
51. Bi, C., Wang, J., Duan, Y., Fu, B., Kang, J.-R., Shi, Y.: Mobilenet based apple leaf diseases identification. In: Mobile Networks and Applications, pp. 1–9 (2022)
52. Sai Reddy, B., Neeraja, S.: Plant leaf disease classification and damage detection system using deep learning models. Multimedia Tools Appl. **81**(17), 24021–24040 (2022)
53. Sankareshwaran, S.P., Jayaraman, G., Muthukumar, P., Krishnan, A.: Optimizing rice plant disease detection with crossover boosted artificial hummingbird algorithm based AX-RetinaNet. Environ. Monitor. Assess. **195**(9), 1070 (2023)
54. Ashwinkumar, S., Rajagopal, S., Manimaran, V., Jegajothi, B.: Automated plant leaf disease detection and classification using optimal MobileNet based convolutional neural networks. Mater. Today: Proc. **51**, 480–487 (2022)
55. Huang, X., Chen, A., Zhou, G., Zhang, X., Wang, J., Peng, N., Yan, N., Jiang, C.: Tomato leaf disease detection system based on FC-SNDPN. Multimedia Tools Appl. **82**(2), 2121–2144 (2023)
56. Hari, P., Singh, M.P., Singh, A.K.: An improved federated deep learning for plant leaf disease detection. Multimedia Tools Appl. 1–21 (2024)
57. Nawaz, M., Nazir, T., Javed, A., Amin, S.T., Jeribi, F., Tahir, A.: Coffeenet: a deep learning approach for coffee plant leaves diseases recognition. Expert Syst. Appl. **237**, 121481 (2024)
58. Eunice, J., Popescu, D.E., Chowdary, M.K., Hemanth, J.: Deep learning-based leaf disease detection in crops using images for agricultural applications. Agronomy **12**(10), 2395 (2022)
59. Harakannanavar, S.S., Rudagi, J.M., Puranikmath, V.I., Siddiqua, A., Pramodhini, R.: Plant leaf disease detection using computer vision and machine learning algorithms. Glob. Transit. Proc. **3**(1), 305–310 (2022)
60. Khan, A.I., Quadri, S., Banday, S., Shah, J.L.: Deep diagnosis: a real-time apple leaf disease detection system based on deep learning. Comput. Electron. Agric. **198**, 107093 (2022)
61. Singh, P., Singh, P., Farooq, U., Khurana, S.S., Verma, J.K., Kumar, M.: Cottonleafnet: cotton plant leaf disease detection using deep neural networks. Multimedia Tools Appl. **82**(24), 37151–37176 (2023)
62. Sharma, R., Kukreja, V., Kaushal, R.K., Bansal, A., Kaur, A.: Rice leaf blight disease detection using multi-classification deep learning model. In: 2022 10th International Conference on Reliability, Infocom Technologies and Optimization (trends and Future directions) (ICRITO), pp. 1–5. IEEE (2022)
63. Hasan, S., Jahan, S., Islam, M.I.: Disease detection of apple leaf with combination of color segmentation and modified dwt. J. King Saud Univ.-Comput. Inf. Sci. **34**(9), 7212–7224 (2022)
64. Pandian, J.A., Kanchanadevi, K., Kumar, V.D., Jasińska, E., Goňo, R., Leonowicz, Z., Jasiński, M.: A five convolutional layer deep convolutional neural network for plant leaf disease detection. Electronics **11**(8), 1266 (2022)
65. Mathew, M.P., Mahesh, T.Y.: Leaf-based disease detection in bell pepper plant using yolo v5. Signal Image Video Process 1–7 (2022)
66. Turkoglu, M., Yanikoğlu, B., Hanbay, D.: Plantdiseasenet: convolutional neural network ensemble for plant disease and pest detection. Signal Image Video Process. **16**(2), 301–309 (2022)
67. Moussafir, M., Chaibi, H., Saadane, R., Chehri, A., Rharras, A.E., Jeon, G.: Design of efficient techniques for tomato leaf disease detection using genetic algorithm-based and deep neural networks. Plant Soil **479**(1), 251–266 (2022)
68. Sharma, V., Tripathi, A.K., Mittal, H.: DLMC-net: deeper lightweight multi-class classification model for plant leaf disease detection. Ecol. Inform. **75**, 102025 (2023)
69. Seetharaman, K., Mahendran, T.: Leaf disease detection in banana plant using gabor extraction and region-based convolution neural network (RCNN). J. Inst. Eng. (India): Ser. A **103**(2), 501–507 (2022)
70. Chen, Z., Wu, R., Lin, Y., Li, C., Chen, S., Yuan, Z., Chen, S., Zou, X.: Plant disease recognition model based on improved yolov5. Agronomy **12**(2), 365 (2022)
71. Nobel, S.N., Imran, M.A., Bina, N.Z., Kabir, M.M., Safran, M., Alfarhood, S., Mridha, M.F.: Palm leaf health management: a hybrid approach for automated disease detection and therapy enhancement. IEEE Access (2024)
72. Mazumder, M.K.A., Kabir, M.M., Rahman, A., Abdullah-Al-Jubair, M., Mridha, M.: Densenet201plus: cost-effective transfer-learning architecture for rapid leaf disease identification with attention mechanisms. Heliyon (2024)
73. Afifi, A., Alhumam, A., Abdelwahab, A.: Convolutional neural network for automatic identification of plant diseases with limited data. Plants **10**(1), 28 (2020)
74. Shafik, W., Tufail, A., Namoun, A., De Silva, L.C., Apong, R.A.A.H.M.: A systematic literature review on plant disease detection: Motivations, classification techniques, datasets, challenges, and future trends. IEEE Access **11**, 59174–59203 (2023)
75. Tiwari, V., Joshi, R.C., Dutta, M.K.: Dense convolutional neural networks based multiclass plant disease detection and classification using leaf images. Ecol. Inform. **63**, 101289 (2021)
76. Xu, M., Kim, H., Yang, J., Fuentes, A., Meng, Y., Yoon, S., Kim, T., Park, D.S.: Embracing limited and imperfect training datasets: opportunities and challenges in plant disease recognition using deep learning. Front. Plant Sci. **14**, 1225409 (2023)
77. Liu, X., Min, W., Mei, S., Wang, L., Jiang, S.: Plant disease recognition: a large-scale benchmark dataset and a visual region and loss reweighting approach. IEEE Trans. Image Process. **30**, 2003–2015 (2021)
78. Wang, X., Liu, J., Liu, G.: Diseases detection of occlusion and overlapping tomato leaves based on deep learning. Front. Plant Sci. **12**, 792244 (2021)
79. Xia, C., Lee, J.-M., Li, Y., Song, Y.-H., Chung, B.-K., Chon, T.-S.: Plant leaf detection using modified active shape models. Biosyst. Eng. **116**(1), 23–35 (2013)

80. Al-Badri, A.H., Ismail, N.A., Al-Dulaimi, K., Salman, G.A., Khan, A., Al-Sabaawi, A., Salam, M.S.H.: Classification of weed using machine learning techniques: a review-challenges, current and future potential techniques. J. Plant Dis. Protect. **129**(4), 745–768 (2022)
81. Prodeep, A.R., Hoque, A.M., Kabir, M.M., Rahman, M.S., Mridha, M.: Plant disease identification from leaf images using deep CNN's efficientnet. In: 2022 International Conference on Decision Aid Sciences and Applications (DASA), pp. 523–527. IEEE (2022)
82. Yağ, İ, Altan, A.: Artificial intelligence-based robust hybrid algorithm design and implementation for real-time detection of plant diseases in agricultural environments. Biology **11**(12), 1732 (2022)
83. Hassan, S.M., Maji, A.K.: Plant disease identification using a novel convolutional neural network. IEEE Access **10**, 5390–5401 (2022)

Plant Leaf Disease Detection Using Deep Stacking: Integrating CNNs and Gradient Boosting for Enhanced Classification Accuracy

Md. Sabbir Hossain, Mostafijur Rahman, Md. Golam Rabbani Abir, Jannatul Maua, and Ashifur Rahman

Abstract

This paper presents a novel approach to detecting plant leaf diseases by employing a deep stacking method that integrates the advantages of deep learning (DL) and traditional machine learning (ML) techniques. Convolutional Neural Networks (CNNs) serve as base learners, capturing intricate features from leaf images, while Gradient Boosting acts as a meta-learner to refine and enhance classification accuracy. The model completed training and assessment on a comprehensive dataset of plant leaf pictures, achieving impressive results getting an overall accuracy of 99.57% and significant improvements in validation loss across training epochs. The approach is validated through detailed evaluation metrics, such as precision, recall, and F1-scores, across multiple plant disease categories. The proposed method demonstrates high effectiveness in identifying and classifying various plant diseases, offering the potential for real-world agricultural applications in disease monitoring and prevention.

Keywords

Deep learning • Computer vision • Plant leaf disease detection • Artificial intelligence in agriculture • Convolutional neural networks • Gradient boosting

1 Introduction

Plant diseases pose a significant threat to global agricultural production, leading to substantial economic losses and food security challenges [1]. Early and accurate detection of plant diseases is crucial to mitigate these effects, enabling timely interventions and reducing the spread of infections [2]. Traditional methods of disease identification, which often rely on expert manual inspection, are time-consuming, labor-intensive, and prone to errors [3]. Hence, the requirement for automation, reliable, as well as scalable solutions for the identification of plant diseases has become more pressing than ever.

The emergence of computer vision and machine learning (ML) techniques has created new possibilities for automating plant disease detection. Convolutional Neural Networks (CNNs), a type of deep learning (DL) model, have demonstrated considerable success in image classification tasks, making them well suited for identifying disease patterns in plant leaves [4, 5]. However, despite the promising results, single-model approaches such as CNNs may still suffer from limitations in generalization and robustness, particularly when faced with complex or noisy data. To address these challenges, ensemble learning techniques, combining multiple models' strengths, have gained attention [6]. Combining ensemble learning and deep learning in order to get the best output is currently being worked on [7]. The deep stacking method, which integrates the outputs of multiple base learners through a meta-learner, has shown the potential to improve classification accuracy and robustness [8].

This study holds importance for various reasons. First, it contributes to the expanding knowledge of applying deep learning and ensemble methods in agricultural contexts. By integrating CNNs with traditional machine learning models like Gradient Boosting [9, 10], this work offers a novel approach that enhances the accuracy and reliability of plant disease detection systems. Second, the outcomes of this study have practical consequences for farmers, agronomists,

Md. S. Hossain
Western Illinois University, Macomb, Illinois, USA

M. Rahman
Washington University of Science and Technology, Alexandria, USA

Md. G. R. Abir · A. Rahman
Bangladesh University of Business and Technology, Dhaka, Bangladesh

J. Maua (✉)
Advanced Machine Intelligence Research Lab, Dhaka, Bangladesh
e-mail: janmauakorobi@gmail.com

M. F. Mridha and N. Dey (eds.), *Machine Vision in Plant Leaf Disease Detection for Sustainable Agriculture*, Studies in Computational Intelligence 1202, https://doi.org/10.1007/978-981-96-4520-6_2

and agricultural technology developers, providing them with tools that can facilitate early disease detection and ultimately improve crop yields and food security. Finally, this study tackles the limitations of single-model approaches by demonstrating the effectiveness of ensemble techniques in a real-world application, thereby advancing the leading-edge advancements in plant disease detection.

This paper is motivated by the need to overcome the limitations of current plant disease detection methods. While CNNs have proven effective in identifying diseases from images, their performance can be hindered by data variability, such as differences in lighting, angles, and leaf conditions. By leveraging the deep stacking method, this research seeks to combine the strengths of multiple CNN models, each capturing different aspects of the data, and enhance them further with a powerful meta-learner, Gradient Boosting. The goal is to create a more robust more precise and reliable detection system that can handle diverse and complex datasets, thereby improving the practical use and relevance of these systems in real-world agricultural settings. Key contributions of this paper are summarized below:

- Proposing a novel leaf disease detection framework using deep stacking, combining CNNs as base learners and Gradient Boosting as a meta-learner.
- Demonstrating the performance of the proposed approach through extensive experiments on a diverse dataset, achieving a high classification accuracy and improved validation loss.
- Providing detailed performance metrics across multiple plant disease categories to validate the method's robustness.
- Examining the practical applications of the proposed system for early disease detection, highlighting its potential to enhance agricultural productivity and support global food security.

2 Related Works

Traditional machine learning (ML) models have been used for a long time in the field of identification of plant diseases [11]. Early approaches primarily focused on feature extraction methods such as texture, shape, and color analysis, followed by classification using algorithms like Support Vector Machines (SVM), Random Forests (RF), and k-Nearest Neighbors (k-NN). In a paper, Das et al. [12] used traditional machine learning algorithms such as SVM, RF, and Logistic Regression (LR) to classify as well as identify different types of leaf diseases. They used feature extraction techniques to make the classification more accurate. From these machine learning techniques, SVM performed better with good accuracy. The setback of this paper is that the researchers used the majority of the image dataset for training as well as testing which reduces the speed of the system.

Hatuwal et al. [13] presented some crucial information in their paper regarding traditional machine learning and CNN. In their paper, they compared CNN against Random Forest, SVM, and K-NN. They used the plant village dataset with 16 classes. The result was a landslide win for CNN achieving 97.8% accuracy where the Random Forest model attained 87.436%, while SVM achieved 78.61% and K-NN scored 76.969%.

Padol et al. [14] researched the segmentation of grape leaf disease. They used k-means clustering for the image segmentation. Main features like shape, color, and texture were selected, and a Linear SVM was used to classify the grape leaf disease. The images they classified were categorized into two classes, Downy and Powderly, based on extracted texture and color features. These features are crucial for the segmentation of all three parts of a single-leaf image.

Rice leaf diseases are really dangerous. They can destroy an entire paddy field. To make sure they cannot spread, the disease needs to be detected at an early stage. Pothen et al. [15] discussed this issue in their paper where for segmentation Otsu's method was used. Along with that, Local Binary patterns (LBP) and Histogram of oriented gradients (HOG) were implemented to separate various features from it. Using SVM, the classified data achieved a high-level accuracy of 94.6%. For this, a polynomial kernel was used. Similarly, using traditional machine learning algorithms such as the K-means clustering algorithm, Reddy et al. [16] have detected fainted or diseased parts of the leaf with great effectiveness. They also used K-NN and SVM to compare accuracies between them where SVM has 95% and K-NN has 85% accuracies.

With the advent of DL, particularly CNNs, the area of plant disease detection has seen significant advancements. CNNs can independently perform automatic tasks to learn hierarchical extract features directly from raw images, minimizing the reliance on manual feature extraction. Sujatha et al. [17] compared machine learning and deep learning models to detect citrus plant leaf disease. The authors used SVM, RF, and Stochastic Gradient Descent (SGD) for ML models and Inception-v3, VGG-16, and VGG-19 for deep learning models. They achieved really impressive results in deep learning compared to machine learning. In their experiment, RF acquired 76.8% accuracy, SGD 86.5%, and SVM with the highest accuracy among ML models with 87%. Compared to that VGG-19 achieved 87.4% accuracy; Inception-v3, 89% accuracy; and VGG-16, 89.5% accuracy. This proves that DL performs superior to ML in detecting citrus plant diseases. Pradeep et al. [18] have highlighted the need for detecting plant disease effectively. They used their own dataset from various online sources containing leaf images of ten different plant species. In this study, the authors compared various famous CNN designs. They discovered that EfficientNet outperforms other CNN architectures in multi-label plant disease classification tasks. The study also found that incorporating features like skip connections, spatial convolutions, and optimized network layers enhances the precision and effectiveness of leaf disease classification, providing significant benefits for agricultural management and phytopathology. Mohanty et al. [19] developed a deep convolutional neural network model using the GoogLeNet architecture, which achieved a remarkable accuracy of 99.35% in classifying 14 crop species and 26 diseases from a dataset of 54,306 images. This model, trained with transfer learning on color images, underscores the feasibility of employing deep learning for large-scale, smartphone-assisted disease diagnosis. Such technological advancements are crucial for improving food security by facilitating early and precise detection of crop diseases. Advancements in DL have greatly enhanced the accuracy and effectiveness of plant disease diagnosis. Kabir et al. [20] explored multi-label classification methods using convolutional neural networks (CNNs) for diagnosing diseases across six plant species. In their study, several CNN architectures were evaluated, with Xception and DenseNet169 models demonstrating the optimal F1-scores of 97.38% and 97.36%, respectively. These results highlight the possibility of CNNs accurately recognizing multiple tree diseases simultaneously, providing a robust solution for agricultural automation.

Ensemble learning, which merges several models to enhance prediction accuracy, has gained attention as a way to address the limitations of single-model approaches. Techniques such as bagging, boosting, and stacking have been explored in this context. Enkvetchakul et al. [21] proposed an approach to enhance the precision of plant disease identification by leveraging a stacking ensemble of CNN models, including InceptionResNetV2, NASNetMobile, MobileNetV2, and EfficientNetB1. The study evaluated these models on three datasets: PlantDoc, Crop-PlantDoc, and iCassava2019, demonstrating that EfficientNetB1 consistently provided the highest accuracy. By employing ensemble learning and stacking these CNN outputs with classifiers like Random Forest, the method achieved significant improvements in disease recognition accuracy, with the best results obtained by combining EfficientNetB1 with Random Forest, particularly on the Crop-PlantDoc and iCassava2019 datasets. The study concludes that stacking ensembles of CNNs offers a robust approach to enhancing plant disease detection systems. In recent advancements, Asad et al. [22] proposed an ensemble learning framework leveraging homogeneous stacking of base models trained on diverse agricultural datasets. By integrating these models through a UNET meta-architecture, the framework significantly enhanced detection performance for Canola crops and Kochia weeds in uncontrolled field conditions. Their approach resulted in an mIOU improvement of 12% for Canola and 6% for Kochia, with the MUNE model achieving an mIOU of 90.40% and an IOU of 81.59% for Canola, highlighting the potential of ensemble learning in improving model generalization across varied field conditions. Mazumder et al. [23] developed a modified DenseNet201 architecture, termed DenseNet201Plus, which integrates advanced attention mechanisms and data augmentation for enhanced leaf disease detection. The model was tested on Banana and Black Gram Leaf Disease datasets, achieving accuracies of 90.12% and 99.50%, respectively. This study highlights DenseNet201Plus as a superior model in efficiently detecting plant diseases across diverse conditions, surpassing traditional models like MobileNetV2 and ResNet152V2.

Deep stacking is a sophisticated ensemble learning method that integrates the predictions of various base models using a meta-learner, allowing for more complex and accurate decision-making. Although deep stacking has been effectively used in multiple fields, such as image recognition and natural language processing, its application in plant disease detection remains relatively unexplored. In an attempt to identify plant leaf diseases, Jha et al. [24] proposed a unique approach that mixes techniques like deep learning and image processing. Their main focus is on a complex deep neural network ensemble model combining Residual Network, MobileNet, and Inception models. This model is mainly built for maximizing accuracy and robustness in plant leaf disease detection. The model had a really spectacular result with 98.86% accuracy on the potato leaf dataset with three classes belonging to healthy and infected potato leaves. Yilma et al. [25] proposed attention augmented CNN with AAR deep stacking for fine-grained plant disease classification and visualization. The model improved the detailed categorization of plant diseases and provided better visualization through attention mechanisms. Their multi-head self-attention augmented CNN (AAC) network achieved a 99.56% accuracy using stochastic transformed training samples. With geometrically transformed samples, they acquired 98.01 and 97.34% without transformation. Nobel et al. [26] introduced a cutting-edge hybrid model combining Efficient Channel Attention Network (ECA-Net) with ResNet50 and DenseNet201 for palm leaf disease detection. The model attained a noteworthy training accuracy of 99.54% and validation accuracy of 98.67%, significantly outperforming existing approaches. The study underscores the model's effectiveness in accurately identifying diseases caused by the Dubas bug, marking a substantial advancement in automated agricultural disease management.

The combination of IoT and machine learning has unlocked new possibilities in precision agriculture. Islam et al. [27] developed an IoT-enabled system that employs CatBoost and Random Forest algorithms to monitor soil nutrients and recommend appropriate crops and fertilizers. Their system achieved high precision, with the CatBoost model having 97.5% accuracy in crop recommendations and the Random Forest model achieving 99.56% accuracy in fertilizer prediction. This method highlights the capability of machine learning to enhance agricultural decision-making and maximize resource efficiency, supporting sustainable farming practices.

While existing research has made significant strides in improving plant disease detection through traditional machine and deep learning approaches, several gaps remain. First, many studies rely on single-model architectures, which may not generalize well across diverse datasets. Second, while ensemble learning has been explored, the potential of deep stacking methods combining deep learning and traditional machine learning models remains underutilized. Finally, most existing works focus on accuracy as the primary metric, with less attention given to other important aspects such as robustness, interpretability, and computational efficiency.

The paper aims to address the aforementioned lackings by proposing a deep stacking framework that integrates CNNs as base learners with Gradient Boosting as a meta-learner for plant disease detection. This approach not only improves precision but also enhances the robustness of the model against data variability. By harnessing the capabilities of both DL and traditional ML, our method provides a more comprehensive approach to overcoming the challenges of plant disease detection.

3 Methodology

3.1 Proposed Method

This research introduces a plant leaf disease detection framework using a deep stacking approach. The methodology combines CNNs as base learners and Gradient Boosting as a meta-learner. The primary goal is to utilize the feature extraction capabilities of CNNs and the robust classification abilities of Gradient Boosting to achieve high accuracy and generalization in plant disease detection. The overall architecture is designed to process input leaf images, extract meaningful features, and output the predicted disease class. Figure 1 represents the architecture of the proposed method.

3.2 Dataset and Preprocessing

The deep learning model needs a large dataset to train properly [28]. Here employed a dataset that consists of a large collection of labeled images representing various plant diseases and healthy leaves. The images were sourced from openly available datasets, such as the PlantVillage dataset, which has high-resolution images of different plant species and their associated diseases.

3.2.1 Image Augmentation

To improve the model's robustness and mitigate overfitting image augmentation techniques were implemented during the preprocessing stage. These techniques included random rotations, flips, zooms, and shifts. Formally, given an image a, the augmented image a' is obtained through a transformation function G:

$$a' = G(a) \tag{1}$$

where G represents the random combination of rotation, flip, zoom, and shift transformations.

3.2.2 Normalization

All images were adjusted to a consistent resolution, such as 224×224 pixels, and their pixel values were normalized to fall within a specific range [0, 1]. Mathematically, the normalization process can be expressed as

$$a_{\text{norm}} = \frac{a - a_{\min}}{a_{\max} - a_{\min}} \tag{2}$$

where a is the original pixel value, a_{norm} is the normalized pixel value, and $a_{\min}$ and $a_{\max}$ represent the minimum and maximum pixel values, respectively.

3.3 Base Learners: Convolutional Neural Networks (CNNs)

CNNs were chosen as the base learners in the deep stacking framework because of their demonstrated success in tasks involving image classification. Multiple CNN architectures are used to capture diverse features from the input images. The specific architectures used in this study include VGG16, ResNet50, and InceptionV3.

The output of each CNN can be expressed as

$$q_i = \text{CNN}_i(a) \tag{3}$$

where q_i represents the output probability vector from the i-th CNN model and a is the input image.

3.4 Meta-Learner: Gradient Boosting

The outputs from the base CNN models were combined using a Gradient Boosting classifier as the meta-learner. Gradient Boosting iteratively fits decision trees to the residuals of previous models to minimize the classification error. The loss function for Gradient Boosting can be expressed as

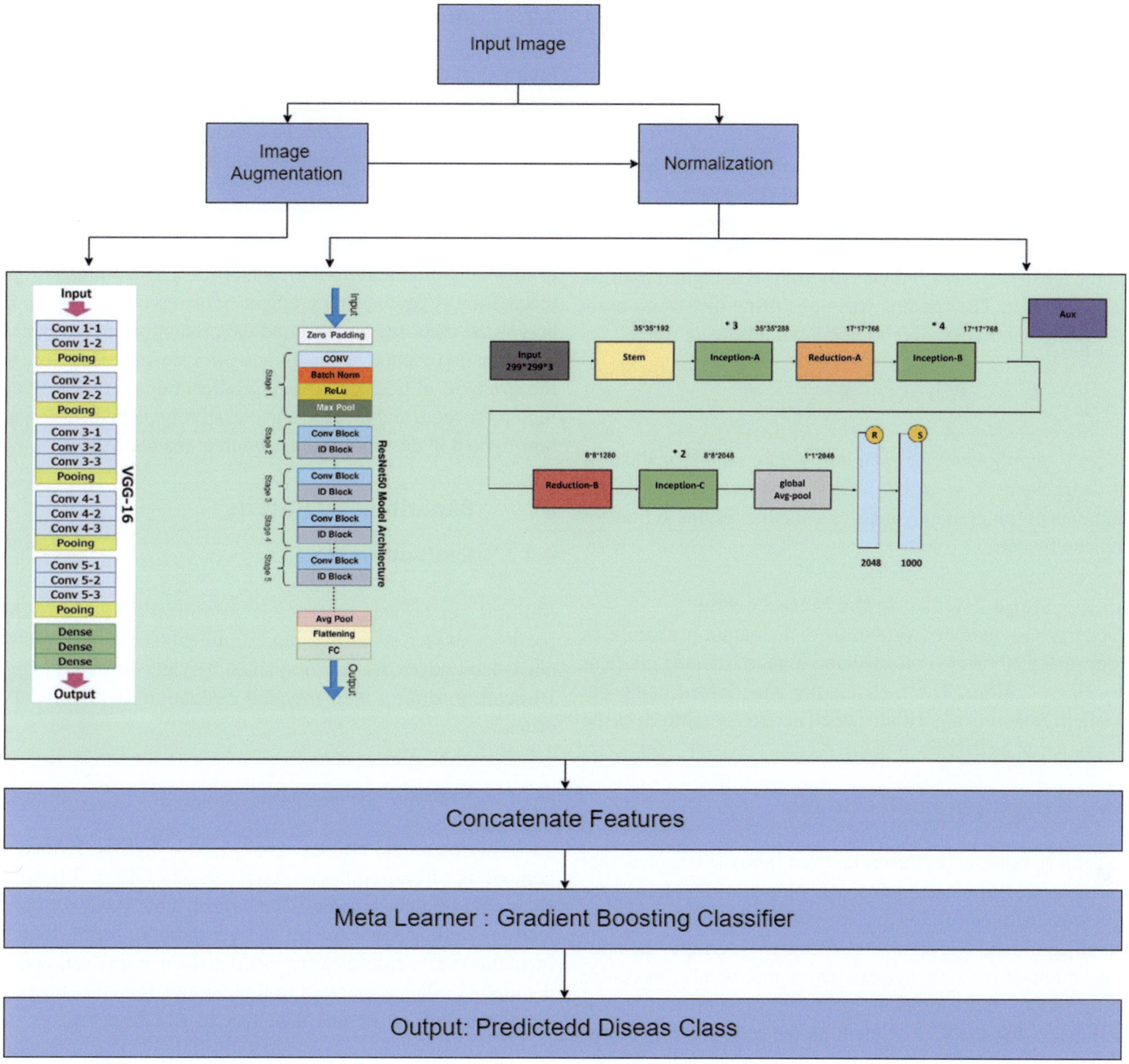

Fig. 1 Architecture DeepStack model

$$L(x, \hat{x}) = \sum_{i=1}^{n} \left(x_i - \hat{x}_i\right)^2 \tag{4}$$

where x_j is the true label, $\hat{x}_j$ is the predicted label, and n is the total number of samples.

3.4.1 Feature Extraction for Meta-Learner

The predicted probabilities from each CNN model were concatenated to form a feature vector for each input image. This feature vector, denoted as w, is given by

$$w = [q_1, q_2, \ldots, q_n] \tag{5}$$

where $q_1, q_2, \ldots, q_n$ are the output probability vectors from the n CNN models.

3.4.2 Training the Meta-Learner

The Gradient Boosting classifier was trained on the concatenated feature vectors using cross-entropy as the loss function, defined as

$$L_{CE} = -\sum_{i=1}^{n} \left(x_i \log(\hat{x}_i) + (1 - x_i) \log(1 - \hat{x}_i)\right) \tag{6}$$

where x_i is the true binary label and $\hat{x}_i$ represents the predicted probability for each class 1.

3.5 Training Procedure

The training process was carried out in two stages.

3.5.1 Stage 1: Training the Base Learners

Each CNN model was individually trained using the preprocessed dataset. The training was done using the Adam optimizer, which updates weights using the following rule:

$$\theta_{t+1} = \theta_t - \eta \frac{g_t}{\sqrt{h_t} + \epsilon} \tag{7}$$

where θ_t symbolizes the weights at time step t, η denotes the learning rate, g_t and h_t denote first- and second-order moments of the gradient, and ϵ is a small constant to prevent division by zero.

3.5.2 Stage 2: Training the Meta-Learner

Once the base learners were trained, the output probabilities from each CNN were collected and used to train the gradient boosting classifier. The meta-learner was trained using the Gradient Boosting algorithm, which iteratively minimizes the loss function as described in Eq. 4.

3.6 Hyperparameter Tuning

To achieve optimal performance from both the base learners and the meta-learner, an extensive hyperparameter tuning process was conducted. This process involved a systematic grid search combined with cross-validation to determine the optimal combination of hyperparameters that would maximize both training and validation accuracies. The critical hyperparameters for the CNNs, such as the learning rate, batch size, and number of layers, were carefully adjusted to enhance the model's generalization capabilities and minimize overfitting. Similarly, for the Gradient Boosting (GB) meta-learner, parameters like the learning rate, number of estimators, and maximum depth were fine-tuned to optimize its effectiveness in integrating the outputs of the base models.

The tuning process involved testing various configurations of these hyperparameters, and the resulting training and validation accuracies for different hyperparameter settings are presented in the table below. The results demonstrate the sensitivity of the model's performance to changes in learning rate and batch size, highlighting the importance of careful hyperparameter selection in building robust deep learning models.

3.7 Model Architecture

Figure 1 illustrates the proposed deep stacking framework for plant leaf disease detection. The process begins with the input images undergoing image augmentation and normalization to ensure robust preprocessing. Two pre-trained CNNs, VGG-16 and Inception-v4, are employed as base learners to extract intricate and diverse features from the leaf images. The extracted features from these networks are then concatenated to form a unified feature representation. This combined representation is passed to a gradient boosting classifier, which acts as the meta-learner to refine the predictions and enhance classification accuracy. The framework is designed to leverage the strengths of both CNNs and gradient boosting, achieving high robustness and precision in classifying plant diseases, as highlighted in the study's experimental results.

4 Experimental Results

4.1 Evaluation Metrics

Multiple performance metrics were utilized to evaluate the proposed model. Accuracy, representing the ratio of correctly classified images, was accompanied by precision, recall, and F1-score to offer a well-rounded evaluation of the model's efficacy.

4.2 Experimental Setup

To train this model efficiently we need a high-performance computing cluster equipped with multiple GPUs. The software environment included TensorFlow and Keras for deep learning and Scikit-learn for the Gradient Boosting implementation. The results were evaluated on a test set, and a detailed classification report was generated to analyze the performance across different plant species and diseases.

4.3 Implementation Details

The proposed method was implemented in Python, leveraging popular libraries such as TensorFlow, Keras, and Scikit-learn. The code was structured to allow easy replication of the experiments, and the trained models were saved for future use. The entire workflow, from data preprocessing to model training and evaluation, was automated using custom scripts.

4.4 Dataset Details

The dataset used in this study originates from the PlantVillage repository, comprising over 50,000 high-quality images of plant leaves. It encompasses 38 distinct categories, showcasing both healthy and diseased leaves from various plant species like apples, grapes, corn, and tomatoes. The data was divided into training (70%), validation (15%), and testing (15%) sets. Images were resized to dimensions of 224 × 224 pixels and prepared according to the procedures described in the methodology section. Figure 2 presents some examples from the dataset.

4.5 Training and Validation Accuracy for Different Hyperparameter Settings

The proposed model was trained and validated under different hyperparameter settings to determine the optimal configuration. The following table demonstrates the training and validation accuracies for various learning rates and batch sizes, with the best results highlighted.

Table 1 illustrates how different learning rates affect the model's performance, as well as the influence of batch sizes on both the training and validation accuracies of the CNN models when integrated with the gradient boosting meta-learner. A 0.001 learning rate with a 32 batch size yielded the highest validation accuracy at 99.56%, indicating that this configuration strikes the best balance between learning speed and model generalization. Conversely, reducing the learning rate to 0.0001, even with the same batch size, resulted in lower training and validation accuracies, which suggests that the model may have converged more slowly or potentially underfitted the data. These findings emphasize the critical role of hyperparameter optimization in improving the predictive accuracy of DL models for detecting plant diseases.

4.6 Classification Report

The classification report for the top-performing model on the test set is presented in Figs. 3, 4, and 5. The classification metrics in the provided graphs offer valuable details about the model's effectiveness for different plant diseases. Precision in Fig. 3 shows remarkably high accuracy across most classes, with several diseases such as "Apple—Black Rot" and "Grape—Leaf Blight" achieving a perfect score of 100%, indicating that nearly all positive predictions are correct. Recall in Fig. 4 is also high but slightly lower for certain classes like "Corn—Cercospora Leaf Spot" (88%) and "Tomato—Late Blight" (99%). This implies that although the model is very accurate, it occasionally misses some true positive cases in these categories. The F1-score in Fig. 5, which balances precision and recall, further highlights the model's robustness, with scores close to or at 100% for most classes, reflecting the model's overall effectiveness. The model's overall accuracy is impressive, with a 100% precision and F1-score, and a 99% recall, indicating it performs exceptionally well in both identifying diseases and minimizing false predictions.

4.7 Comparison with Existing Work

The comparison of multiple DL models shown in Table 2 utilizing deep stacking techniques for plant disease detection demonstrates significant advancements in accuracy across different datasets. Enkvetchakul et al. [21] employed a Stacking Ensemble model combining EfficientNetB1 with Random Forest, achieving accuracies of 90.71% on the Crop-PlantDoc dataset and 91.87% on iCassava2019, highlighting the effectiveness of stacking ensembles in diverse environments. Jha et al. [24] implemented a stacking ensemble with transfer learning on potato leaf images, resulting in a high accuracy of 98.86%, showing the power of transfer learning when integrated with deep stacking.

Yilma et al. [25] utilized Attention Augmented Residual (AAR) deep stacking residual modules, reaching an impressive 99.56% accuracy on plant disease images, underscoring the benefits of attention mechanisms in improving model performance. Asad et al. [22] developed a diverse data ensemble with deep stacking, achieving a mean Intersection Over Union (mIOU) of 90.40% and an Intersection Over Union (IOU) of 81.59% on crop and weed images, which illustrates the model's robustness in handling varied agricultural data. Our proposed model, which integrates Convolutional Neural Networks (CNN) with gradient boosting through deep stacking, achieved a leading accuracy of 99.56% on the PlantVillage dataset, demonstrating the superior performance of this hybrid approach in plant disease detection tasks. This comparison highlights the key role of deep stacking techniques in advancing the precision and effectiveness of plant disease detection models.

5 Discussion

The primary goal of this study was to develop a reliable and effective plant leaf disease detection system using a deep stacking framework that combines CNNs as base learners with gradient boosting as a meta-learner. The experimental results demonstrate the efficacy of this approach, achieving an impressive overall accuracy of 99.57% on the test set. The results consistently improve classification accuracy across different plant species and disease categories, outperforming several cutting-edge methods. These findings validate the hypothesis that integrating DL and traditional ML models through stacking can enhance the generalization and robustness of plant disease detection systems.

Fig. 2 Dataset examples

Table 1 Performance metrics for various hyperparameter configurations

Model	LR	Batch size	Training accuracy (%)	Val. accuracy (%)
CNN + GB	0.001	32	99.57	99.56
CNN + GB	0.001	64	98.23	98.50
CNN + GB	0.0005	32	98.45	98.60
CNN + GB	0.0001	32	97.80	97.92
CNN + GB	0.0001	64	96.75	96.88

The success of the deep stacking approach in this study can be attributed to several key factors. First, the CNNs used as base learners capture diverse features from the input images, each model specializing in different aspects of the data. For instance, VGG16 excels at learning fine-grained features, ResNet50 effectively handles deeper feature representations through residual connections, and InceptionV3 captures multi-scale features using inception modules. Combining these complementary strengths through a meta-learner like gradient boosting benefits the final model from a richer feature set, leading to improved classification accuracy.

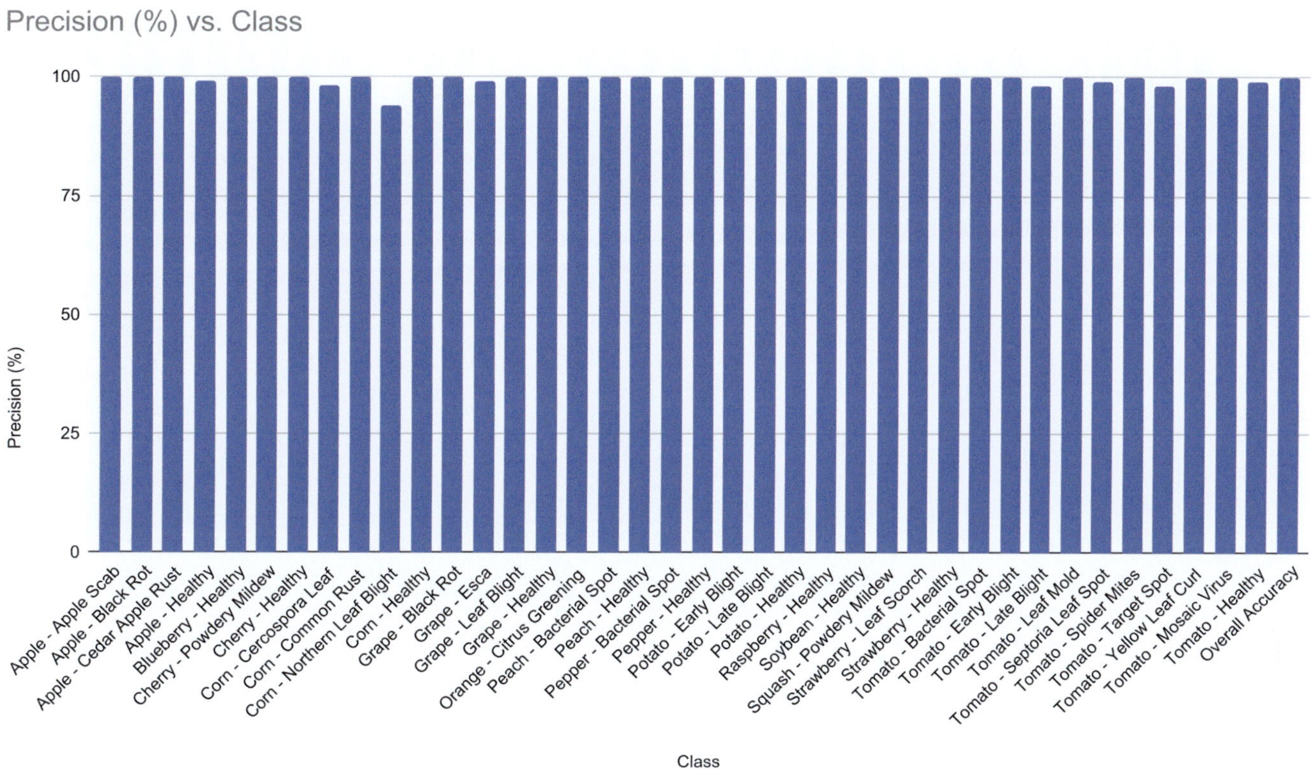

Fig. 3 Precision of the model according to the dataset

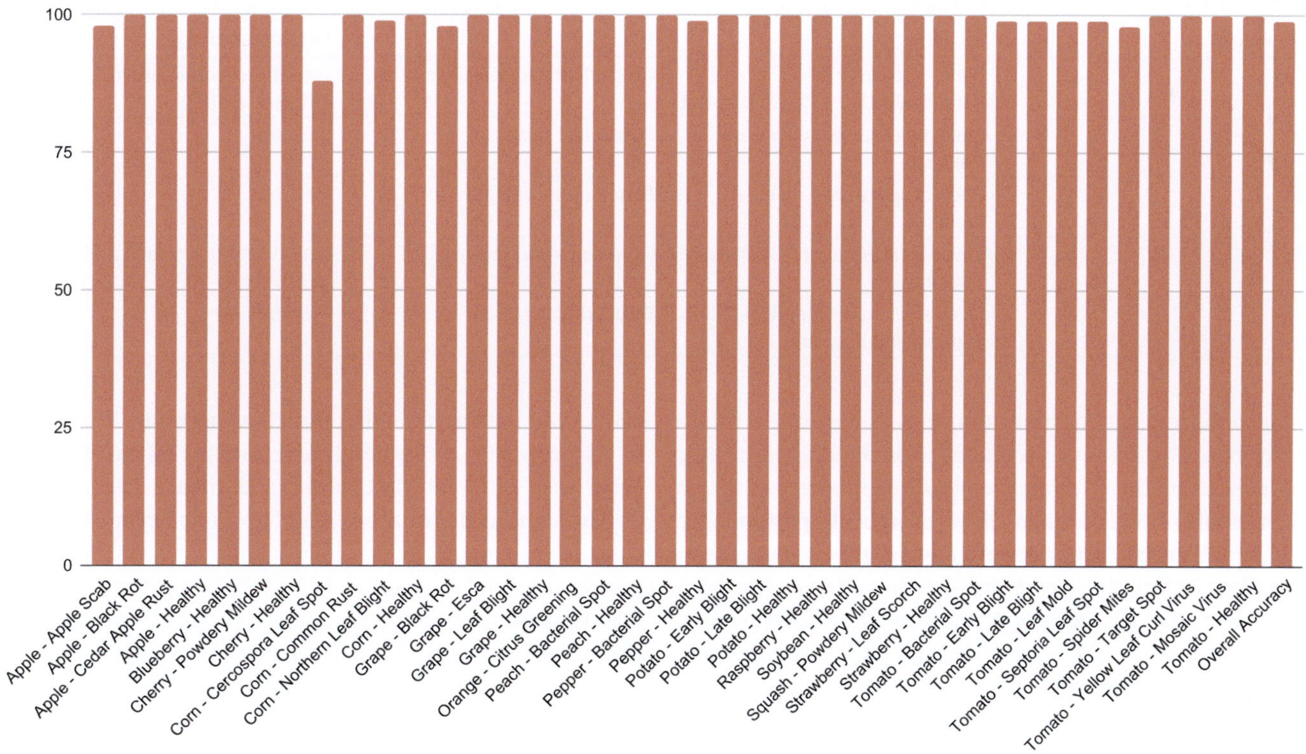

Fig. 4 Recall of the model according to the dataset

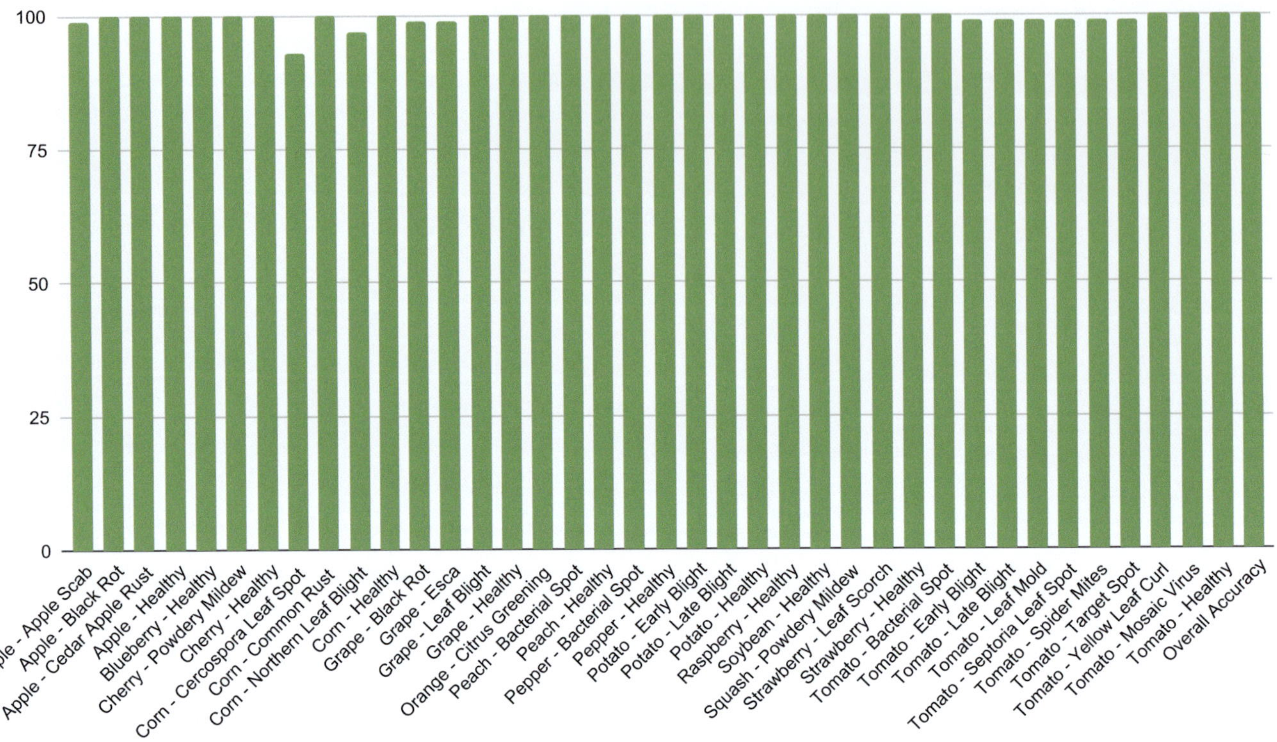

Fig. 5 F1-score of the model according to the dataset

Table 2 State-of-the-art results from five papers using deep stacking

References	Model	Dataset	Result (accuracy)
[21]	Stacking ensemble (EfficientNetB1 + Random Forest)	Crop-PlantDoc, iCassava2019	90.71% (Crop-PlantDoc), 91.87% (iCassava2019)
[24]	Stacking ensemble with transfer learning	Potato leaf images	98.86%
[25]	AAR deep stacking residual modules	Plant disease images	99.56%
[22]	Diverse data ensemble with deep stacking	Crop and weed images	mIOU of 90.40% and an IOU of 81.59%
Ours	Deep stacking integrating CNN and gradient boosting	PlantVillage dataset	99.56%

Second, gradient boosting, as a meta-learner, offers robust performance by iteratively refining the predictions of the base learners. Its ability to minimize classification errors through an ensemble of decision trees allows the model's performance in handling unseen data more effectively. This is particularly important in agricultural settings where variations in lighting, angles, and leaf conditions can introduce noise into the data. The findings of this study suggest that the deep stacking approach is more resilient to such variations, resulting in higher accuracy compared to single-model architectures.

A key advantage of our approach is its capacity to strike a balance between bias and variance, outperforming existing methods. While single CNN models may suffer from high variance due to their complexity, the deep stacking method effectively reduces variance by integrating multiple models and leveraging the strengths of gradient boosting. This results in a more stable and generalizable model that excels across varieties of different plant species and disease categories.

The significance of this research extends beyond the technical advancements in machine learning. From a practical perspective, the proposed plant leaf disease detection system could significantly impact agriculture. Timely detection of plant diseases is important for minimizing crop losses and safeguarding food security, particularly in areas with limited access to expert diagnostics. By offering an automated and precise approach to disease detection, this study advances scalable solutions that can be implemented in real-world agricultural settings.

Furthermore, the use of a deep stacking framework opens new avenues for future research in other domains where multi-modal data or complex classification tasks are involved. The flexibility of the stacking approach allows for the integration of various base models, not limited to CNNs, and can be extended to other applications such as medical image analysis, environmental monitoring, and industrial inspection.

While the results of this study are promising, several challenges and limitations should be acknowledged. The main challenges encountered during the research were the computational cost associated with training multiple CNN models and the meta-learner. The deep stacking framework requires substantial computational requirements, particularly when handling large-scale datasets and complex model architectures. This may limit the accessibility of the proposed method for small-scale or resource-constrained applications. Future research could explore methods to optimize the training process, such as model pruning or quantization, to reduce the computational burden.

Another limitation is the reliance on high-quality, labeled datasets. The success of the proposed method relies significantly on the presence of extensive, accurately labeled datasets such as PlantVillage. In practice, obtaining such datasets may not always be feasible, especially in regions where data collection infrastructure is lacking. A possible solution to this limitation could be to investigate semi-supervised or unsupervised learning techniques to lessen the reliance on labeled data.

The findings of this study suggest several possible avenues for future research. One possibility is to explore the incorporation of additional data types, like hyperspectral or multispectral imaging, into the deep stacking framework. These imaging methods can provide more intricate details about plant health and disease signs, which could enhance the model's accuracy.

An alternative approach is to explore the use of transfer learning (TL) to improve the performance of base CNN models. TL allows models that have been trained on large, general datasets to be adapted to smaller, domain-specific datasets, reducing the requirement for large amounts of labeled training data and computational power.

Additionally, future research could explore real-time implementation of the proposed method in field conditions. Deploying the model on mobile devices or drones equipped with cameras could enable on-the-fly disease detection, providing farmers with immediate feedback on crop health. This would require optimizing the model for efficiency and robustness in dynamic environments.

Finally, exploring the interpretability and explainability of the model's predictions is another critical direction for future work. Understanding the decision-making process of the model is essential for gaining the trust of end-users, particularly in high-stakes applications such as agriculture. Techniques such as saliency maps or Grad-CAM could be employed to visualize which parts of the image the model is focusing on when making a prediction.

To sum up, this research has highlighted the effectiveness of a deep stacking method for detecting plant leaf diseases, delivering top-tier performance, and providing practical solutions to address real-world agricultural issues. The integration of CNNs and gradient boosting as base and meta-learners, respectively, has proven to be a powerful combination, enhancing the robustness and generalization of the model. While challenges remain, the contributions of this research lay the foundation for future advancements in automated plant disease detection and beyond.

6 Conclusions

This paper proposed an advanced deep stacking framework that combines CNNs as base learners with gradient boosting as a meta-learner for plant leaf disease detection. The experimental results indicated that this approach reached an overall accuracy of 99.57%, surpassing current leading methods. The integration of multiple CNN architectures and the robust classification capabilities of gradient boosting allowed the model to effectively generalize across a diverse set of plant species and disease categories. This research not only advances the field of automated detection of plant disease but also practical implications for improving agricultural productivity and food security. Future research could focus on optimizing this framework for real-time applications and broadening its use to other areas that involve complex image classification tasks.

References

1. Strange, R.N., Scott, P.R.: Plant disease: a threat to global food security. Annu. Rev. Phytopathol. **43**(1), 83–116 (2005)
2. John, M.A., Bankole, I., Ajayi-Moses, O., Ijila, T., Jeje, O., Lalit, P.: Relevance of advanced plant disease detection techniques in disease and pest management for ensuring food security and their implication: a review. Am. J. Plant Sci. **14**(11), 1260–1295 (2023)
3. Bock, C., Poole, G., Parker, P., Gottwald, T.: Plant disease severity estimated visually, by digital photography and image analysis, and by hyperspectral imaging. Crit. Rev. Plant Sci. **29**(2), 59–107 (2010)
4. Sladojevic, S., Arsenovic, M., Anderla, A., Culibrk, D., Stefanovic, D.: Deep neural networks based recognition of plant diseases by leaf image classification. Comput. Intell. Neurosci. **2016**(1), 3289801 (2016)
5. Amara, J., Bouaziz, B., Algergawy, A.: A deep learning-based approach for banana leaf diseases classification (2017)
6. Kunapuli, G.: Ensemble Methods for Machine Learning. Simon and Schuster (2023)
7. Ganaie, M.A., Hu, M., Malik, A.K., Tanveer, M., Suganthan, P.N.: Ensemble deep learning: a review. Eng. Appl. Artif. Intell. **115**, 105151 (2022)
8. Low, C.-Y., Park, J., Teoh, A.B.-J.: Stacking-based deep neural network: deep analytic network for pattern classification. IEEE Trans. Cybern. **50**(12), 5021–5034 (2019)

9. Chen, T., Guestrin, C.: Xgboost: a scalable tree boosting system. In: Proceedings of the 22nd ACM Sigkdd International Conference on Knowledge Discovery and Data Mining, pp. 785–794 (2016)
10. Ke, G., Meng, Q., Finley, T., Wang, T., Chen, W., Ma, W., Ye, Q., Liu, T.-Y.: Lightgbm: a highly efficient gradient boosting decision tree. In: Advances in Neural Information Processing Systems, vol. 30 (2017)
11. Anam, I., Arafat, N., Hafiz, M.S., Jim, J.R., Kabir, M.M., Mridha, M.: A systematic review of UAV and AI integration for targeted disease detection, weed management, and pest control in precision agriculture. Smart Agric. Technol. 100647 (2024)
12. Das, D., Singh, M., Mohanty, S.S., Chakravarty, S.: Leaf disease detection using support vector machine. In: 2020 International Conference on Communication and Signal Processing (ICCSP), pp. 1036–1040 (2020). IEEE
13. Hatuwal, B.K., Shakya, A., Joshi, B.: Plant leaf disease recognition using random forest, KNN, SVM and CNN. Polibits **62**, 13–19 (2020)
14. Padol, P.B., Yadav, A.A.: SVM classifier based grape leaf disease detection. In: 2016 Conference on Advances in Signal Processing (CASP), pp. 175–179. IEEE (2016)
15. Pothen, M.E., Pai, M.L.: Detection of rice leaf diseases using image processing. In: 2020 Fourth International Conference on Computing Methodologies and Communication (ICCMC), pp. 424–430. IEEE (2020)
16. Reddy, J.N., Vinod, K., Ajai, A.R.: Analysis of classification algorithms for plant leaf disease detection. In: 2019 IEEE International Conference on Electrical, Computer and Communication Technologies (ICECCT), pp. 1–6. IEEE (2019)
17. Sujatha, R., Chatterjee, J.M., Jhanjhi, N., Brohi, S.N.: Performance of deep learning versus machine learning in plant leaf disease detection. Microprocess. Microsyst. **80**, 103615 (2021)
18. Prodeep, A.R., Hoque, A.M., Kabir, M.M., Rahman, M.S., Mridha, M.: Plant disease identification from leaf images using deep CNN's efficientnet. In: 2022 International Conference on Decision Aid Sciences and Applications (DASA), pp. 523–527. IEEE (2022)
19. Mohanty, S.P., Hughes, D.P., Salathé, M.: Using deep learning for image-based plant disease detection. Front. Plant Sci. **7**, 1419 (2016)
20. Kabir, M.M., Ohi, A.Q., Mridha, M.F.: A multi-plant disease diagnosis method using convolutional neural network. In: Computer Vision and Machine Learning in Agriculture, pp. 99–111 (2021)
21. Enkvetchakul, P., Surinta, O.: Stacking ensemble of lightweight convolutional neural networks for plant leaf disease recognition. ICIC Express Lett. **16**(5), 521–528 (2022)
22. Asad, M.H., Anwar, S., Bais, A.: Improved crop and weed detection with diverse data ensemble learning. In: Proceedings of the IEEE/CVF Conference on Computer Vision and Pattern Recognition, pp. 5336–5345 (2024)
23. Mazumder, M.K.A., Kabir, M.M., Rahman, A., Abdullah-Al-Jubair, M., Mridha, M.: Densenet201plus: Cost-effective transfer-learning architecture for rapid leaf disease identification with attention mechanisms. Heliyon (2024)
24. Jha, P., Dembla, D., Dubey, W.: Deep learning models for enhancing potato leaf disease prediction: implementation of transfer learning based stacking ensemble model. Multimedia Tools Appl. **83**(13), 37839–37858 (2024)
25. Yilma, G., Qin, Z., Assefa, M., Alemu, G., Ayalew, M.: Attention augmented convolutional neural network for fine-grained plant disease classification and visualization using stochastic sample transformations. In: Proceedings of the 5th International Conference on Advances in Image Processing, pp. 13–19 (2021)
26. Nobel, S.N., Imran, M.A., Bina, N.Z., Kabir, M.M., Safran, M., Alfarhood, S., Mridha, M.F.: Palm leaf health management: a hybrid approach for automated disease detection and therapy enhancement. IEEE Access (2024)
27. Islam, M.R., Oliullah, K., Kabir, M.M., Alom, M., Mridha, M.: Machine learning enabled IoT system for soil nutrients monitoring and crop recommendation. J. Agric. Food Res. **14**, 100880 (2023)
28. Kabir, M.M., Hafiz, M.S., Bandyopadhyaa, S., Jim, J.R., Mridha, M.: Tea leaf age quality: age-stratified tea leaf quality classification dataset. Data Brief **54**, 110462 (2024)

Revolutionizing Agriculture: Deep Learning Models for Crop Pest and Disease Analysis

Arifa Akter Eva, Tasmia Khan, Tabassoom Rahman, Ahmed Taj Noor, and Nour Elhouda Ben Saadi

Abstract

Crop diseases significantly impact agricultural production, causing substantial losses in quantity and quality. The early detection of diseases through regular monitoring is vital for minimizing these losses. Integrating advanced technologies, such as image processing, machine learning (ML), and deep learning (DL), has led to the development of innovative agriculture solutions. Researchers have utilized these techniques to create automatic crop disease identification systems, offering an innovative approach to address the challenges associated with crop health. The paper used different lightweight convolutional neural networks (CNNs) with modifications. These are good and efficient for classifying images specifically for identifying crop diseases by analyzing crop-leaf images.This cutting-edge approach promises to provide an effective and cost-effective solution for automatically categorizing crop diseases. The performance of the model is assessed using a comprehensive evaluation framework applied to the CCMT dataset that collectively encompasses various crop varieties. The dataset consists of 22 classes and 24,881 rows of images of four different types of crops. The experimental results highlight the superiority of the model over recent DL approaches in the field of crop disease, achieving an impressive accuracy of 99.73% on the VGG16 model. This research contributes to advancing automated crop pest disease categorization, offering a promising solution for practical and flexible agriculture.

A. A. Eva (✉) · T. Khan
Computer Science and Engineering, American International University Bangladesh, Dhaka, Bangladesh
e-mail: arifaeva2019@gmail.com

T. Rahman
Electrical and Computer Engineering, North South University, Dhaka, Bangladesh

A. T. Noor
Computer Science and Engineering, Southeast University, Dhaka, Bangladesh

N. E. Ben Saadi
Faculty of Informatics, Eötvös Loránd University, Budapest, Hungary

Keywords

Crop pest disease categorization • Deep learning • Agriculture • Convolutional neural networks (CNNs) • Image classification

1 Introduction

Agriculture is the fundamental element in the economies of numerous developing nations, serving as both the primary source of sustenance and a crucial revenue generator. This essential sector supports approximately 61% of the global population, representing a vital aspect of livelihoods and subsistence for millions around the world [1]. Within the agricultural field of Bangladesh, tomatoes, maize, cassava, etc., emerge as pivotal and commercially significant crops. The majority of the people rely on agriculture for money and survival. Notably, tomatoes are the fourth most extensively cultivated vegetable in the country, with approximately 4.26 lakh tons of tomatoes being produced in 2023 [2]. The economic prowess of Bangladesh receives a substantial boost through tomato cultivation, fostering positive financial outcomes [2]. Notably, the year 2021 witnessed tomato exports from Bangladesh amounting to approximately one hundred million dollars, underscoring the economic significance of this agricultural sector. Millions of individuals find employment within this growing domain, further emphasizing its role in sustaining livelihoods [3].

Crop health is a crucial component of agricultural output, and the agriculture industry is essential to maintaining global food security [4]. Diseases and pests affecting crops pose severe risks and cause global production losses. In the field of addressing crop disease identification, researchers have

M. F. Mridha and N. Dey (eds.), *Machine Vision in Plant Leaf Disease Detection for Sustainable Agriculture*, Studies in Computational Intelligence 1202, https://doi.org/10.1007/978-981-96-4520-6_3

proposed various solutions by applying image processing, machine learning (ML), and deep learning (DL) methodologies [5]. Recent advancements primarily lean toward ML and mostly DL, showcasing impressive performance attributed to the abundance of data, enhanced computing resources, and the sophisticated learning capabilities of both algorithms [6, 7]. Noteworthy among the various approaches are Support Vector Machines (SVM) [8], k-nearest Neighbors (K-NN) [9], Multilayer Layer Perception (MLP) [10], Fisher Linear Discriminant (FLD) [11], and Random Forest (RF) classifiers [12], which have emerged significantly in the field of crop disease identification. Further, the landscape of computer vision has witnessed a transformative impact with the growth of DL, proving to be an effective tool across a spectrum of challenges. Deep Convolutional Neural Networks (DCNNs) [13], Transfer Learning [14], and CNN [15] have demonstrated remarkable efficacy in the field of crop disease identification and categorization.

Deploying DCNN models for real-time, on-the-spot crop disease categorization using resource-constrained devices presents significant difficulties. The conventional deep CNN architectures, characterized by their substantial memory requirements and intensive computational demands, may present challenges in this context. Given this, there is an increasing interest in researching lightweight CNN architectures with changed layers, with the goal of striking a compromise between computing economy and keeping a comparable level of precision in illness categorization. Recently, CNN has developed some lightweight models that are good for image categorization and other computer vision issues. EfficientNet [16], InceptionNet [17], ShuffleNet [18], GhostNet [19], MobileNet [20], and ResNet [21] are lightweight and efficient models for image categorization. Despite their overall effectiveness, existing models have shown inconsistent performance across diverse crop disease datasets. In this research article, we are using the CNN approaches for classifying diseases by analyzing crop and leaf images. This method aims to assist farmers in identifying crop diseases and leaf issues independently, eliminating the need to consult plant expertise. By enabling timely treatment of crop diseases, our approach is expected to enhance both the quality and quantity of crop yields, ultimately boosting the profits of the farmers. For our experiments, we utilized the CCMT-crop pest diseases dataset. Using this dataset, we developed CNNs model to classify the images. We evaluated the performance using various metrics, including different accuracies and the number of trainable parameters compared to a pre-trained model.

Our models evaluate the performance of recently introduced DL models across the CCMT publicly available dataset. The current work brings forth several noteworthy contributions, outlined as follows:

- This thorough examination explores CNN models, offering a comprehensive exposition of the operational methodologies. The analysis provides depth clarification of the inherent strengths and weaknesses associated with each model, providing valuable knowledge of their respective capabilities and limitations.
- The assessment of the performance of the models is carried out on widely recognized agricultural datasets, with a particular focus on the CCMT dataset. This precise evaluation measures their effectiveness and accuracy in pattern recognition tasks, supported by the presentation of crucial graphs to substantiate the findings.
- Developed automatic crop disease identification systems reflect a commitment to smart agriculture solutions. These systems can enhance efficiency, reduce labor-intensive processes, and provide timely responses to disease outbreaks, contributing to improved overall agricultural productivity.

The paper is structured as follows: Section 2 goes into recent research findings on the topic, providing a comprehensive overview. In Sect. 3, the proposed method for model development is elaborated upon in detail. The outcomes of the proposed models and the experiment of our research are presented in Sect. 4. The study concludes with significant findings summarized in Sect. 5.

2 Related Work

A multitude of agricultural [22], food security [23], economic [24], and other aspects depend on the accurate evaluation of crops. Farmers can reduce the effects of a disease by acting immediately and appropriately when the condition is discovered early. The diagnosis of a crop disease can help with early classification, identification and invention, crop productivity and quality, food security, economic impact, environmental impact, global and market access, and technical innovation [25]. Prompt action can minimize agricultural losses and prevent the disease from spreading [26]. A disease can significantly reduce both agricultural yield and the quality of produced items. By protecting their crops early on, farmers may boost yields and enhance the quality of the produce they grow. An extensive examination of the existing literature highlights that the automated categorization of crop diseases and pests is predominantly conducted through the application of DL-based techniques. The subsequent sections provide an in-depth review of recent works on this topic.

Mondal et al. utilized DL to identify tomato leaf diseases. They used convolutional recurrent neural network (CRNN) with a gated-recurrent unit employed for tomato leaves, achieving a 99.62% success rate, while transfer learning was applied for rice leaves, resulting in a 91.63% success rate [27]. Meena et al. introduced a methodology for crop pest, weed, and disease detection using DL. They used a UAV-captured image dataset that contained 400 types of segmented weed occurrences. The fine-tuned Inception-V3 model achieved the highest accuracy of 87.85% in classifying pests, weeds, and diseases [28]. Further, Ullah et al. presented DeepPestNet, a novel DL model achieving 100% accuracy in recognizing and classifying crop datasets. They used a Pest Dataset, and it attained 98.92% accuracy. DeepPestNet offers potential aid to specialists and farmers, reducing economic losses [29]. Moreover, Sanghavi et al. proposed a Hunger Games search-based deep CNN (HGS-DCNN) model to efficiently detect crop pests. Their proposed model achieved impressive accuracy, precision, F-score, sensitivity, and specificity of 99.1%, 97.80%, 97.80%, 97.82%, and 99.43%, respectively [30]. Additionally, Kasinathan et al. addressed challenges in crop insect detection, proposing an algorithm using CNN for Wang, Xie, Deng, and IP102 datasets. They achieved a high identification rate of 91.5 and 90% for 9 and 24 insect classes, respectively. The algorithm enhances detection performance with reduced computation time and it demonstrates significant improvements in accuracy and efficiency, offering potential applications in early-stage insect recognition for improved crop yield and quality in agriculture [31].

Islam et al. developed CNN-based models CNN, VGG16, VGG19, and ResNet50. They achieved high accuracies of 98.60%, 92.39%, 96.15%, and 98.98%, respectively. ResNet50 was chosen as a web application for real-time prediction. They contributed to reducing disease impact, using DL for identification, and suggesting an efficient framework with ResNet50 at its core for crop disease detection [32]. Malathi et al. applied deep CNN and a transfer learning for classifying pests in paddy crops. They used pest images dataset of 3549 and constructed a neural model using various DCNN architectures. The resulting fine-tuned ResNet50 model achieved an impressive accuracy of 95.012%, outperforming other models [33]. Further, Wang et al. introduced a DL-based model for crop disease and pest identification, aiming to address the challenges of slow manual identification. AlexNet model used for experimental parameters, and the results indicate an impressive average recognition accuracy of 96.26%. They contributed significantly to crop disease and pest identification, showcasing high accuracy and efficiency through DL techniques [34]. Moreover, Turkoglu et al. introduced two classification models for detecting plant diseases and pests that use deep feature extraction from pre-trained CNNs. Trained on a dataset featuring 15 classes and 4,447 images, the models achieved a high accuracy of 97.56% [35]. Xin et al. proposed a DCNN-G model, leveraging DL and Google data fusion for precise crop pest and disease identification, and achieved an impressive 95% recognition accuracy. The integration of loss functions enhances boundary quality and segmentation, offering a novel method for assessing image quality in this context [36].

Nandhini et al. addressed the impact of diseases on plantain tree cultivation, emphasizing the global importance of bananas. They proposed a novel gated-recurrent CNN (G-RecConNN) model for early disease detection, combining CNN and RNN and used real-time datasets from Tamil Nadu, India, are used for experiments, showcasing advantages such as reduced data preprocessing and online performance evaluation [42]. Thakur et al. proposed a lightweight CNN, VGG-ICNN, for disease identification of crop using plant-leaf images. They achieved 99.16% accuracy on the PlantVillage dataset and consistent performance across five datasets. They offered a promising solution for smart agriculture by enabling early disease detection and reducing crop losses [43]. Kurmi et al. used a CNN model for crop disease detection using leaf images, incorporating a novel color-based localization method. Their model outperformed existing methods with a mean accuracy of 95.35% and contributes to precision agriculture by employing artificial intelligence for visual feature extraction and classification [44]. Additionally, Chodey et al. implemented a hybrid DL model for real-time pest detection in plants. They include denoising, image enhancement, semantic segmentation with ResNet, and pest detection using CNN. They demonstrated suitability for real-time monitoring and detection, and results show high effectiveness with a structural similarity index measure (SSIM) of 99%, mean absolute error (MAE) below 2%, and average precision (AP) of 89.67% [45].

According to the survey, a significant section of the global population relies on agriculture for their food. Because they can reduce the availability of essential crops, crop diseases have the potential to compromise food security [46]. Early detection and categorization help to maintain a stable and secure food supply. For farmers and the agricultural sector as a whole, crop infections can have catastrophic financial effects [47]. Using a subset of CNN and DL architecture, several ML models are utilized to diagnose agricultural illnesses, including AlexNet, GoogleNet, ResNet-101, CNN, DCNN, and lightCNN models, which have been frequently used in previous years. Table 1 presents more existing work on crop pest diseases.

3 Methodology

In this chapter, the implemented methodology for our work briefly. Firstly we collected the dataset then we preprocessed the dataset and split it into three folders: train, validate, and test. Based on the dataset, we selected three suitable and appropriate DL-based CNN models. In addition, pre-trained

Table 1 The following table presents more existing work on crop pest diseases

References	Description
Sourav et al. [37]	Introduced an intelligent model utilizing transfer learning and DCNNs for accurate identification of jute pests. They achieved a high accuracy of 95.86% on four major pest classes. Additionally, their model was integrated into practical Android and iOS applications for widespread use
Adem et al. [38]	Employed faster R-CNN, SSD, VGG16, and YOLOv4 DL models, along with a hybrid approach integrating YOLOv4 with image processing, to automatically detect and classify leaf spot disease severity in sugar beet plants and achieved a classification accuracy of 96.47%, outperforming individual DL models. Their approach reduced diagnosis time, eliminated human error, and facilitated timely pesticide application
Parthiban et al. [39]	Focused on detecting diseases in citrus and tomato plants using a CNN. With data augmentation, the model is trained on 18,392 images, achieving a high validation accuracy of 97.02%. Their approach offered promise for accurate disease identification, potentially benefiting real-time agricultural systems and reducing economic losses for farmers
Usha Ruby et al. [40]	Introduced a DL approach for the precise classification of wheat leaf diseases by modifying the ResNet50 architecture with additional layers, their proposed method achieved superior performance compared to other models, with an identification accuracy of 98.44%
Singh et al. [41]	Presented an optimized DL approach for leaf type classification and disease detection in agriculture. They employed anisotropic filtering and Mask R-CNN for preprocessing and segmentation, respectively. SqueezeNet with adaptive snake optimizer (ASO) achieved high accuracy in leaf type classification 92.4%, while deep Qnet with ASO performed well in disease detection 91.9%

models including Xception, ResNet50, VGG16, and others, Inception-V3 and ResNet50 are particularly popular for transfer learning implementations, and their structures and speed make them ideal for this purpose. For the paper, the first model we used for training and testing was VGG16, the second one was Inception-V3, and the final model we implemented was ResNet50. We used these three models VGG16, Inception-V3, and ResNet50 for their strong performance in image classification and more efficient than other models. The working method of our study is shown in Fig. 1. The process begins with data collection, followed by partitioning into two segments: typically, 80% for training and 20% for validation. Performance metrics are then employed for image classification (identifying specific crop diseases and pests), and, finally, visualization techniques are utilized for image classification.

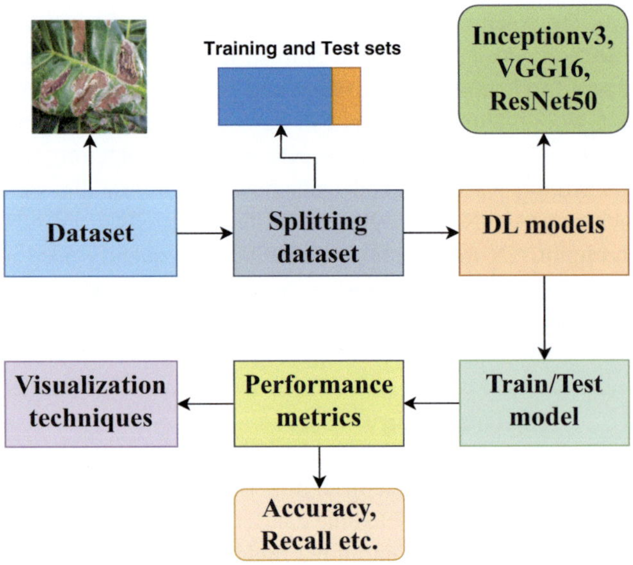

Fig. 1 The figure presents workflow of crop pest disease classification using DL models

3.1 Description of Our Training Models

3.1.1 Inception-V3

Inception-V3 is a classification-based CNN model. It is a part of the DL inception series of Google. A Google team of researchers created Inception-V3. Szeged et al. were key contributors to the Inception-V3 architecture. They detailed the changes and enhancements introduced into the Inception-V3 model in 2015 [17].

The primary uses of InceptionNets are classification, identification, and feature extraction. Because of the effectiveness of its architecture, it is frequently used in transfer learning applications to adapt and fine-tune pre-trained models for particular image-related tasks, making it an increasingly common choice in computer vision applications. Mainly InceptionNet has four different types and they are different because of their parameters and computational efficiency. The architectures include Inception-V1 [48], Inception-V2 [49], Inception-V3 [17], Inception-V4 [50], and Inception-ResNet [50]. Inception-V1 is mainly popular as GoogLeNet [48].

The Base Architecture of Inception-V3 Model

- Input Layer: Image data as input.
- Initial Convolutional Layers: Several convolutional layers are utilized at the start to extract low-level information from the input image.
- Inception Modules: The main of the Inception-V3 architecture is made up of numerous Inception modules stacked

together. Each Inception module consists of convolutional layers with various filter sizes (1 × 1, 3 × 3, 5 × 5) and pooling procedures.
- Global Average Pooling Layer: It is applied to the output feature maps to reduce spatial dimensions while maintaining critical information.
- Fully Connected Layers: The output of the global average pooling layer is flattened into a vector and routed via one or more fully linked layers.
- Softmax Layer and Final layer: The last layer is a softmax layer, which generates class probabilities based on the outputs of the fully connected layers. The softmax function normalizes the output scores into probabilities that represent the likelihood of each class.

The Customized Architecture of Inception-V3 Model

In our research, we employed the Inception-V3 architecture, which is divided into two primary sections: feature extraction and classification. However, we made adaptations to suit our specific task of MRI brain tumor image classification into two classes: healthy and tumor cases. The way we customized Inception-V3 for our work is shown in Fig. 2:

- Firstly, we removed the fully connected layers from the original Inception-V3 model and replaced them with a customized fully connected part tailored to our two classes.
- The convolutional part of the model, which remains unchanged from the original Inception-V3, consists of five convolutional layers, each followed by batch normalization, 2 pooling layers, and 11 inception modules.
- Then for the fully connected layer, we used a flattened layer, dense layer, and dropout layer to enhance the robustness of the model.
- Finally, we used the softmax layer to generate normalized probabilities for the output classes: "healthy" or "infected crop."

The advantages and disadvantages of the Inception-V3 model are

- **Advantage**: Several major advantages distinguish Inception-V3 for its novel inception modules, which are meant to capture a wide range of information at varied scales. Its multi-branch architecture enables fast information aggregation via concurrent convolutions of varying sizes, allowing the network to learn both fine-grained and high-level characteristics at the same time. This design decision improves its ability to recognize subtle patterns in photos while retaining computing efficiency.
- **Disadvantage**: While Inception-V3 is a powerful architecture for computer vision tasks, its complexity poses issues, resulting in greater computing demands during training and inference. Although its sophisticated modular design is excellent for recording varied features at multiple dimensions, it can be difficult to comprehend or adjust for specific applications without significant experience. Furthermore, the presence of numerous parallel routes within the architecture may result in increased memory usage and optimization complexity when compared to simpler architectures such as VGG or ResNet.

3.1.2 VGG16

The VGG16 is a CNN design proposed by the University of Oxford's visual graphics group (VGG). Simonyan et al. sig-

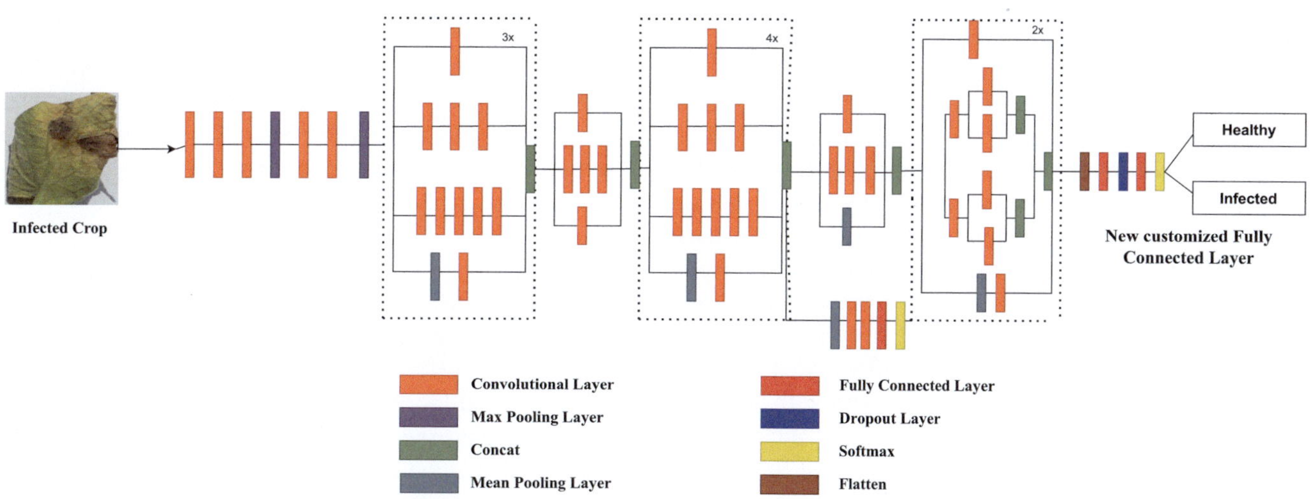

Fig. 2 The figure presents the customized architecture of the Inception-V3 model

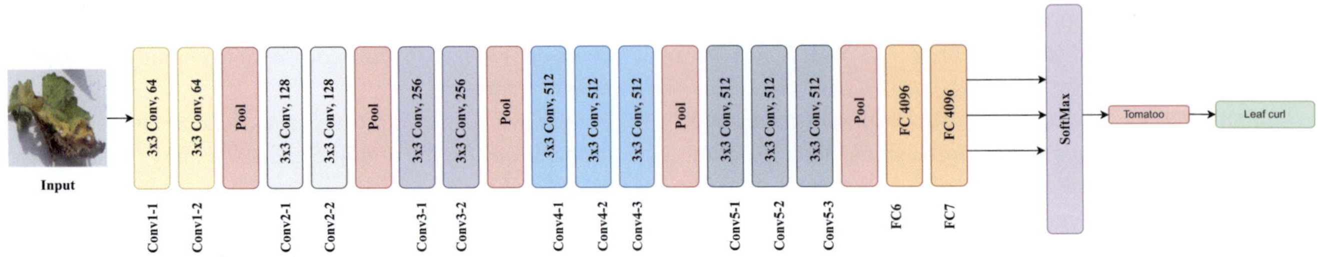

Fig. 3 The figure presents the customized architecture of VGG16 model

nificantly contributed to the VGG network in 2014 [51]. It is popular for its simplicity, interpretability, and high accuracy.

VGG16 is largely used in computer vision for image categorization problems. Its deep convolutional architecture improves at accurately recognizing and categorizing items within images. VGG16 is extensively used by researchers for a pre-trained model for transfer learning, where its learned features are used in a variety of applications such as object detection [52], image segmentation [53], and feature extraction [53] in neural networks [54]. Because of its simplicity and effectiveness, it has become a standard for studying and experimenting with deep convolutional networks in image recognition and classification applications.

VGG is available in various depths or layer configurations, and some typical versions are (1) VGG11: 11 convolutional layers. (2) VGG13: 13 convolutional layers. (3) VGG16: The most popular VGG model, with 16 convolutional layers. (4) VGG19: VGG19 is comparable to VGG16, except it contains 19 convolutional layers. It is significantly more precise than VGG16, but slightly more computationally expensive. (5) VGG-A to E: These are several VGG network configurations, with A being the shallowest and E being the deepest. They are less often used than VGG16 and VGG19, but they can be effective for jobs requiring a different combination of accuracy and processing efficiency.

The Base Architecture of VGG16 Model

VGGNet, with 138 million parameters, can be difficult to manage. Transfer learning can handle this by altering pre-trained parameters to improve accuracy. VGG16 consists of 13 convolutional layers and three fully linked layers, beginning with input dimensions of 224 × 224 × 3. It is divided into five blocks, each including convolutional layers followed by max-pooling layers for downsampling. The fully connected layers are made up of two layers of 4096 neurons with ReLU activation, followed by a final layer of 1000 neurons representing ImageNet classes with softmax activation. VGG16's strength is its consistent architecture and simple but effective layout, making it useful for picture classification tasks.

The Customized Architecture of VGG16 Model

The way we employed the VGG16 architecture is shown in Fig. 3.

- Firstly, we removed the fully connected layers from the original VGG16 model and replaced them with a fully connected layer of 1024 neurons long.
- Then we used the softmax layer for our data classification of two classes.

The advantages and disadvantages of the VGG16 model are

- **Advantage**: VGG16 is a straightforward design, built with stacked 3 × 3 convolutional layers, making it popular and easy to understand and train. Its depth allowed it to capture intricate image features, resulting in strong classification performance. While newer models such as ResNet and Inception surpassed it in accuracy and efficiency with innovations such as skip connections, VGG16 remains a crucial baseline for CNN development in image recognition tasks.
- **Disadvantage**: VGG16's deep architecture and large number of parameters result in a high computational cost and memory requirement, making it less suitable for deployment on resource-constrained devices or real-time applications. Furthermore, its uniform architecture lacks explicit methods like skip connections, making training deeper networks more difficult and limiting its ability to achieve state-of-the-art performance when compared to more recent architectures such as ResNet or EfficientNet.

3.1.3 ResNet50

ResNet50 is a subset of the residual network family of CNN architectures. It was first proposed in 2015 by He et al. ResNet50 has 50 layers and is noted for its deep structure, which aids in constructing very deep networks while avoiding the vanishing gradient problem [21].

The ResNet50 neural network is mostly used for image identification, categorization, and computer vision tasks. Its

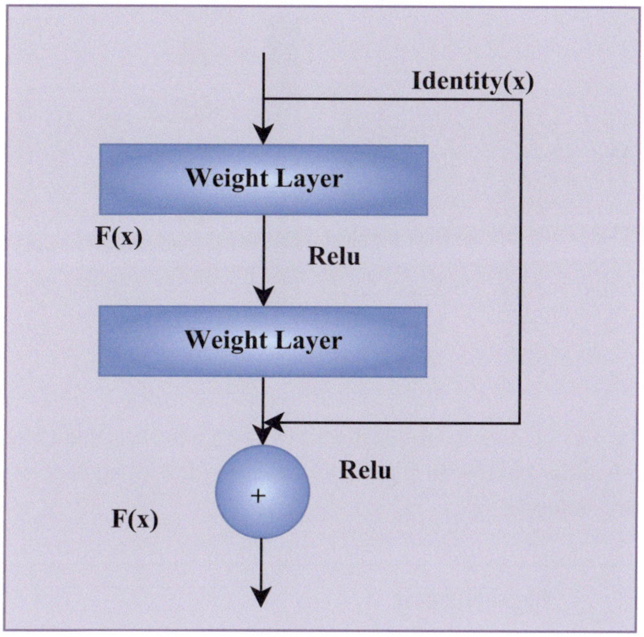

Fig. 4 The figure presents residual learning block and skip connections

deep design with residual connections allows it to do highly accurate picture classification, object detection, and image segmentation [55]. With 50 layers, it balances computational efficiency and depth, making it appropriate for a variety of real-world uses. ResNet50 has been widely used in a variety of industries for visual data processing tasks, autonomous vehicles for scene understanding, healthcare for diagnostic analysis, and other fields due to its proven performance in large-scale image datasets like ImageNet and its capacity to extract complex features from images.

ResNet is available in a variety of depths or layer configurations, as well as some typical versions are (1) ResNet-18 is a smaller version compared to ResNet50, consisting of 18 layers. It provides a lighter architecture while keeping some of the advantages of residual connections. (2) ResNet-101 and ResNet-152: These are ResNet variants with 101 and 152 layers, respectively. They provide better performance than ResNet50 but have a higher computational complexity. (3) ResNeXt: This is a version of ResNet that emphasizes a more modular architecture by substituting the conventional construction block with a cardinality parameter, offering greater efficiency by aggregating a group of parallel paths [56]. (4) Wide Residual Networks (WRN): WRNs attempt to enhance performance by increasing the network's breadth (number of channels) rather than its depth, resulting in improved generalization [57].

The Base Architecture of ResNet50 Model

ResNet50's architecture consists of convolutional layers, identity blocks, convolutional blocks, and fully connected layers. Then convolutional layers extract features including edges and textures. Identity blocks process input and add it back, facilitating optimal mapping. Bottleneck blocks reduce filter numbers before applying convolution. Fully connected layers perform the final classification, with softmax activation yielding class probabilities. The overall structure of ResNet50 is shown in Fig. 5. Skip connections in ResNet50 address vanishing gradients in deep neural networks, allowing the network to learn more effectively in deeper architectures. Figure 4 presents the skip connections of ResNet.

The advantages and disadvantages of the ResNet50 model are

- **Advantage**: ResNet50s architecture, featuring residual connections, solves the vanishing gradient problem, enabling the training of deep networks without degradation. With 50 layers, it balances performance and efficiency. Pre-trained models on datasets including ImageNet allow for effective transfer learning, faster convergence, and reduced data requirements. ResNet50 has become a favored choice, excelling in various vision tasks and proving adaptable for applications such as detection, segmentation, and feature extraction.
- **Disadvantage**: ResNet50, while highly effective, has some drawbacks. Due to its depth and complexity, training can be computationally demanding and resource-intensive, requiring substantial memory and computational capacity. Additionally, the skip connections, while beneficial for mitigating the vanishing gradient problem, may present optimization challenges, making fine-tuning and training on smaller datasets more difficult. Moreover, without proper regularization and data augmentation, the increased depth of ResNet50 may lead to over-fitting issues.

3.2 Dataset

The dataset is core to DL research. For this chapter, we used a new crop dataset named "CCMT Dataset" [58]. In this dataset, there are four different types of crops available. Every crop has five to seven different sub-classes. The dataset has two folders, one for raw data and the next one for augmented data. Raw data consists of 24,881 images and augmented data consists of 102,976 images. 22 classes are available in this dataset. Table 2 contains the details of the dataset and classes are available for training and testing.

3.2.1 Data Prepossessing

The CCMT dataset consists of RGB photos of varying sizes. Images obtained via the Internet come in various formats, resolutions, and quality. Preprocessing is essential for accurate

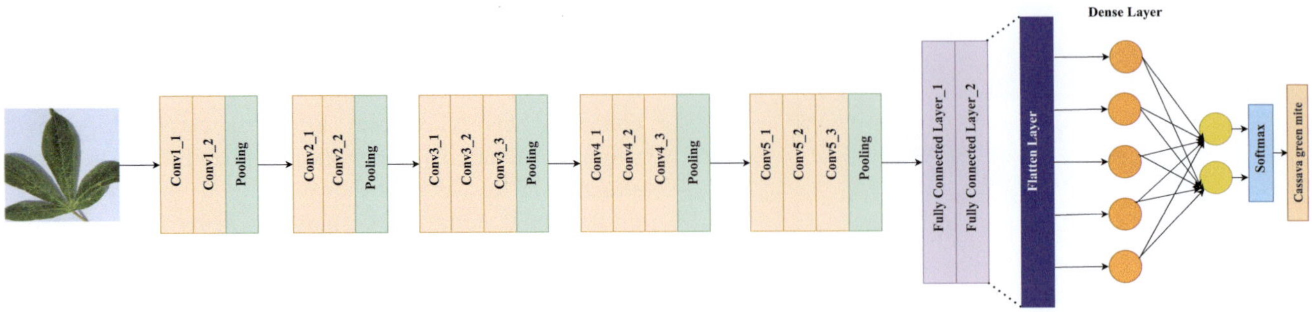

Fig. 5 The figure presents the architecture of the ResNet50 model

Table 2 The table presents the number of trains and test images and total classes of image in CCMT dataset

Plant	Disease type	Training	Testing
Cashew	Anthracnose	600	120
	Gumosis	600	120
	Healthy	600	120
	Leaf miner	600	120
	Red rust	600	120
Cassava	Bacterial blight	600	120
	Brown spot	600	120
	Green mite	600	120
	Healthy	600	120
	Mosaic	600	120
Maize	Fall armyworm	600	120
	Grasshopper	600	120
	Healthy	600	120
	Leaf beetle	600	120
	Leaf blight	600	120
	Leaf spot	600	120
	Streak virus	600	120
Tomato	Verticulium wilt	600	120
	Septoria leaf spot	600	120
	Leaf curl	600	120
	Leaf blight	600	120
	Healthy	600	120
Total		13,200	2,640

feature extraction in all collected photos. Expert verification is critical, especially for photographs obtained from the Internet or taken in the field and labeled data. Ensuring dataset accuracy is critical for training and validation, as accurately identified images are required. Following human labeling and verification, data resizing is used for resizing images and 224 × 224 pixels to make standardized dimensions for subsequent processing. Then we removed the duplication of images. However, the training image dataset is subjected to various random augmentations such as rotation, shearing, zooming, and flipping. These enhancements are intended to reduce overfitting and improve ability. The input is divided into distinct classes that depict various disorders, with one class dedicated to healthy photos. In Fig. 6, we can see a few images of our used dataset.

4 Experiment

For our experiment, we used Google Collaboratory [59], to perform the experiment we needed a GPU, and the GPU we used was Tasla T4, the RAM was 48 GB, and the hard disk was 78 GB. Multiple libraries such as TensorFlow, Keras, Matplotlib, ImageDataGenerator, pandas, and others were used for training the models and processing the dataset. For our experiment, we added modified layers on VGG16, Inception-V3, and ResNet50 and the batch size was 16. We used Adam optimizer and softmax for the activation function. The input shapes for both models were 224 × 224 × 3. The investigation extended over 40 training epochs, a learning rate was set 0.0001 and kernel size of 3 × 3 were incorporated into the model.

From the split dataset, with 80% remarked for training data and the 20% set aside for testing and validation. The training CCMT dataset comprised 13,200 images, while the test and validate datasets consisted of 2,640 images each. This meticulous distribution of data aimed to guarantee a thorough and representative collection of samples for the experiment. You can find the detailed data distribution in Table 2.

4.1 Result Analysis and Discussion

In our chapter, we used different measures to check how well our model performed. These include accuracy (ACC), F1-score, recall (R), and precision (P). Accuracy (ACC) tells us the percentage of correctly identified images among all images, providing us with a general idea of how well the model categorized diseases and pests.

The main goal of the study was to conduct a depth examination of the effectiveness, cost, and execution time of VGG16, Inception-V3, and ResNet50 architectures in the context of a crop pest and disease.

Fig. 6 The figure presents few images of the CCMT dataset: the first row presents all maize images (1) fall armyworm, (2) grasshopper, (3) healthy, (4) leaf beetle, (5) leaf blight, (6) leaf spot, (7) streak virus. The second row presents all cashew images (8) anthracnose, (9) gumosis, 10) healthy, (11) leaf miner, (12) red rust. The third row presents all cassava images (13) bacterial blight, (14) brown spot, (15) green mite, (16) healthy, (17) mosaic. The fourth row presents all tomato images (18) healthy, (19) leaf blight, (20) leaf curl, (21) septoria leaf spot, (22) verticillium wilt

Table 3 The following table presents the matrices for the CCMT dataset

Models	Precision (%)	Recall (%)	F1-score (%)
VGG16	98.01	98.40	98.61
Inception-V3	97.65	96.78	98.07
ResNet50	92.47	94.67	95.9

In particular, our investigation revealed that the input size significantly influences achieving peak accuracy levels for VGG16, ResNet50, and Inception-V3. Through the study identified that an input size of 224 × 224 px resulted in the highest accuracy for the three models. This conclusion is substantiated by the precision, F1-score, and recall metrics, meticulously outlined in Table 3 and a graph in Fig. 7.

In our research, dedicated to classifying crop diseases and pests, we have achieved noteworthy success by leveraging advanced DL models. Specifically, the VGG16 model demonstrated exceptional effectiveness in identifying diseases and pests from crops, outperforming an impressive rate of accuracy of 98.50%. Figure 8 presents the training accuracy and loss, validation accuracy and loss of the VGG16 model. Our next trained model was Inception-V3, which showcased strong performance, achieving an accuracy of 94.79% in the same context. Further, the last one we used was ResNet50 and the accuracy we got was 90.70%. Figures 9 and 10 present the training accuracy, training loss, validation accuracy, and validation loss of the Inception-V3 and ResNet50 models, respectively.

The comprehensive details of these significant accomplishments are outlined in Table 4, emphasizing the robustness of our approach. Figure 11 presents our used models' accuracies and validation accuracies. Notably, using graphs as an essential part of our detection and assessment procedure has greatly improved the precision and dependability of our findings. Together, these results highlight how well our approach works to further the field of agricultural picture classification.

Upon completing our experiment, we conducted a comparative analysis with other experimental results. Our findings

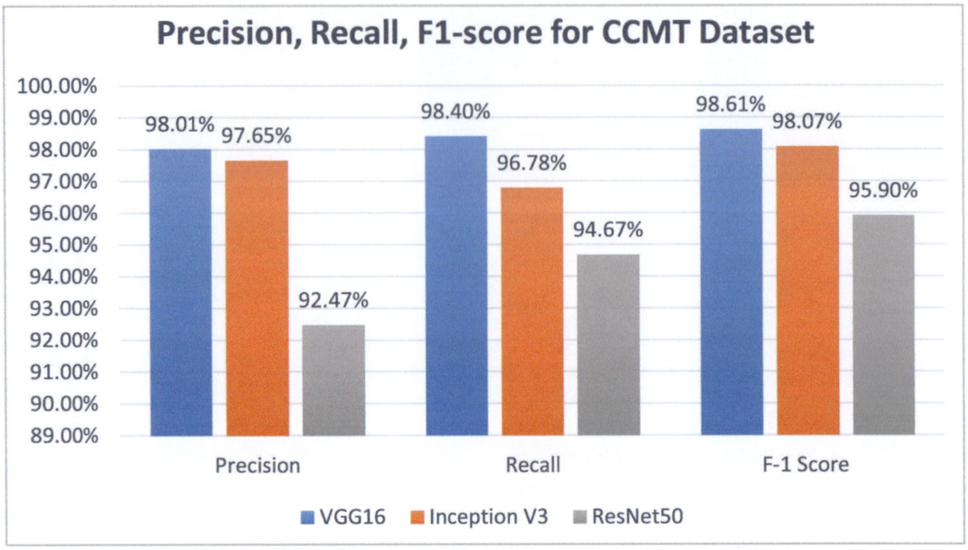

Fig. 7 The graph presents the precision, recall, and F1-score for the CCMT dataset

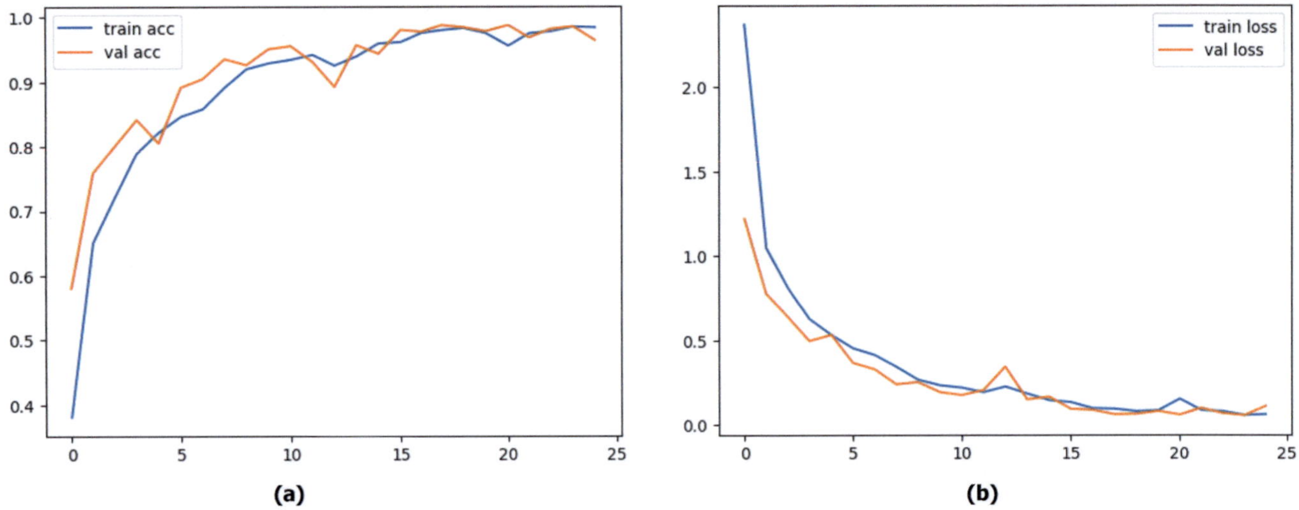

Fig. 8 The graph presents training accuracy and loss, validation accuracy, and loss of the VGG16 model

indicate that our model outperformed another model significantly. Figure 12 demonstrates that our method achieved the highest outcome compared to the alternative model. Moupojou et al. achieved 82.54% accuracy on the classification of plantDoc dataset [60]. Kaur et al. experimented on tomato leaves and achieved 98.92% accuracy [61]. Further, Alirezazadeh et al. achieved 86.89% accuracy on the DiaMOS Plant dataset [62]. Mallick et al. used mung bean dataset for the experiment and achieved 93.65% accuracy [63]. Additionally, Chen et al. achieved 99.61% accuracy on the PlantVillage dataset [64]. Table 5 compares the accuracy with other CNN models based on crop disease datasets.

5 Conclusion

This study introduces a lightweight neural network designed for the resource-intensive domain of agriculture, earning a remarkable accuracy of 99.73% in detecting healthy and infected crop pests and diseases. The framework employs a multichannel ensemble approach, integrating Inception-V3, VGG16, and ResNet50 as feature extractors. Extracted features are fused and processed through fully connected layers to capture optimal patterns, culminating in a robust detection process using a stacked ensemble technique.

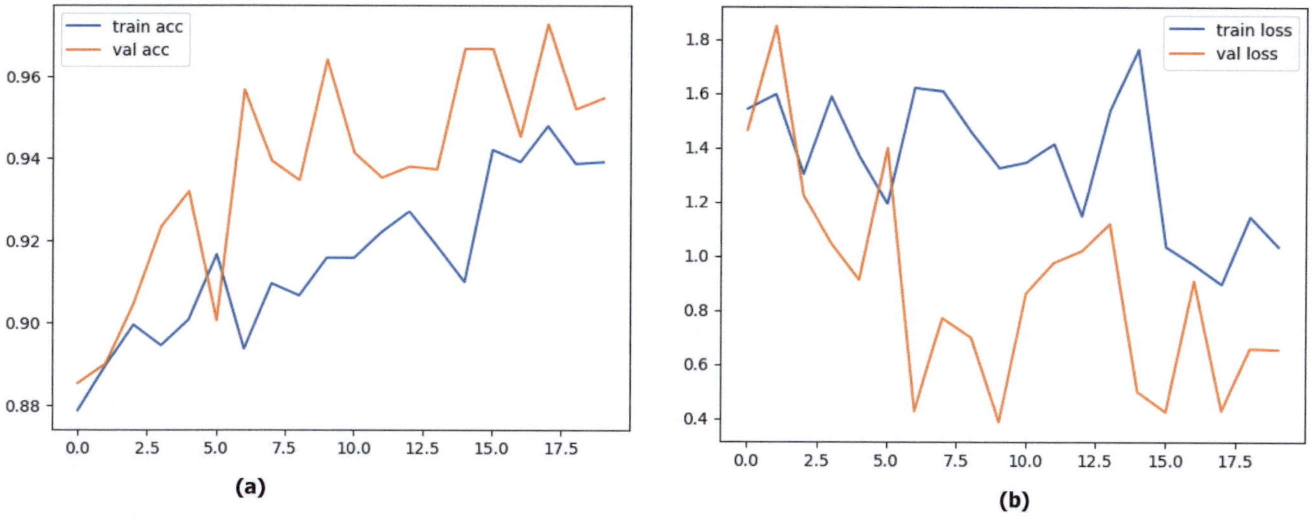

Fig. 9 The graph presents training accuracy and loss, validation accuracy, and loss of the Inception-V3 model

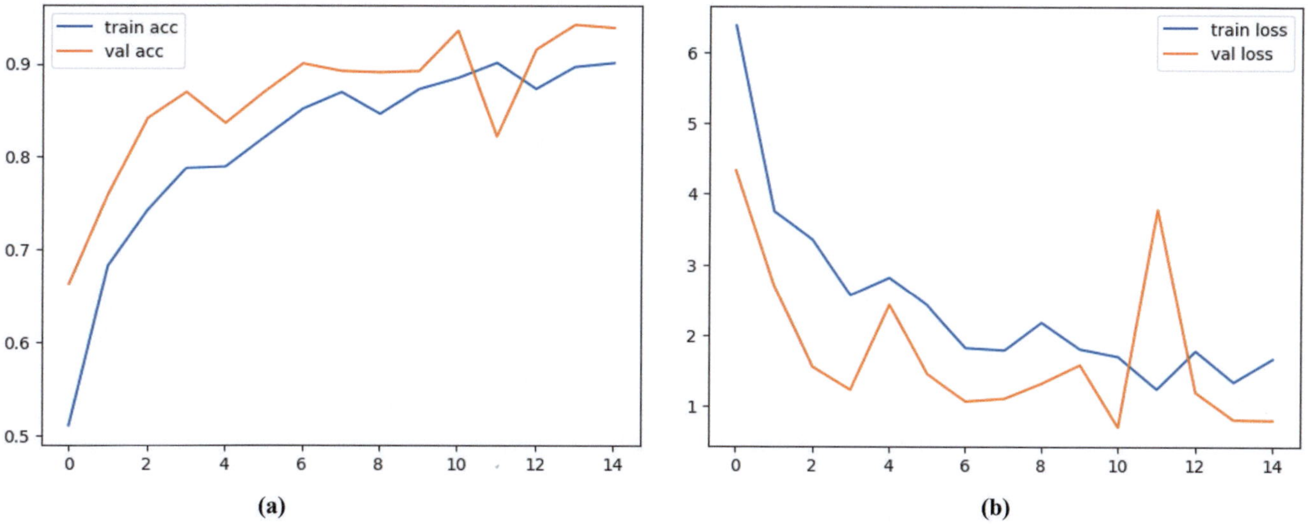

Fig. 10 The graph presents training accuracy and loss, validation accuracy, and loss of the ResNet50 model

The methodology showcases strong robustness and generalization on uncharted datasets, emphasizing its potential for smart agricultural applications. However, further evaluation on real-time datasets is essential to validate its reliability in practical scenarios. Additionally, the model's susceptibility to adversarial challenges in smart agricultural environments necessitates future exploration of advanced techniques to enhance its resilience. Prospective directions include developing hybrid architectures, leveraging label dependencies, and integrating segmentation information to improve adaptability and effectiveness in managing complex agricultural datasets and navigating challenging conditions.

Table 4 The following table presents the training and validation accuracies for the CCMT dataset

Models (%)	Training accuracy (%)	Validation accuracy (%)
VGG16	98.50	99.73
Inception-V3	94.79	97.27
ResNet50	90.70	94.27

Fig. 11 The graph presents the accuracy of validation and training of VGG16, Inception-V3, and ResNet50 models

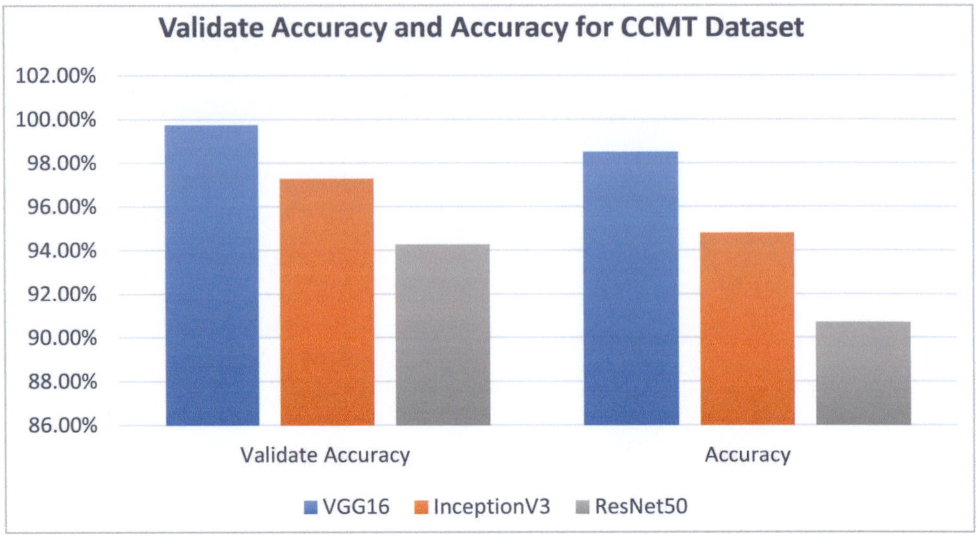

Fig. 12 The graph presents a comparison between various models vs. accuracy in percentage (%) with our model

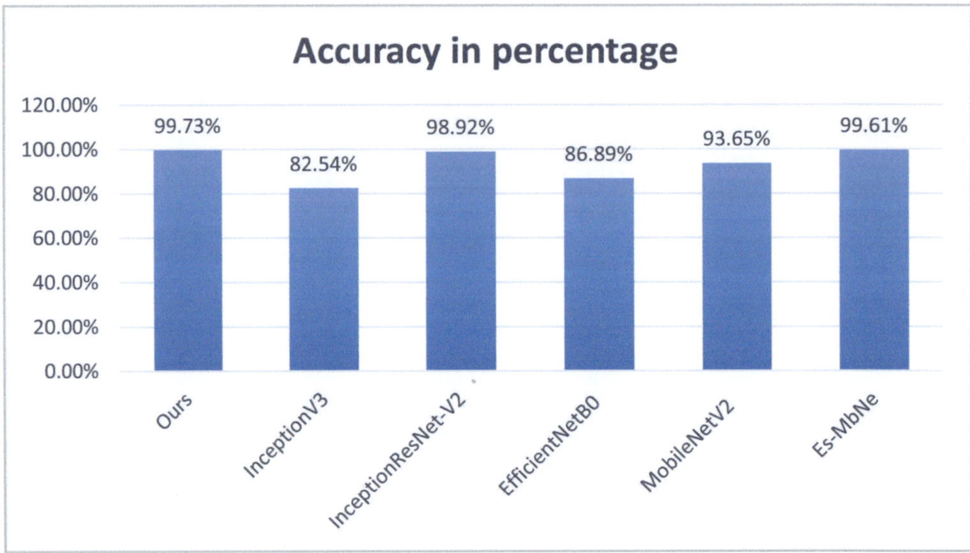

Table 5 The following table presents comparison of our results with other models based on crop disease datasets

References	Dataset	Model	Accuracy (%)
Moupojou et al. [60]	plantDoc	Inception-V3	82.54
Kaur et al. [61]	Tomato leaves	Inception-ResNet-V2	98.92
Alirezazadeh et al. [62]	DiaMOS plant	EfficientNetB0+CBAM	86.89
Mallick et al. [63]	Mung bean	MobileNet V2	93.65
Chen et al. [64]	PlantVillage	Ensemble learning	99.61
Thakur et al. [43]	PlantVillage	Lightweight CNN	99.16
Islam et al. [32]	Crop image	VGG16	92.39
Turkoglu et al. [35]	4,447 plant images	ResNet50	97.56

References

1. Singh, V., Misra, A.K.: Detection of plant leaf diseases using image segmentation and soft computing techniques. Inf. Process. Agric. **4**(1), 41–49 (2017)
2. Dipankar Roy, J.C.: Tomato farmers face peak season plight. https://www.thedailystar.net/business/economy/news/tomato-farmers-face-peak-season-plight-3252416 (2023)
3. Fu, X., Ma, Q., Yang, F., Zhang, C., Zhao, X., Chang, F., Han, L.: Crop pest image recognition based on the improved ViT method. Inf. Process. Agric. **11**(2), 249–259 (2024)
4. Ngugi, H.N., Ezugwu, A.E., Akinyelu, A.A., Abualigah, L.: Revolutionizing crop disease detection with computational deep learning: a comprehensive review. Environ. Monitor. Assess. **196**(3), 302 (2024)
5. Chen, J., Yin, H., Zhang, D.: A self-adaptive classification method for plant disease detection using GMDH-logistic model. Sustain. Comput.: Inform. Syst. **28**, 100415 (2020)
6. Kaur, P., Pannu, H.S., Malhi, A.K.: Plant disease recognition using fractional-order Zernike moments and SVM classifier. Neural Comput. Appl. **31**, 8749–8768 (2019)
7. Islam, M.R., Oliullah, K., Kabir, M.M., Alom, M., Mridha, M.: Machine learning enabled IoT system for soil nutrients monitoring and crop recommendation. J. Agric. Food Res. **14**, 100880 (2023)
8. Meyer, D., Wien, F.: Support vector machines. Interface Libsvm Package e1071 **28**(20), 597 (2015)
9. Peterson, L.E.: K-nearest neighbor. Scholarpedia **4**(2), 1883 (2009)
10. Tang, J., Deng, C., Huang, G.-B.: Extreme learning machine for multilayer perceptron. IEEE Trans. Neural Netw. Learn. Syst. **27**(4), 809–821 (2015)
11. Daqi, G., Jun, D., Changming, Z.: Integrated fisher linear discriminants: an empirical study. Pattern Recogn. **47**(2), 789–805 (2014)
12. Pal, M.: Random forest classifier for remote sensing classification. Int. J. Remote Sens. **26**(1), 217–222 (2005)
13. Patil, N., Ali, R., Wankhedkar, V., Nayak, D.: Crop disease detection using deep convolutional neural networks. Int. J. Eng. Res. Technol. (IJERT) **8**(03), 2278–2281 (2019)
14. Abbas, A., Jain, S., Gour, M., Vankudothu, S.: Tomato plant disease detection using transfer learning with C-GAN synthetic images. Comput. Electron. Agric. **187**, 106279 (2021)
15. Agarwal, M., Gupta, S.K., Biswas, K.: Development of efficient CNN model for tomato crop disease identification. Sustain. Comput.: Inform. Syst. **28**, 100407 (2020)
16. Tan, M., Le, Q.: Efficientnet: rethinking model scaling for convolutional neural networks. In: International Conference on Machine Learning, pp. 6105–6114. PMLR (2019)
17. Szegedy, C., Vanhoucke, V., Ioffe, S., Shlens, J., Wojna, Z.: Rethinking the inception architecture for computer vision. In: Proceedings of the IEEE Conference on Computer Vision and Pattern Recognition, pp. 2818–2826 (2016)
18. Ma, N., Zhang, X., Zheng, H.-T., Sun, J.: Shufflenet v2: Practical guidelines for efficient CNN architecture design. In: Proceedings of the European Conference on Computer Vision (ECCV), pp. 116–131 (2018)
19. Han, K., Wang, Y., Tian, Q., Guo, J., Xu, C., Xu, C.: Ghostnet: more features from cheap operations. In: Proceedings of the IEEE/CVF Conference on Computer Vision and Pattern Recognition, pp. 1580–1589 (2020)
20. Sandler, M., Howard, A., Zhu, M., Zhmoginov, A., Chen, L.-C.: Mobilenetv2: inverted residuals and linear bottlenecks. In: Proceedings of the IEEE Conference on Computer Vision and Pattern Recognition, pp. 4510–4520 (2018)
21. He, K., Zhang, X., Ren, S., Sun, J.: Deep residual learning for image recognition. In: Proceedings of the IEEE Conference on Computer Vision and Pattern Recognition, pp. 770–778 (2016)
22. Kabir, M.M., Hafiz, M.S., Bandyopadhyaa, S., Jim, J.R., Mridha, M.: Tea leaf age quality: age-stratified tea leaf quality classification dataset. Data Brief **54**, 110462 (2024)
23. Mbow, C., Rosenzweig, C.E., Barioni, L.G., Benton, T.G., Herrero, M., Krishnapillai, M., Ruane, A.C., Liwenga, E., Pradhan, P., Rivera-Ferre, M.G., et al.: Food security. Technical report, IPCC (2020)
24. Lewis, W.A.: Theory of Economic Growth. Routledge (2013)
25. Nobel, S.N., Imran, M.A., Bina, N.Z., Kabir, M.M., Safran, M., Alfarhood, S., Mridha, M.: Palm leaf health management: a hybrid approach for automated disease detection and therapy enhancement. IEEE Access (2024)
26. Ben Ayed, R., Hanana, M.: Artificial intelligence to improve the food and agriculture sector. J. Food Qual. **2021**, 1–7 (2021)
27. Mondal, D., Roy, K., Pal, D., Kole, D.K.: Deep learning-based approach to detect and classify signs of crop leaf diseases and pest damage. SN Comput. Sci. **3**(6), 433 (2022)
28. Meena, S.D., Susank, M., Guttula, T., Chandana, S.H., Sheela, J.: Crop yield improvement with weeds, pest and disease detection. Procedia Comput. Sci. **218**, 2369–2382 (2023)
29. Ullah, N., Khan, J.A., Alharbi, L.A., Raza, A., Khan, W., Ahmad, I.: An efficient approach for crops pests recognition and classification based on novel deeppestnet deep learning model. IEEE Access **10**, 73019–73032 (2022)
30. Sanghavi, V.B., Bhadka, H., Dubey, V.: Hunger games search based deep convolutional neural network for crop pest identification and classification with transfer learning. Evol. Syst. **14**(4), 649–671 (2023)
31. Kasinathan, T., Singaraju, D., Uyyala, S.R.: Insect classification and detection in field crops using modern machine learning techniques. Inf. Process. Agric. **8**(3), 446–457 (2021)
32. Islam, M.M., Adil, M.A.A., Talukder, M.A., Ahamed, M.K.U., Uddin, M.A., Hasan, M.K., Sharmin, S., Rahman, M.M., Debnath, S.K.: Deepcrop: deep learning-based crop disease prediction with web application. J. Agric. Food Res. **14**, 100764 (2023)
33. Malathi, V., Gopinath, M.: Classification of pest detection in paddy crop based on transfer learning approach. Acta Agriculturae Scandinavica, Sect. B-Soil Plant Sci. **71**(7), 552–559 (2021)
34. Wang, B.: Identification of crop diseases and insect pests based on deep learning. Sci. Program. **2022**, 1–10 (2022)
35. Turkoglu, M., Yanikoğlu, B., Hanbay, D.: Plantdiseasenet: convolutional neural network ensemble for plant disease and pest detection. Signal Image Video Process. **16**(2), 301–309 (2022)
36. Xin, M., Wang, Y.: Image recognition of crop diseases and insect pests based on deep learning. Wirel. Commun. Mob. Comput. **2021**, 1–15 (2021)
37. Sourav, M.S.U., Wang, H.: Intelligent identification of jute pests based on transfer learning and deep convolutional neural networks. Neural Process. Lett. **55**(3), 2193–2210 (2023)
38. Adem, K., Ozguven, M.M., Altas, Z.: A sugar beet leaf disease classification method based on image processing and deep learning. Multimedia Tools Appl. **82**(8), 12577–12594 (2023)
39. Parthiban, S., Moorthy, S., Sabanayagam, S., Shanmugasundaram, S., Naganathan, A., Annamalai, M., Balasubramanian, S.: Deep learning based recognition of plant diseases. In: Computer Vision and Machine Intelligence Paradigms for SDGs: select Proceedings of ICRTAC-CVMIP 2021, pp. 83–93. Springer, Singapore (2023)
40. Usha Ruby, A., George Chellin Chandran, J., Chaithanya, B., Swasthika Jain, T., Patil, R.: Wheat leaf disease classification using modified resnet50 convolutional neural network model. Multimedia Tools Appl. 1–19 (2024)
41. Singh, V., Kaushik, V.D.: Adaptive snake optimization-enabled deep learning-based multi-classification using leaf images. Signal Image Video Process. 1–10 (2024)
42. Nandhini, M., Kala, K., Thangadarshini, M., Verma, S.M.: Deep learning model of sequential image classifier for crop disease detection in plantain tree cultivation. Comput. Electron. Agric. **197**, 106915 (2022)

43. Thakur, P.S., Sheorey, T., Ojha, A.: VGG-ICNN: a lightweight CNN model for crop disease identification. Multimedia Tools Appl. **82**(1), 497–520 (2023)
44. Kurmi, Y., Saxena, P., Kirar, B.S., Gangwar, S., Chaurasia, V., Goel, A.: Deep CNN model for crops' diseases detection using leaf images. Multidimens. Syst. Signal Process. **33**(3), 981–1000 (2022)
45. Chodey, M.D., Noorullah Shariff, C.: Hybrid deep learning model for in-field pest detection on real-time field monitoring. J. Plant Dis. Protect. **129**(3), 635–650 (2022)
46. Chen, X., Shuai, C., Wu, Y., Zhang, Y.: Analysis on the carbon emission peaks of china's industrial, building, transport, and agricultural sectors. Sci. Total Environ. **709**, 135768 (2020)
47. Li, L., Zhang, S., Wang, B.: Plant disease detection and classification by deep learning-a review. IEEE Access **9**, 56683–56698 (2021)
48. Szegedy, C., Liu, W., Jia, Y., Sermanet, P., Reed, S., Anguelov, D., Erhan, D., Vanhoucke, V., Rabinovich, A.: Going deeper with convolutions. In: Proceedings of the IEEE Conference on Computer Vision and Pattern Recognition, pp. 1–9 (2015)
49. Ioffe, S., Szegedy, C.: Batch normalization: accelerating deep network training by reducing internal covariate shift. In: International Conference on Machine Learning, pp. 448–456. PMLR (2015)
50. Szegedy, C., Ioffe, S., Vanhoucke, V., Alemi, A.: Inception-v4, inception-resnet and the impact of residual connections on learning. In: Proceedings of the AAAI Conference on Artificial Intelligence, vol. 31 (2017)
51. Simonyan, K., Zisserman, A.: Very deep convolutional networks for large-scale image recognition (2014). arXiv:1409.1556
52. Sarker, S., Tushar, S.N.B., Chen, H.: High accuracy keyway angle identification using vgg16-based learning method. J. Manuf. Process. **98**, 223–233 (2023)
53. Donepudi, S., Palagani, S.C., Pramod, P.S.N., Kumar, Y.R., Karthikeya, S., Praveen, S.P.: Brain metastasis tumor detection using image segmentation and vgg16 architecture. In: 2023 5th International Conference on Smart Systems and Inventive Technology (ICSSIT), pp. 150–155. IEEE (2023)
54. Sharma, S., Guleria, K.: A deep learning based model for the detection of pneumonia from chest X-ray images using VGG-16 and neural networks. Procedia Comput. Sci. **218**, 357–366 (2023)
55. Koonce, B., Koonce, B.: Resnet 50. Convolutional Neural Networks with Swift for Tensorflow: image Recognition and Dataset Categorization, pp. 63–72 (2021)
56. Xie, S., Girshick, R., Dollár, P., Tu, Z., He, K.: Aggregated residual transformations for deep neural networks. In: Proceedings of the IEEE Conference on Computer Vision and Pattern Recognition, pp. 1492–1500 (2017)
57. Zagoruyko, S., Komodakis, N.: Wide residual networks (2016). arXiv:1605.07146
58. Mensah, P.K., Akoto-Adjepong, V., Adu, K., Ayidzoe, M.A., Bediako, E.A., Nyarko-Boateng, O., Boateng, S., Donkor, E.F., Bawah, F.U., Awarayi, N.S., et al.: CCMT: dataset for crop pest and disease detection. In: Data in Brief, p. 109306 (2023)
59. https://research.google.com/colaboratory/
60. Moupojou, E., Tagne, A., Retraint, F., Tadonkemwa, A., Wilfried, D., Tapamo, H., Nkenlifack, M.: Fieldplant: a dataset of field plant images for plant disease detection and classification with deep learning. IEEE Access **11**, 35398–35410 (2023)
61. Kaur, P., Harnal, S., Gautam, V., Singh, M.P., Singh, S.P.: A novel transfer deep learning method for detection and classification of plant leaf disease. J. Ambient Intell. Hum. Comput. **14**(9), 12407–12424 (2023)
62. Alirezazadeh, P., Schirrmann, M., Stolzenburg, F.: Improving deep learning-based plant disease classification with attention mechanism. Gesunde Pflanzen **75**(1), 49–59 (2023)
63. Mallick, M.T., Biswas, S., Das, A.K., Saha, H.N., Chakrabarti, A., Deb, N.: Deep learning based automated disease detection and pest classification in Indian mung bean. Multimedia Tools Appl. **82**(8), 12017–12041 (2023)
64. Chen, J., Zeb, A., Nanehkaran, Y., Zhang, D.: Stacking ensemble model of deep learning for plant disease recognition. J. Ambient Intell. Hum. Comput. **14**(9), 12359–12372 (2023)

BananaLeaf-Net: A Proposed Deep Learning Model for Accurate Classification of Banana Leaf Diseases

Anika Saba Ibte Sum, Anik Kumar Saha, Kamruddin Md. Nur, and Khandaker Tabin Hasan

Abstract

A serious problem for agriculture, banana leaf disease is brought on by bacterial, viral, or fungal infections and can cause large crop losses. Worldwide farmers are seriously threatened by the recurrent incidence of diseases like Banana Bunchy Top Virus, Panama disease, and Black Sigatoka, which lower yields and compromise food security. In order to prevent extensive harm to banana crops, rapid handling and correct diagnosis of these diseases are essential. In order to protect farmers' livelihoods, guarantee healthy crop production, and reduce financial losses in the agriculture industry, timely identification techniques are essential. This paper showcases a comparative analysis with other state-of-the-art models for Banana Leaf disease. Using a banana leaf disease image dataset, this research compares the proposed BananaLeaf-Net model to a wide range of CNN-based architectures, such as EfficientNet B0, EfficientNet B1, EfficientNet B2, EfficientNet B3, MobileNetV2,ResNet50, ResNet101, ResNet152V, and DenseNet201. This study examines the most recent Banana Leaf Disease classification algorithms in this context to increase classification accuracy; our proposed BananaLeaf-Net model outperforms base models with an exceptional accuracy of 97%.

A. S. I. Sum · A. K. Saha (✉) · K. Md. Nur · K. T. Hasan
Department of Computer Science, American International University-Bangladesh, Dhaka, Bangladesh
e-mail: anikkumarsaha08@gmail.com

K. Md. Nur
e-mail: kamruddin@aiub.edu

K. T. Hasan
e-mail: tabin@aiub.edu

Keywords

Convolutional neural networks (CNNs) • Artificial intelligence (AI) • Deep learning • Banana leaf disease

1 Introduction

1.1 Background and Motivations

Agriculture is an art and a science, and it is the backbone of human civilization as well as a vital sector of the global economy [1]. Tilling the land, raising cattle, and growing crops are the ways that agriculture generates the resources required for human sustenance [1]. Being the primary industry and a major force behind the national economy, it is far more significant in developing countries. Increasing agricultural output has become a key goal in the effort to ensure food security and boost the economy. For optimal growth, crops require a precise balance of 16 critical elements; any imbalance leads to stunted growth and, in extreme cases, substantial crop loss [1]. Farmers must be able to visually assess these nutrient shortages in order to take prompt action to increase yields and improve plant health. Bananas, in particular, are extremely significant due to their broad use and economic importance. According to the Food and Agriculture Organization (FAO), bananas are the second largest fruit crop in the world. Their production and sale have a major economic impact on agriculture, especially in tropical and subtropical regions [2]. Nowadays, most banana leaf diseases are manually detected by trained farmers using visual cues to assess the health of the plant; nevertheless, these methods are often prone to errors, increasing the challenge of maintaining crop yields [2].

In terms of economics, planting bananas is important, especially in countries like India where the sector greatly increases GDP. In the 2019 fiscal year, bananas, a significant crop in India, contributed 346 billion Indian rupees to the GDP [3]. However, there are challenges in the banana industry as well.

M. F. Mridha and N. Dey (eds.), *Machine Vision in Plant Leaf Disease Detection for Sustainable Agriculture*, Studies in Computational Intelligence 1202, https://doi.org/10.1007/978-981-96-4520-6_4

Banana crops are vulnerable to diseases like Mosaic, Panama wilt, and Black Sigatoka, which reduce yields and negatively impact farmers' and the agricultural industry's profitability [3]. The spread of these diseases means that improved methods of disease identification and management are desperately needed. Bananas are a vital component of agribusiness in developing nations where agriculture remains the primary source of revenue. China, India, and Indonesia are global leaders in banana farming and contribute significantly to agricultural production globally [4]. According to agribusiness data, countries like Ecuador, Argentina, and the Philippines are significant banana producers, accounting for over 20% of global agricultural output. Despite the large-scale nature of banana production, an estimated 85% of farms globally are run by smallholders, many of whom lack access to advanced agricultural tools or procedures [4]. This makes managing banana diseases more challenging.

Bananas are crops that originated in Southeast Asia and are now farmed in over 130 countries, primarily in tropical and subtropical climates. As a highly sought-after staple food, they account for about 16% of global fruit production and are the second largest fruit crop after citrus [5]. The fact that bananas are the fifth most traded food crop in the world, behind coffee, cereals, sugar, and cocoa, makes them significant [5]. There are two main types of bananas: plantains, also referred to as kitchen bananas since they can be eaten either raw or cooked, and sweet bananas, sometimes referred to as dessert bananas. Bananas are rich in phytosterols, carotenoids, biogenic amines, and phenolics, among other bioactive chemicals, all of which offer a number of health benefits, including antioxidant properties. Bananas require a well-balanced diet because of these characteristics. Eating bananas is widely encouraged to preserve good health, as they were once used to treat long-term degenerative disorders. The World Health Organization's recommendation that consumers consume 400 g of fruits and vegetables each day [5] further emphasizes the significance of bananas in the global diet. With a total production of 128,778,738 tons from 5,517,027 ha worldwide, India is the world's largest producer of bananas, contributing 27% of the crop [5]. The importance of bananas for the global economy and diet is demonstrated by the fact that Bangladesh, another significant producer, grows 833,309 tons of bananas annually from 48,850 ha of cropland [5].

Though diseases like Mosaic, Panama wilt, and Black Sigatoka pose serious threats to banana crops and may place a strain on smallholder farmers' economics, banana crops are essential to the world's food supply and the economy [2]. For plant disease identification, these farmers frequently rely on outdated visual techniques, which are prone to inaccuracy and can cause treatment delays and further crop loss. Technology advancements are desperately needed because of the increasing need for more precise and effective disease management. With the ability of precise and early identification of diseases, automated disease detection systems, driven by deep learning and advanced image processing, provide an attractive option. By implementing such practices in place, farmers may better protect their crops, resulting in increased production resilience for bananas and preserving the livelihoods of individuals who depend on this vital agricultural product. The demand for scalable, technologically advanced solutions that can revolutionize disease management and boost productivity, particularly in developing countries where agriculture is essential to economic growth, is what motivated our research.

1.2 Related Works

Recent years indicate significant progress in the study of banana leaf disease detection due to the creation of innovative machine learning models and deep learning algorithms. The denoising of banana leaf pictures using the K-scale VisuShrink algorithm (KVA) when combined with the Ghost ResNeSt-Attention RReLU-Swish Net (GR-ARNet) has made a notable contribution as well to this field. Using 13,021 images, the KVA achieved an impressive accuracy of 95.98% and precision of 89.31%. Even though the study was successful, it was restricted to three diseases of banana leaves: Sigatoka, Cordana, and Pestalotiopsis. It also only used image data, which meant that other diseases and data types could not be explored as much [2]. Furthermore, the BananaLSD dataset which focuses on leaf spot diseases was initially created as a large collection of 937 photos and eventually expanded to 1,600. The dataset was collected in the real world with the intention of assisting with early disease identification using 95.25% accurate models such as BananaSqueezeNet with precision of 96.53% and recall of 96.25%. The dataset's focus on just three diseases, among other drawbacks, will force future research to broaden the scope of disease categories and improve the generalizability of the model [5, 6]. Across a range of architectures, deep learning-based efforts to improve disease detection have produced encouraging results. For instance, a study that used a modified version of the ResNet50 model to identify Fusarium Wilt in banana leaves had astounding results, exceeding previous models by 0.07 in accuracy with 94% accuracy, precision (88%), recall (91%), and F1-score (89%). The 600 photos in the collection, however, are a limitation, highlighting the necessity for larger datasets to reflect real-world situations and enable real-time monitoring systems for disease management [1, 7] more accurately. In comparison, the Faster R-CNN model was used to detect leaf diseases, and it performed faster and more accurately than the usual CNN, R-CNN, and SVM models. However, due to its high computing requirements, it is not as applicable in real-world scenarios. Scholars are urged to maximize computational effectiveness and investigate scalability in various crop types [8].

Furthermore, pre-processing and picture segmentation methods have made a substantial contribution to the accuracy of disease classification. In one study, disease diagnosis accuracy was above 92% using genetic algorithm-based segmentation and ensemble machine learning models, outperforming individual classifiers such as SVM, Naïve Bayes, and K-NN [9]. A different approach that used the k-means clustering algorithm showed good image quality retention with a high PSNR of 57.41, low RMSE of 0.0013, and SSIM of 0.9982, but adaptability was limited due to its reliance on publicly available image datasets. Future research could investigate the integration of sophisticated machine learning algorithms, diversify the dataset, and extend the methodology to other plant species for wider applicability to overcome these concerns [10, 11]. In the field of hybrid techniques, a study that combined Support Vector Machines (SVMs) and Convolutional Neural Networks (CNNs) produced a maximum average accuracy of 94% in the classification of banana leaf disease. Although this method worked well, more research into real-time applications and wider dataset validation is necessary because it depends on high-quality labeled data and computational resources [12]. These investigations have yielded several applications, including models being used for real-time disease diagnosis in mobile applications. A noteworthy instance is the Leaf Guard application, which utilizes deep learning and image processing methods to identify illnesses such as Panama Wilt and Black Sigatoka and offers customized chemical suggestions for disease control [13]. For real-world deployment, it is still difficult to extend the model's detection capabilities to include more disease classes and strengthen its resilience to changing environmental factors [14, 15]. With accuracy rates over 90%, transfer learning techniques like those used in research with VGG16 and MobileNetV2 have also proven useful in real-time applications. These models still have issues, such as dataset size restrictions, but these could be resolved with more development and optimization for use in contexts with limited resources [16].

The detection of disease in banana leaves has been found to be highly effective with the application of deep learning techniques, specifically Convolutional Neural Networks (CNN). ShuffleNet V2 and YOLOv4 were two CNN models that were used in a few research to identify disorders like Panama Disease and Black Sigatoka. In diagnosing Black Sigatoka, for example, a ShuffleNet V2-based CNN achieved 95% accuracy, 96.67% sensitivity, and 93.33% specificity [17]. Similarly, a YOLOv4 algorithm trained on a dataset of 500 photos of infected banana leaves achieved 90% accuracy in identifying Panama disease [18]. To accurately calculate leaf damage percentages, another study used OpenCV and YOLOv4 to detect the same disease and determine leaf degradation by image segmentation [19]. Nonetheless, certain obstacles persist, including the requirement for pictures of excellent quality, the possibility of misidentification, and dependence on datasets that could not accurately represent real-world scenarios. Future research should concentrate on establishing real-time monitoring systems to improve disease identification and accuracy, as well as increasing datasets to include new disease classes, to solve these constraints [17–19]. Other methods have combined CNNs with machine learning strategies like Support Vector Machines (SVM) to create hybrid models that increase classification accuracy. A hybrid CNN-SVM model, for instance, was able to classify different banana leaf diseases with 90% accuracy, although it was limited by large processing demands and dataset limits [20]. The potential of federated learning in agricultural disease detection was demonstrated by a cooperative learning technique that used CNNs and federated learning to produce promising results, including excellent accuracy, precision, and recall across clients representing different disease severity levels [21]. Like this, it was discovered that picture pre-processing and segmentation techniques like k-means were very successful in maintaining image quality during analysis, obtaining a low RMSE (0.0013) and high SSIM (0.9982) [22]. While Modified Generative Adversarial Networks (MGAN) and Modified Faster Region-based Convolutional Neural Networks (MFRCNN) achieved even higher performance metrics of 96% accuracy, 94% precision, and 95% F1-score [23], hybrid approaches, such as combining CNNs with SVM, reached an average accuracy of 94% [24]. Thoughts should be directed toward incorporating cutting-edge machine learning techniques, growing datasets, and applying these approaches to a wider range of plant species and agricultural diseases in the future as the use of datasets or image sources may limit the generalizability of these models [20–25].

The potential for realistic field deployment has also been demonstrated by developments in real-time disease detection technologies and mobile applications. To improve crop health in Sri Lanka, for example, the Leaf Guard app uses deep learning and image processing to analyze photos of banana leaves and delivers precise disease identification and customized chemical recommendations [26]. Furthermore, early diagnosis of diseases like Xanthomonas wilt and Sigatoka leaf spot has been made easier by the creation of CNN-based models that use pre-trained architectures like VGG16, VGG19, and MobileNet, with accuracy rates surpassing 80% [27, 28]. Convolutional neural networks (CNNs) were used to diagnose Sigatoka leaf spot disease using an Indian dataset [25]. The usefulness of hybrid models was demonstrated by CNNs and SVMs combined to obtain a maximum accuracy of 94% [24]. However, adoption in resource-constrained situations is hampered by issues including scarce computational resources, diverse datasets, and sophisticated models. In order to improve model interpretability and precision, future research could investigate refining these models for real-time deployment, adding more extensive and varied datasets, and utilizing sophisticated methods such as Shapley values [26–29]. Although deep learning, and CNNs in particular, has made significant progress recently in the diagnosis of banana leaf disease, there are still issues with processing demands and dataset constraints. Expanding datasets, improving models for real-time application, and incorporating cutting-edge methods for enhanced generalizability and deployment should be the main goals of future research.

2 Research Gaps, Main Contributions, and Novelties

Accurate banana leaf identification is essential especially for farmers and need a good success in agriculture. Researchers can help to implement automated classification techniques which can significantly enhance disease management, helping farmers protect their crops, improve yields, and sustain the livelihoods of those who rely on this critical agricultural resource. Reliable classification is challenging due to implementation of automated advanced classification.

(a) **Research Gaps**

- **Limited Disease Coverage**: A small number of banana leaf diseases, including Sigatoka, Cordana, and Pestalotiopsis, are within the scope of several research. For wider applicability, the scope needs to be expanded to encompass more diseases such as Panama Disease, Xanthomonas Wilt, and others.
- **Geographic and Dataset Restrictions**: A lot of the datasets utilized in these research (such as images from Bangladesh, Assam, and Karnataka) are geographically limited and might not be universally applicable to areas of the world where bananas are grown. To increase the resilience of the model, more different datasets are needed.
- **Manual Image Pre-processing**: A lot of research uses biased manual image pre-processing methods, such as resizing and cropping. To standardize the input data, it is necessary to investigate advanced image pre-processing techniques and automate this process.
- **Dependency on only Image Data**: The majority of research uses just image-based data. Accurate and practical identification of diseases may be enhanced by including additional data types, such as genetic information or environmental parameters (weather, soil conditions).
- **Model Complexity and Resource Requirements**: Several of the models that have been suggested, including Faster R-CNN, are computationally intensive and thus less suitable for real-time, in-the-field applications. It is essential to create compact, effective models for use in contexts with limited resources.
- **Dataset Size and Augmentation**: Analyses usually start with relatively small datasets, even though many apply data augmentation to enhance dataset size. Establishing larger datasets will improve performance, especially when training deep learning models.
- **Lighting and Environmental Factors**: It has been determined that variations in lighting, image quality, and background noise can have an impact on how well a model performs. More effective image capture and data augmentation techniques should be used in additional research to solve these problems.
- **Real-time Disease Detection and Scalability**: A number of research have weaknesses in this area. Managing banana diseases on an extensive basis may be made possible by integrating real-time monitoring systems with applications for mobile devices or drones.
- **Model Generalizability**: Regional specificity and dataset size are two factors that restrict the high accuracy of models such as ResNet50 and VGG19. The generalizability of these models to various geographical locations, seasons, and disease stages should be investigated in future studies.

(b) **Contributions**

As research into banana leaf diseases and treatments advances, the development and implementation of robust classification algorithms presents a significant opportunity for transformation. Through improved banana leaf disease classification due to new understanding obtained from this collaborative, multidisciplinary research work, the aim is to deliver individualized farmer concern to all the farming community worldwide. The overall contributions of the paper are as follows:

- Application of a task-specific lightweight advanced deep learning model that has been developed for the classification of banana leaf disease.
- Applied advanced image pre-processing methods, including resizing, cropping, pixel normalization, and color space adjustments, to enhance image quality and consistency.
- Developing a strong and robust BananaLeaf-Net model for image classification by reducing overfitting and performance problems.
- Improving safety and accuracy for classification tasks on images implementing proposed model.
- Assessing proposed model with similarity of several advanced models and multiple classification task performance metrics.

(c) **Novelties**

- **BananaLeaf-Net Model Presenting**: The BananaLeaf-Net model is a significant advancement in agriculture area for banana leaf classification, owing to its highly accurate outcomes.

- **Customized Augmentation Process**: To improve the model's generalizability across a range of real-world situations in banana leaf images, the model integrates a specific augmentation technique, including rotations, flips, and brightness modifications.
- **Adaptive Resizing and Cropping**: To preserve pertinent details and standardize input dimensions, images are cropped and scaled adaptively, maximizing feature extraction at a range of resolutions.
- **Color and Contrast Enhancement**: Advanced color adjustment and contrast enhancement techniques are part of pre-processing, which makes leaf details more visible and clear. These features are essential for accurately detecting and classifying diseases.

2.1 Paper Organization

The structure of the paper follows the Introduction: The methodology is provided in Sect. 3. Subsequently, Sect. 4 contains the analysis of the results, whereas Sect. 5 contains the discussion. The Conclusion is finally presented in Sect. 6.

3 Methodology

3.1 General Overview of the Method

The main concentration of this study is the classification of banana leaf diseases by comparing various CNN models. Ensuring data accessibility, converting images to JPG format, and resizing and padding them to standardize their size are the initial steps in the method's data pre-processing. Divide a dataset into testing, validation, and training sets to separate it. Augmentation techniques like brightness adjustment, rescaling, flipping, random rotation, and translation are used to increase dataset diversity and improve model generalization. The model architecture is the suggested model with preloaded ImageNet weights. After the top layer has been removed, further levels are added using the Functional API. These layers consist of batch normalization, dropout layers for regularization, linear layers with ReLU activation and kernel regularization, linear layers for multi-class prediction, and linear layers with ReLU activation. The next layers are then unfrozen in order to adjust the system to changing weights. A categorical cross-entropy loss function and an Adam optimizer are employed for multi-class classification. Early stopping functions and learning rate scheduling are used to reduce overfitting and maximize model training. In simple terms, this approach targets to maximize model performance through architectural adjustments and data augmentation, while simultaneously effectively classifying banana leaf diseases using deep learning approaches.

3.2 Dataset Description

For Convolutional Neural Networks (CNNs) to be trained, validated, and tested in deep learning applications and to achieve high accuracy in identification and prediction tasks, a large, high-quality dataset is required. A specialized image dataset of banana leaves from plants was gathered from the Southern Nations, Nationalities, and Peoples' Regional State (SNNPRS) in Ethiopia for the purpose of our research on the identification of banana diseases. More precisely, the data came from the Omolante kebele in Mierab Abaya Woreda and the Lante and Chano kebeles in Arbaminch Zuria Woreda. These regions are noted for their significant banana agriculture and high rates of Sigatoka leaf spot and Xanthomonas wilt. Images from four separate farms, each covering one hectare, are included in the dataset. Every day during the data collecting period, banana leaves were captured. The pictures were first classified as "healthy" or "infected" depending on whether or not disease symptoms could be seen. A second expert examined and validated the plant pathologist's first classification to assure the labeling process' accuracy as well as reliability. Xanthomonas-infected leaves, Sigatoka-infected leaves, and healthy banana leaves were the final divisions into which the dataset was divided. Despite the difficulties in gathering large numbers of categorized images in agricultural contexts, 1,288 carefully selected images make up this dataset, which is a valuable resource for developing machine learning models targeted at the early identification of these serious banana diseases. This dataset has the potential to advance automated disease detection system's development and enhance agricultural disease management. The example images for these three categories are shown in Fig. 1. Additionally, Table 1 presents the image distribution for the several classes of trains, tests, and validations. The images were split into 70%-15%-15% ratios for training, validation, and testing, accordingly.

3.3 Proposed Model

The benchmarking models are trained using the banana leaf image dataset, as seen in Fig. 2. The MobileNetV2 architecture is used as the base for the modified model, BananaLeaf-Net, which has been modified to enhance its classification performance for banana leaf diseases. Through its lightweight architecture, MobileNetV2 is well suited for embedded and mobile applications without sacrificing accuracy. Generalized feature representations from millions of images are utilized by the base model through the use of ImageNet-pre-trained

Fig. 1 Banana leaf disease sample images

weights. Every layer of the convolutional backbone (features) is unfrozen to make the model task-specific, allowing for fine-tuning on the banana leaf dataset. By making this modification, the model is better able to identify and categorize diseases since it can now learn domain-specific characteristics unique to banana leaves. In order to achieve greater regularization during training and lessen overfitting, the Adam optimizer with weight decay and CrossEntropyLoss are used for multi-class classification. To improve convergence and training stability, the OneCycleLR scheduler dynamically modifies the learning rate. Gradient clipping keeps gradients from expanding and ensures that training proceeds without problems. In order to maximize training time without compromising model performance, an early stopping mechanism is included that stops training if validation accuracy stagnates after a predetermined amount of time five epochs.

Thorough pre-processing and augmentation methods are essential for getting the dataset prepared for efficient training. First, all images are standardized to 425 $\times$ 425 pixels by padding with a white border; corrupt or unreadable images are removed. This ensures that all of the input data are aligned and consistent with the specifications of the model. The training images receive a range of changes, including random horizontal flips, rotations within plus or equal 30°, and color adjustments, including brightness, contrast, saturation, and hue, before being resized to 224 $\times$ 224 for augmentation. By

Table 1 Image dataset distribution for each class

Class name	Train	Validation	Test	Total images
Healthy	1299	279	279	1857
Segatoka	2687	576	576	3839
Xamthomonas	6834	1464	1464	9762
Total	10820	2319	2319	15458

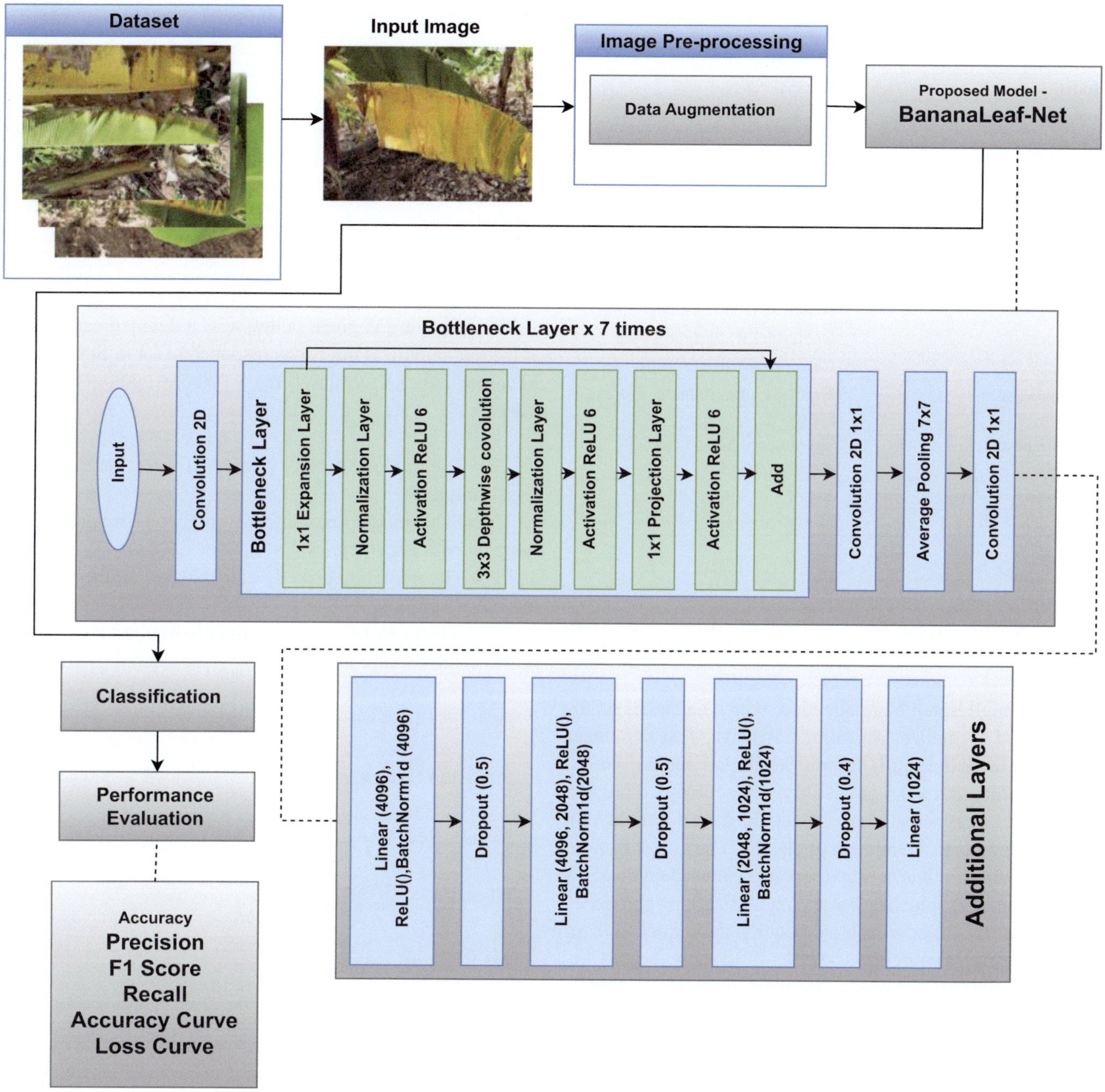

Fig. 2 Overview of proposed BananaLeaf-Net system architecture

simulating real-world differences in lighting and color distribution, these augmentations improve the model's robustness to changes in the environment. Furthermore, the model's capacity to manage different image distortions and alignments is further improved by affine transformations like translation and shear. To make sure that performance evaluations accurately represent the model's generalization to new data, the validation and test datasets only get little augmentation, including resizing and normalization. The convolutional neural network model BananaLeaf-Net was trained with hyper-parameters that reveal information about the optimal configuration for classifying banana leaf diseases which is given in Table 2.

BananaLeaf-Net is designed to have several fully connected layers built on top of the MobileNetV2 feature extractor to improve its learning capability. First, two more linear layers reduce the size to 2048 and then 1024. The linear layer that increases the dimensions from 1280 to 4096 is the initial of the new levels. After every linear layer, non-linearity is introduced through ReLU activations, and training is sta-

Table 2 Hyperparameters for the proposed BananaLeaf-Net model

Hyperparameters	Hyperparameter value
Model	BananaLeaf-Net
Convolution size	3 × 3
Layers	7
Training resolution	(224, 224)
Number of channels	32 to 1280
Number of classes	3
Scaling	Resize (425, 425)
Optimizer	Adam
Learning rate	Adaptive, 0.0001 (at beginning)
Scheduler	OneCycleLR, EarlyStopping
Loss	Categorical CrossEntropy
Augmentation	Normalization, rotation, flipping, zooming
Regularization	Dropout = 0.5, 0.4 | Kernel Regularizer = L2
Activation	ReLU
Max epoch	50
patience	5
Batch	64
Classifier	Linear

bilized through Batch Normalization, which enhances gradient flow and normalizes activations. After the first two linear layers and after the third, dropout layers are applied with a 0.5 probability and a 0.4 probability, respectively, to prevent overfitting. In the banana leaf dataset, the final output layer's size corresponds to the number of classes. The model is more capable to differentiate between separate banana leaf diseases due to the several hidden layers and regularization approaches of this deep classifier, which also equips the model to learn complex feature hierarchies.

Overall, the study provides a thorough method for enhancing Banana leaf diseases through the use of several convolutional neural network models: EfficientNet B0, EfficientNet B1, EfficientNet B2, EfficientNet B3, MobileNetV2, ResNet50, ResNet101, ResNet152V, and DenseNet201. The research initiative focuses on combining BananaLeaf-Net with additional layers to improve representation utilizing the dataset. The proposed method highlights the hyperparameter tuning for BananaLeaf-Net for the classification of banana leaf diseases and applies deep learning to the image dataset.

3.4 Evaluation Metrics

Several criteria for assessment, including Accuracy (Acc), Precision (Pre), Recall (Rec), F1-score, and Matthews Correlation Coefficient score (MCC), were applied to evaluate the performance of our implemented models and compare them with our proposed Bananaleaf-Net. Besides a performance indicator in the most restrictive way, constructed a confusion matrix, which is the source for other measures. Visual representation of the difference between the projected labels and the ground truth labels is provided by the confusion matrix. In the confusion matrix, examples in a predicted class are defined by rows, whereas cases in an actual class are described by columns. True Positive (TP), False Negative (FN), False Positive (FP), and True Negative (TN) are terms that are dependent on the confusion matrix.

The following is given below with a detailed explanation of the mathematical intuitions responsible for it: Specifically, Accuracy (Acc), Precision (Pre), F1-Score, and Recall (Rec), where

TP = The model correctly calculated a number of positive class samples.

TN = The model calculated a number of negative class samples with accuracy.

FP = The model incorrectly calculated a number of negative class samples.

FN = The model failed to accurately forecast a number of positive class samples.

$$\text{Accuracy (Acc)} = \frac{TP+TN}{TP+TN+FP+FN}$$

$$\text{Precision (Pre)} = \frac{TP}{TP+FP}$$

$$\text{Recall (Rec)} = \frac{TP}{TP+FN}$$

$$\text{F1-score} = \frac{2\times \text{Precision}\times \text{Recall}}{\text{Precision}+\text{Recall}}$$

$$\text{MCC} = \frac{(TP\times TN)-(FP\times FN)}{\sqrt{(TP+FP)(TP+FN)(TN+FP)(TN+FN)}}$$

4 Result Analysis

4.1 Environment Setup

A comprehensive overview of the setup of the environment for training various CNN models can be found in Table 3. Significant information such as Model Name, GPU Name, Batch Size, Optimizer, Learning Rate, Epoch, Activation, and Data Augmentation are included in table rows that are linked to specific models. Remarkably, every model on the NVIDIA TESLA T4 GPU was trained for a maximum of 10 epochs with a constant batch size of 224 × 224. The activation function ReLU was uniformly applied to each model in order to introduce non-linearity into the network. The proposed

Table 3 The environment setup summary for the model's applications

Model name	GPU name	Batch size	Optimizer, learning rate	Epoch	Activation function	Data augmentation
EfficientNetB0	NVIDIA TESLA T4	224 × 224	Adam, lr:0.0001	10	ReLU	Applied
EfficientNetB1	NVIDIA TESLA T4	224 × 224	Adam, lr:0.0001	10	ReLU	Applied
EfficientNetB2	NVIDIA TESLA T4	224 × 224	Adam, lr:0.0001	10	ReLU	Applied
EfficientNetB3	NVIDIA TESLA T4	224 × 224	Adam, lr:0.0001	10	ReLU	Applied
MobileNetV2	NVIDIA TESLA T4	224 × 224	Adam, lr:0.0001	10	ReLU	Applied
ResNet50	NVIDIA TESLA T4	224 × 224	Adam, lr:0.0001	10	ReLU	Applied
ResNet101	NVIDIA TESLA T4	224 × 224	Adam, lr:0.0001	10	ReLU	Applied
ResNet152V2	NVIDIA TESLA T4	224 × 224	Adam, lr:0.0001	10	ReLU	Applied
DenseNet201	NVIDIA TESLA T4	224 × 224	Adam, lr:0.0001	10	ReLU	Applied
Proposed model (BananaLeaf-Net)	NVIDIA TESLA T4	224 × 224	Adam, lr:0.0001	50	ReLU	Applied

BananaLeaf-Net model uses Adam with an adaptive learning rate that starts at 0.0001, while some other models used the Adam optimizer with a fixed learning rate of 0.0001. Interestingly, in order to increase the model's robustness and generalization performance, random alterations were incorporated throughout the training phase of the entire model through the use of data augmentation. This configuration provides vital insights into the typical environment for CNN model training in the context of banana leaf disease classification, which helps with reproducibility and comparison across various experiments and architectures.

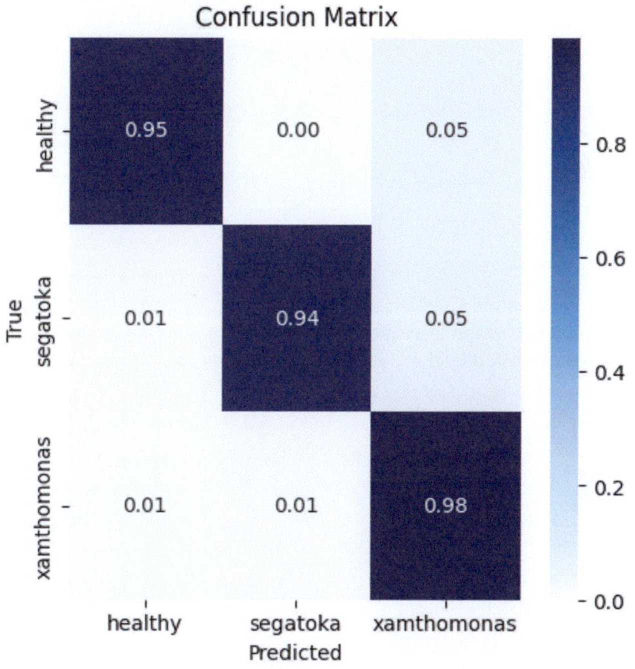

Fig. 3 Confusion matrix of proposed BananaLeaf-Net model

4.2 Result Analysis of Proposed Model

Analyzing the confusion matrix from Fig. 3, the proposed model BananaLeaf-Net outperformed the different base models, including CNN model's EfficientNet B0, EfficientNet B1, EfficientNet B2, EfficientNet B3, MobileNetV2,ResNet50, ResNet101, ResNet152V, and DenseNet201. However, every model that was altered produced positive results. In according to overall performance, the BananaLeaf-Net model outperformed. When comparing the percentage of accurately predicted values that the model produced to the original truth value, the diagonal blue hues in the confusion matrix diagram indicate this. To provide more insight into our model, Figs. 4 and 5 show the accuracy and loss curve outputs. Also, the proposed model's ROC curve is shown in Fig. 6.

4.3 Comparison Analysis with Base Models and Existing Models

The accuracy of various deep learning models for identifying diseases of banana leaves is compared in Table 4. All significant metrics show significant improvements in performance when comparing the proposed BananaLeaf-Net model to base models like EfficientNet, MobileNet, ResNet, and DenseNet. With a testing accuracy of 97.00% and a validation accuracy of 97.24%, BananaLeaf-Net performs much better than ResNet101, the next best model, which only achieves a testing accuracy of 70.22%. With a precision score of 96.33%, BananaLeaf-Net outperforms both ResNet50 and ResNet101, which have accuracy values of 75.13% and 64.78%, respectively, in accurately determining true positives with few false positives. While other approaches, such as DenseNet201 and ResNet152V2, only can identify 55.58% and 58.77% of true positive examples, BananaLeaf-Net leads in recall with 95.66%, demonstrating its power in detecting the majority of true positive instances. With an F1-score of 96.00%, BananaLeaf-Net outperforms other models in terms of preci-

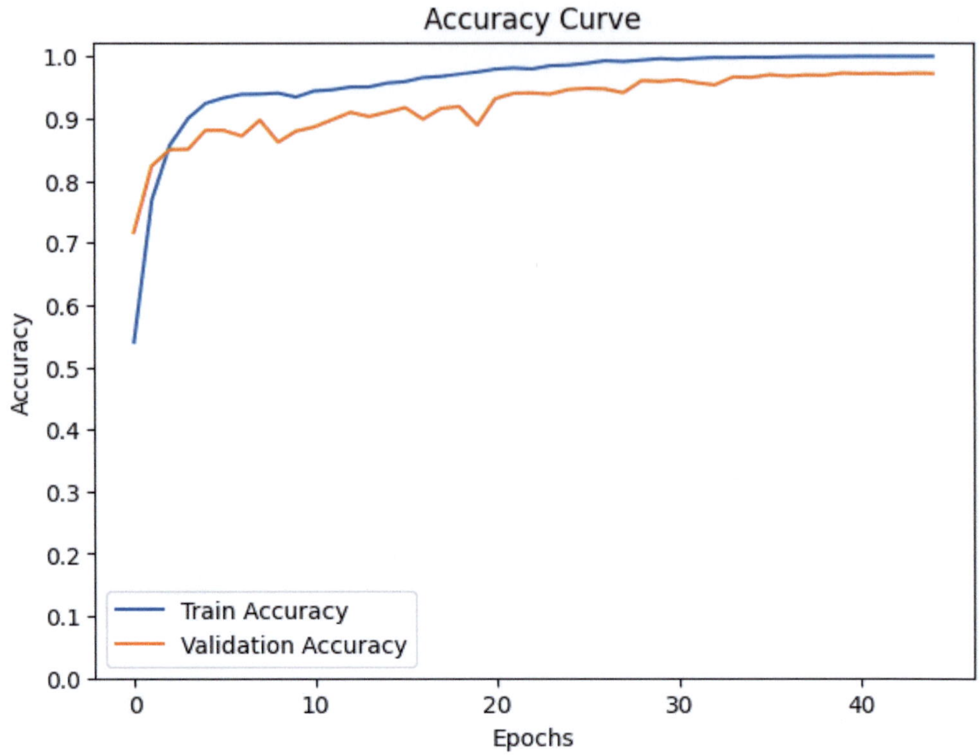

Fig. 4 The training and validation accuracy graph of proposed BananaLeaf-Net model

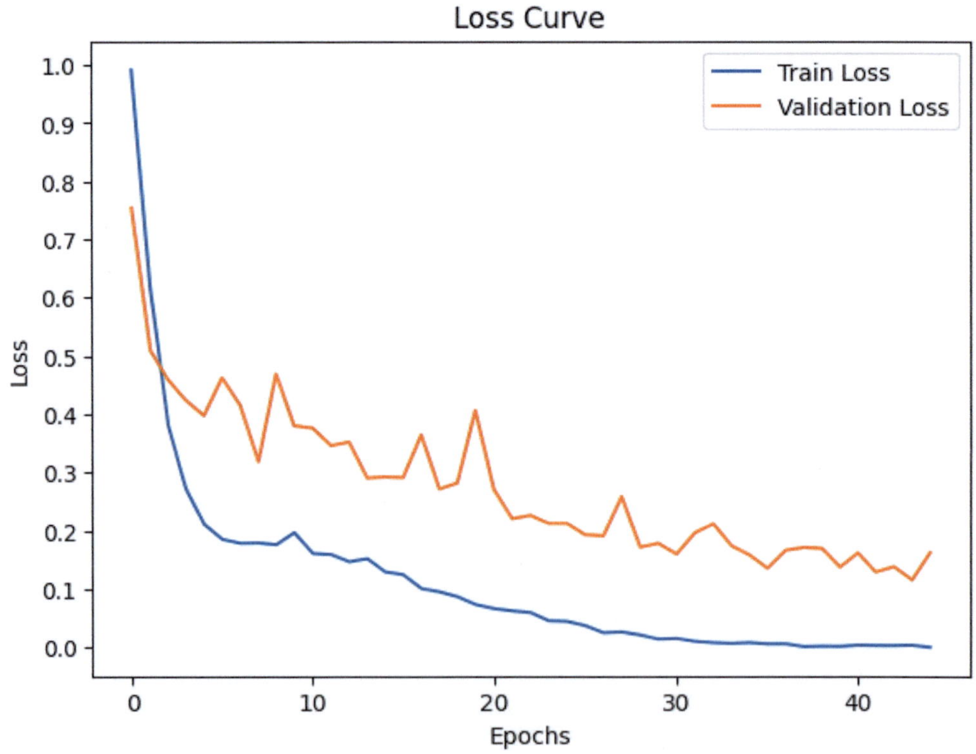

Fig. 5 The training and validation loss graph of proposed BananaLeaf-Net model

sion and recall, with ResNet101 scoring the highest at 50.47%. Additionally, the MCC score for BananaLeaf-Net is at 95.00, demonstrating near-perfect classification, compared to the next highest MCC score of 39.06 from ResNet101. These findings clearly demonstrate that BananaLeaf-Net provides greater precision and overall performance than the other models in the comparison, giving it an exceptional and highly reliable solution for classifying banana leaf diseases.

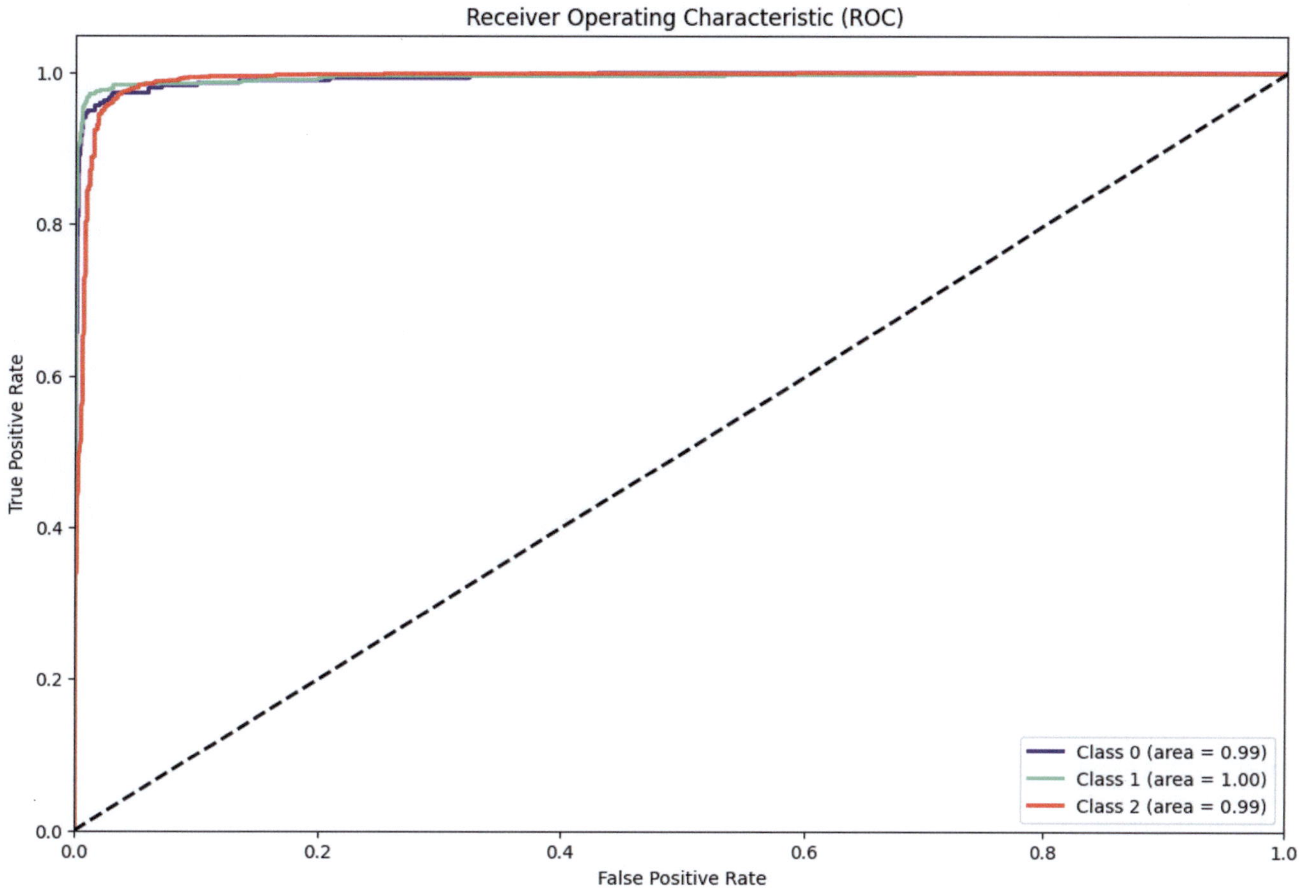

Fig. 6 The ROC curve of proposed BananaLeaf-Net model

Table 4 Model accuracy comparison for banana leaf disease classification using image dataset with proposed BananaLeaf-Net model

Model name	Validation accuracy	Testing accuracy	Precision	Recall	F1-score	MCC score
EfficientNetB0	61.69	60.82	43.11	39.12	38.44	13.98
EfficientNetB1	57.12	57.16	41.25	40.66	40.80	14.72
EfficientNetB2	60.40	60.86	44.18	39.77	39.77	12.88
EfficientNetB3	61.82	62.93	37.43	33.76	28.65	02.20
MobileNetV2	65.57	64.09	51.96	37.63	34.77	15.62
ResNet50	70.02	68.97	75.13	45.23	44.66	33.20
ResNet101	69.59	70.22	64.78	51.04	50.47	39.06
ResNet152V2	64.93	67.41	65.41	58.77	56.74	36.87
DenseNet201	68.59	68.19	62.28	55.58	57.02	36.26
Proposed model (BananaLeaf-Net)	97.24	97.00	96.33	95.66	96.00	95.00

The BananaLeaf-Net model provides important insights into the efficacy of deep learning architectures in medical imaging applications when compared with conventional base models. The proposed model represents its potential for improved accuracy by outperforming the base models in terms of accuracy, F1-score, recall, precision, and MCC (Matthews Correlation Coefficient) score. Because it can capture complex patterns and variables connected to banana leaf disease, the above benefit suggests that the BananaLeaf-Net model has better sensitivity and specificity than other comparable models. More verification that the proposed model may robustly generalize to unknown data comes from the analysis of training and validation loss over epochs, which demonstrates that it not only has improved convergence but also has the abil-

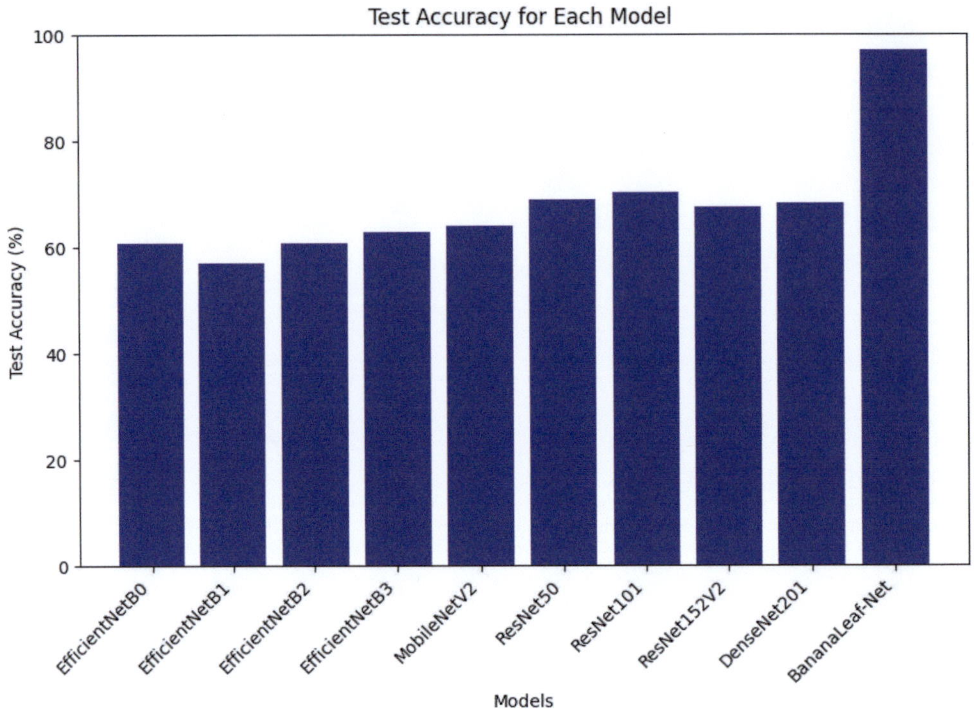

Fig. 7 The accuracy graph for all models including BananaLeaf-Net model

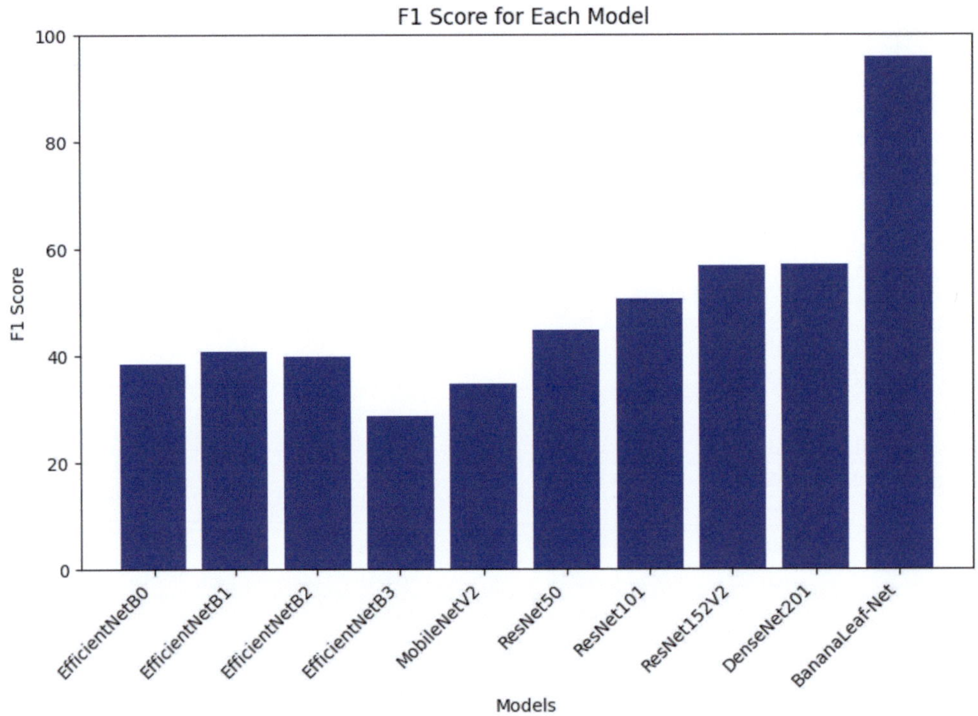

Fig. 8 The F1-score graph for all models including BananaLeaf-Net model

ity to lessen overfitting. The results provide how customized deep learning models may help increase the accuracy and consistency of banana leaf disease.

The performance evaluation of the BananaLeaf-Net model against a collection of well-known base model emphasizes how crucial architectural design choices are to optimizing the accuracy of banana leaf disease classification. This targeted approach not only produces improved performance metrics but also highlights the importance of domain-specific model development for medical imaging applications. The BananaLeaf-Net model has demonstrated superior performance compared to its competitors in various domains such

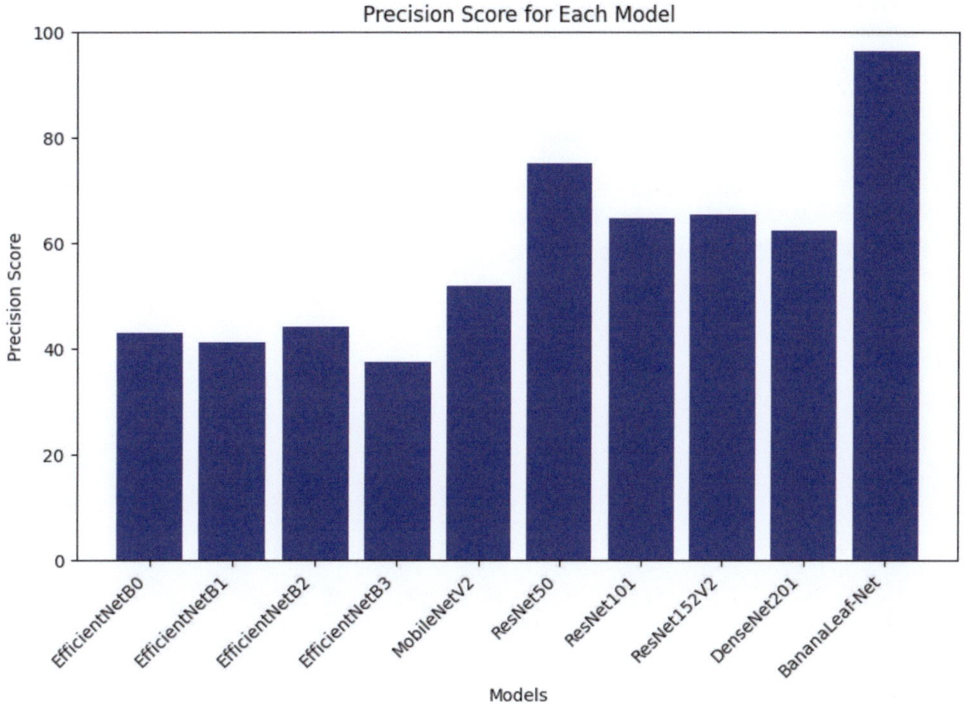

Fig. 9 The precision graph for all models including BananaLeaf-Net model

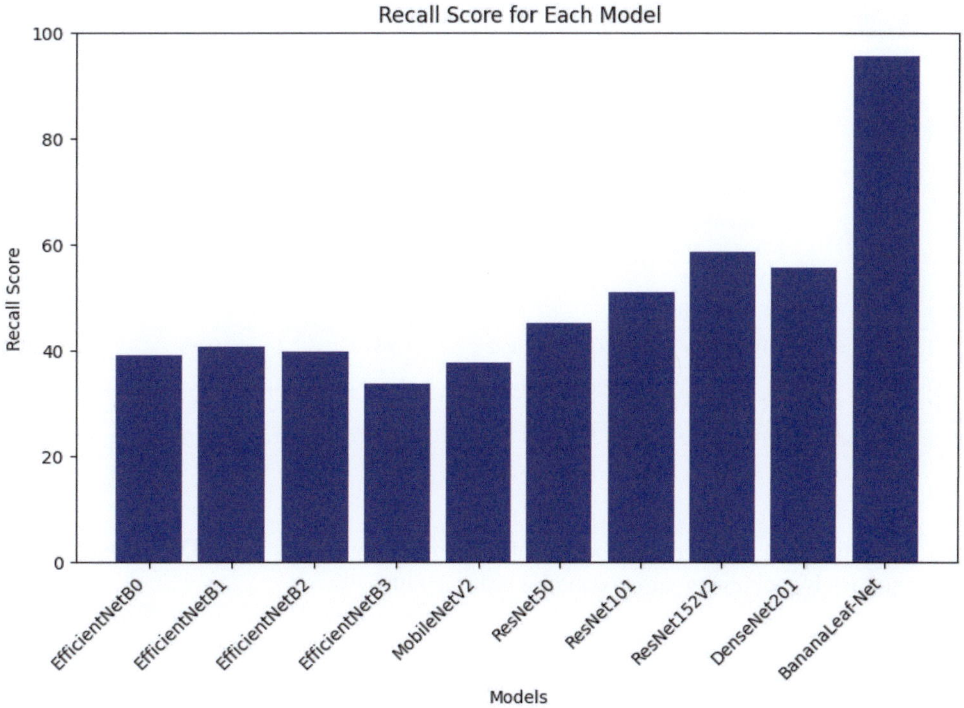

Fig. 10 The recall graph for all models including BananaLeaf-Net model

as accuracy, F1-score, recall, precision, and MCC score. This highlights the significance of further research and innovation focused on enhancing these models for implementation in clinical settings. In addition, it validates the potential of deep learning techniques to revolutionize the field of disease classification. The Accuracy, F1-score, Precision, Recall, and MCC graphs in comparison to other models are shown in Figs. 7, 8, 9, 10, and 11. The outputs to further elucidate our model.

An comprehensive review of model accuracies and performance metrics for image classification tasks is given in Table 5. It includes several well-known models from previous studies as well as the proposed BananaLeaf-Net model.

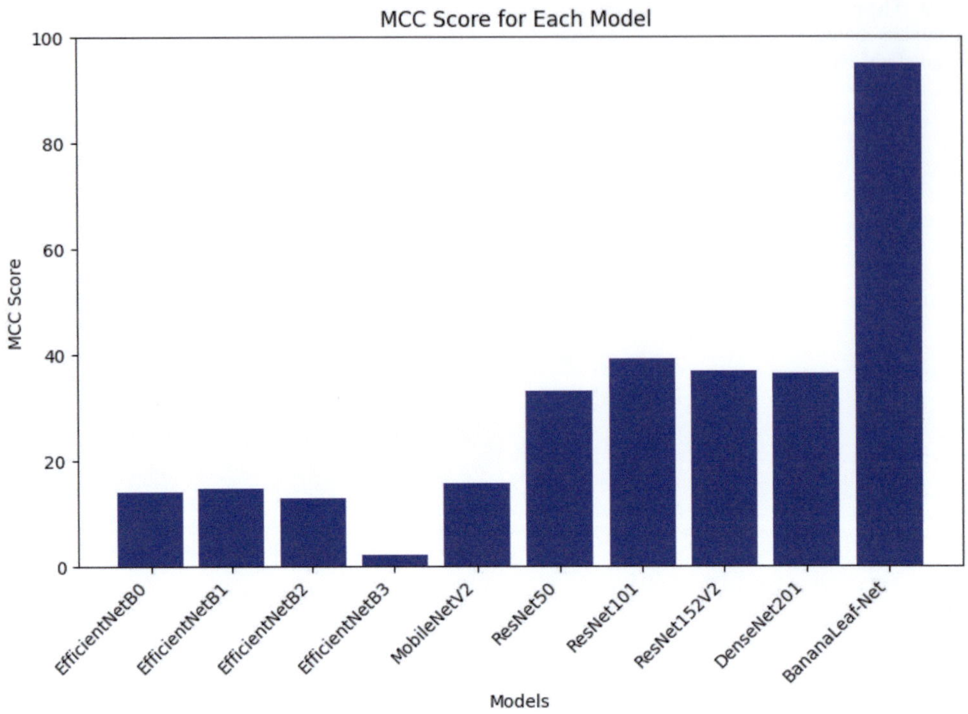

Fig. 11 The MCC graph for all models including BananaLeaf-Net model

Table 5 Model accuracy comparison with existing researches with proposed BananaLeaf-Net model

Author	Algorithm	Test accuracy (%)	Precision	Recall	F1-score
[2]	KVA	95.98	89.31%	–	–
[5]	BananaSqueezeNet	95.25	96.53%	95.25%	–
[1]	Modified ResNet50V5	94	88%	91%	89%
[9]	Ensemble models (SVM, Naïve Bayes, and K-NN)	92	–	–	–
[18]	YOLOV4	90	–	–	–
[17]	ShuffleNet V2	95	–	–	–
[23]	Ensemble models (MGAN and MFRCNN)	96	94%	95%	–
[24]	CNNs and SVMs	94	–	–	–
Proposed model	BananaLeaf-Net	97	96.33%	95.66%	96%

It is interesting that the proposed model outperforms the previous studies by achieving an outstanding accuracy of 97%, while the past works yielded acceptable accuracies between 90 and 95.5%. Furthermore, the proposed model performs outstandingly in terms of accuracy, recall, and F1-score, with 96.33% precision, 96.66% recall, and 96% F1-score. This demonstrates the efficiency and progress of the BananaLeaf-Net model, making it a superior choice for image classification issues.

5 Discussion

Using a modified MobileNetV2 architecture, the proposed BananaLeaf-Net model obtains a high accuracy of 97% in banana leaf disease classification. The model benefits from features acquired on a large-scale dataset since it is started with pre-trained ImageNet weights. Most importantly, the backbone's whole set of convolutional layers is unfrozen, allowing for task-specific feature learning and network optimization for various types of banana leaf diseases. Additional fully connected (linear) layers are added to the system, improving upon MobileNetV2's 1280 final output channel. To ensure the model acquires rich, complex patterns from the input images, these extra layers slowly decrease the dimensionality from 4096 to 2048, then to 1024, and finally to the number of output classes. In order to generate non-linearity, the model further includes ReLU activations after each linear layer. Batch normalization is then used along with this to stabilize training by addressing internal covariate shift. Moreover, dropout layers (probabilities of 0.5 and 0.4) are added to improve the model's generalization capacity by preventing overfitting by randomly zeroing out neuron activations during training. Training is aimed to balance fast convergence and model stability, with AdamW optimizing the weight decay regularization better than normal Adam and OneCycleLR learning rate scheduler providing dynamic control. Gradient clipping is one technique that prevents problems like growing gradients, while early stopping prevents needless overtraining by monitoring validation accuracy and halting training when gains plateau. The BananaLeaf-Net model performs better than other models in the classification of banana leaf diseases through the combination of dropout mechanisms, dynamic learning rate modification, and enhanced regularization.

The ability of the model to recognize varied trends and factors that are important for banana leaf disease highlights how much better identification accuracy and dependability it can provide. With new insights and chances to improve patient outcomes in the management of complicated diseases like banana leaf disease, the ground-breaking results obtained by this model signify the beginning of a new era in precision. Therefore, the BananaLeaf-Net model's performance shows how cutting-edge deep learning models may significantly improve imaging and farmers' satisfaction, clearing the way for further improvements in accuracy and more specialized treatment methods. Furthermore, the results and lack of a significant computational overhead demonstrate the differences between this model and other models, highlighting the strength of our model's ability to produce excellent results while retaining efficiency and using fewer resources. Even while ResNet101 and other models are more efficient, they are still unable to reach the same level of performance.

6 Conclusion

In conclusion, the BananaLeaf-Net model shows impressive effectiveness in identifying diseases of the banana leaf, with a high accuracy of 97%, which is essential for aiding farmers and raising the production of agriculture. Using cutting-edge deep learning models like MobileNetV2, ResNet, DenseNet, and different iterations of EfficientNet (B0, B1, B2, B3), the study emphasizes how crucial it is to analyze a variety of architectures in order to increase classification efficiency and accuracy. Through the provision of useful and dependable solutions for early diagnosis, the comparative comparison of these models highlights the need of applying advanced convolutional neural networks (CNNs) for difficult tasks like disease classification. This achievement provides up exciting opportunities for practical uses by providing farmers with a widely accessible tool to identify diseases early and take quick preventive action, thus improving crop productivity and sustainability in the agriculture field. The effective use of state-of-the-art methods in this study highlights the possibility of additional deep learning innovations resulting in improvements in precision agriculture.

References

1. Sunitha, P., Uma, B., Channakeshava, S., Babu, S.: A fully labelled image dataset of banana leaves deficient in nutrients. Data Brief **48**, 109155 (2023)
2. Deng, J.-S., Huang, W.-Q., Zhou, G.-X., Hu, Y.-H., Li, L.-J., Wang, Y.-F.: Identification of banana leaf disease based on KVA and GR-ARNet1. J. Integr. Agric. (2023)
3. Thiagarajan, J.D., Kulkarni, S.V., Jadhav, S.A., Waghe, A.A., Raja, S., Rajagopal, S., Poddar, H., Subramaniam, S.: Analysis of banana plant health using machine learning techniques. Sci. Rep. **14**(1), 15041 (2024)
4. Rajalakshmi, N., Saravanan, S., Arunpandian, J., Mathivanan, S.K., Jayagopal, P., Mallik, S., Qin, G.: Early detection of banana leaf disease using novel deep convolutional neural network. J. Data Sci. Intell. Syst. (2024)
5. Bhuiyan, M.A.B., Abdullah, H.M., Arman, S.E., Rahman, S.S., Al Mahmud, K.: Bananasqueezenet: a very fast, lightweight convolutional neural network for the diagnosis of three prominent banana leaf diseases. Smart Agric. Technol. **4**, 100214 (2023)
6. Arman, S.E., Bhuiyan, M.A.B., Abdullah, H.M., Islam, S., Chowdhury, T.T., Hossain, M.A.: Bananalsd: a banana leaf images dataset for classification of banana leaf diseases using machine learning. Data Brief **50**, 109608 (2023)
7. Elinisa, C.A., Mduma, N.: Mobile-based convolutional neural network model for the early identification of banana diseases. Smart Agric. Technol. **7**, 100423 (2024)
8. Chaudhari, V., Patil, M.P.: Detection and classification of banana leaf disease using novel segmentation and ensemble machine learning approach. Appl. Comput. Syst. **28**(1), 92–99 (2023)
9. Syihad, I.R., Rizal, M., Sari, Z., Azhar, Y., et al.: CNN method to identify the banana plant diseases based on banana leaf images by giving models of resnet50 and VGG-19. Jurnal RESTI (Rekayasa Sistem dan Teknologi Informasi) **7**(6), 1309–1318 (2023)
10. Sangeetha, R., Logeshwaran, J., Rocher, J., Lloret, J.: An improved agro deep learning model for detection of panama wilts disease in banana leaves. AgriEngineering **5**(2), 660–679 (2023)
11. Yan, K., Shisher, M.K.C., Sun, Y.: A transfer learning-based deep convolutional neural network for detection of fusarium wilt in banana crops. AgriEngineering **5**(4), 2381–2394 (2023)
12. Mohanraj, E., Hariharan, R., Keerthana, S., Cavin, K.: Banana leaf disease detection using advanced convolutional neural network. In: 2023 International Conference on Sustainable Computing and Smart Systems (ICSCSS), pp. 597–603. IEEE (2023)
13. Thomas, D., David, J.M.: Banana plant disease detection-hybrid machine learning approach. In: 2023 Annual International Conference on Emerging Research Areas: international Conference on Intelligent Systems (AICERA/ICIS), pp. 1–5. IEEE (2023)

14. Kakati, J.B., Das, T.K.: Classification of healthy and unhealthy banana leaves using deep learning approach: A comparative assessment. In: 2023 4th International Conference on Computing and Communication Systems (I3CS), pp. 1–5. IEEE (2023)
15. Mathew, D., Kumar, C.S., Cherian, K.A.: Classification of leaf spot diseases in banana using pre-trained convolutional neural networks. In: 2023 International Conference on Control, Communication and Computing (ICCC), pp. 1–5. IEEE (2023)
16. Singh, P., Kumar, A., Chandel, R., Dongre, P., *et al.*: Deep learning insights into banana sigatoka disease: Resnext50 for seriousness classification. In: 2024 IEEE International Conference on Interdisciplinary Approaches in Technology and Management for Social Innovation (IATMSI), vol. 2, pp. 1–6. IEEE (2024)
17. Yumang, A.N., Baguisi, J.M., Buenaventura, B.R.S., Paglinawan, C.C.: Detection of black sigatoka disease on banana leaves using ShuffleNet v2 CNN architecture in comparison to SVM and KNN techniques. In: 2023 15th International Conference on Computer and Automation Engineering (ICCAE), pp. 281–286. IEEE (2023)
18. Ibarra, N.C., Rivera, M.P., Manlises, C.O.: Detection of panama disease on banana leaves using the yolov4 algorithm. In: 2023 15th International Conference on Computer and Automation Engineering (ICCAE), pp. 209–214. IEEE (2023)
19. Ibarra, N.C., Rivera, M.P., Manlises, C.O.: Determination of leaf degradation percentage for banana leaves with panama disease using image segmentation of color spaces and OpenCV. In: 2023 15th International Conference on Computer and Automation Engineering (ICCAE), pp. 269–275. IEEE (2023)
20. Tanwar, V., Sharma, B., Aanand, V.: Implementing a hybrid CNN-SVM model for banana leaf disease classification. In: 2023 International Conference on Research Methodologies in Knowledge Management, Artificial Intelligence and Telecommunication Engineering (RMKMATE), pp. 1–5. IEEE (2023)
21. Suryavanshi, A., Kukreja, V., Shukla, A.N., Mehta, S., Joshi, K.: Farming crop in health analytics: Federated learning cnns in banana leaf disease detection. In: 2024 3rd International Conference for Innovation in Technology (INOCON), pp. 1–6. IEEE (2024)
22. Abisha, A., Bharathi, N.: Impact of image pre-processing in image quality and segmentation for banana leaf disease prediction. In: 2023 International Conference on Sustainable Communication Networks and Application (ICSCNA), pp. 1302–1307. IEEE (2023)
23. Raja, N.B., Rajendran, P.S.: A novel fuzzy-based modified GAN and faster RCNN for classification of banana leaf disease. J. Inst. Eng. (India): Ser. A **104**(3), 529–540 (2023)
24. Banerjee, D., Kukreja, V., Hariharan, S., Sharma, V.: Precision agriculture: classifying banana leaf diseases with hybrid deep learning models. In: 2023 IEEE 8th International Conference for Convergence in Technology (I2CT), pp. 1–5. IEEE (2023)
25. Samridhi, S., Kalpana, M., Parimalarangan, R., VenkatesaPalanichamy, N.: Identification of sigatoka leaf spot disease in banana using convolutional neural network (CNN). Asian J. Agric. Ext. Econ. Sociol. **41**(9), 931–936 (2023)
26. Yathurshan, K., Barathy, S., Karini, G., Sajitha, P., Abeygunawardhna, P., Rajapakse, K.: Leaf guard: detecting the diseases in banana leaf. In: 2023 5th International Conference on Advancements in Computing (ICAC), pp. 514–518. IEEE (2023)
27. Aliff, M., Luqman, M., Yusof, M.I., Sani, N.S., Syafiq, M.U., Sadikan, S.F.N., Mahmud, H.: Utilizing aerial imagery and deep learning techniques for identifying banana plants diseases. In: ITM Web of Conferences, vol. 60, p. 00013. EDP Sciences (2024)
28. Genet, Y.H., Sinshaw, N.T., Assefa, B.G., Mohapatra, S.K.: Sigatoka and Xanthomonas banana leaf disease detection via transfer learning. Scientia Iranica (2024)
29. Freire, A., Roque, R.D.L., Leandro, H.D.S., Fernandes, B.J.: Interpretable diagnosis of banana leaf sigatoka disease using CNNS and Shapley values. In: 2023 IEEE Latin American Conference on Computational Intelligence (LA-CCI), pp. 1–6. IEEE (2023)

Improving Agronomic Disease Detection and Classification: The Superiority of Hybrid Inception-Xception Ensemble Model for Rice Leaf Disease Classification

Md Sadman Hafiz, Asif Anwar, Jungpil Shin, and Jahir Ibna Rafiq

Abstract

Over fifty percent of the global population relies on rice as a fundamental grain, making it crucial for worldwide food security. Its nutritional value, versatility, and adaptability make it indispensable for different diets and agricultural economies. Classification and early detection of rice diseases are crucial for timely action, reducing crop damage, and ensuring food security by maintaining high agricultural productivity. Computer-aided methods to detect rice leaf diseases have been introduced to overcome the limitations of manual detection. They enable faster and more accurate identification and timely intervention to prevent significant crop loss and ensure sustainable agricultural practices. Previous studies have identified diseases in rice plants. In our experiment, we attempted to improve the accuracy disease detections and sub-classified them into mild and severe. It can detect and classify the disease according to the following labels: Bacterial Blight, Blast, Brownspot, and Tungro. We tested different convolutional neural network architectures to determine which architecture provided the best accuracy. We used Inception-V3, Xception, Densenet, MobileNet-V2, VGG-16, VGG-19, Inception Resnet, NASNet Mobile, and an ensemble model combining Xception and Inception-V3. We have seen that the ensemble model outperformed all the other models for this dataset, achieving the best results in all evaluation metrics with an accuracy of 93.33%, which has emerged as the best among all the tested methods for this experiment.

Keyword

Rice leaves disease • Deep learning application • Ensemble model • Inception V3 • Hybrid models • Image processing

1 Introduction

Rice is a significant player in the global food supply network, as the predominant source of nourishment for more than half of the world's population. Its widespread cultivation and consumption highlights its crucial role in maintaining food security in various geographical and cultural settings. Rice's nutritional abundance, versatility in cooking, and capacity to thrive in different agricultural conditions make it a crucial staple in the diets of many regions. Nevertheless, the process of growing rice, like any other agrarian plant, is vulnerable to several diseases which can hurt its productivity and quality. The emergence of illnesses in rice plants, namely in the leaves, crucial for photosynthesis and growth, presents a substantial threat to crop well-being and output.

The agricultural sector faces the dual difficulty of dealing with diseases such as Bacterial Blight, Blast, Brownspot, Tungro effectively and swiftly detecting and identifying them. Conventional approaches to identifying diseases in rice crops mainly depend on visual examinations conducted by farmers or agricultural specialists. However, this strategy has limitations in terms of its scope and efficiency. The technique involves significant physical labor, considerable time investment, and reliance on subjective judgment, potentially resulting in mistakes. Furthermore, delayed identification of infections might permanently harm crops, thereby endangering the

Md S. Hafiz (✉) · A. Anwar
Institute of Information and Communication Technology, Shahjalal University of Science and Technology, Sylhet, Bangladesh
e-mail: hafiz.sust333@gmail.com

A. Anwar
e-mail: asif.anwar.sajid@gmail.com

J. Shin
School of Computer Science and Engineering, The University of Aizu, Aizuwakamatsu, Japan
e-mail: jpshin@u-aizu.ac.jp

J. I. Rafiq
Biomedical Engineering Department, University of Toronto, Toronto, Canada
e-mail: JahirIbna.Rafiq@uhn.ca

M. F. Mridha and N. Dey (eds.), *Machine Vision in Plant Leaf Disease Detection for Sustainable Agriculture*, Studies in Computational Intelligence 1202, https://doi.org/10.1007/978-981-96-4520-6_5

food security and economic stability of regions that rely on rice farming.

The introduction of computer-aided disease detection tools in agriculture represents notable progress in agricultural practices. The application of convolutional neural networks (CNNs) to identify and categorize rice leaf diseases is a revolutionary method for agricultural management. These methods provide a faster, more objective, and more precise way to identify plant diseases than the traditional methods. The combination of deep learning-based image processing and machine learning technologies allows for the accurate identification and classification of different rice leaf diseases, making it easier to implement timely and efficient intervention strategies.

Our research aimed to improve the precision of disease classification and incorporate the ability to categorize diseases into different levels of severity. This study investigates the effectiveness of different CNN architectures, such as Inception-V3, Xception, DenseNet, MobileNet-V2, VGG-16, VGG-19, InceptionResNet, and NASNet Mobile, and a new merging approach that ensembles Xception and Inception-V3. The goal was to ascertain which of these architectures produced the best results.

The rest of the paper is structured as follows: First, the literature review approach is described in Sects. 2 and 3 describes the methodology we have followed. In Sect. 4, the architecture of the proposed model is explained. The evaluation of the model performance and comparison with the other models are presented in Sect. 5. Finally, in Sect. 6, we conclude the paper.

2 Literature Review

Latif et al. used a Deep Convolutional Neural Network (DCNN) transfer learning method in conjunction with drone technology and the Internet of Things (IoT) to propose a new approach for resolving the global problem of rice diseases. The modified VGG19-based transfer learning method showed exceptional accuracy (96.08%) in detecting and classifying six different rice leaf diseases, outperforming existing approaches in the literature. This study proposed the methodology to minimize crop losses and increase agricultural productivity. Dey et al. [1] investigated how well Convolutional Neural Networks and pre-trained deep CNN models can quickly recognize Hispa, brown spot, leaf blight, and NPK deficiency symptoms in rice crops using different image datasets. The VGG19 model achieved the highest accuracy of 91.8%, particularly when trained on a mixture of public and field datasets. This study shows how non-invasive deep-learning techniques can accurately detect different signs of stress in rice crops. This provides farmers with a valuable tool for using mitigation strategies and precision agriculture at the right time. The results also suggest the potential applicability of similar learning algorithms for detecting stress symptoms in economically important crops. A study [2] was proposed by Singh et al. in which a specific Convolutional Neural Network architecture was proposed, which was developed to efficiently detect and classify common diseases in rice plants to combat threats to agricultural productivity. Their proposed model achieved a maximum accuracy of 99.83% using the Adaptive Moment Estimation (Adam) optimization method. This study provides an advanced process for accurate disease detection in rice plants, which could improve food security and help farmers to intervene promptly.

Liu et al. made a significant contribution [3] to identifying rice diseases by addressing the challenges posed by changing environmental conditions. It introduces a novel deep learning-based approach that involves the standardization of datasets and optimization of parameters. The result is an optimized model with a commendable accuracy of 98.64% for detecting rice blasts, rice blight, and bacterial leaf blight. The results provide an advanced and accurate method for identifying common rice diseases, which is essential for effective agricultural management in the face of changing environmental factors. A remarkable contribution [4] was reported by Mohapatra et al. to improve food safety and efficiency in agriculture by introducing an automated system for plant disease diagnosis using computer vision and deep convolutional neural network models. This study focused on detecting four types of rice leaf infections and offers a more efficient alternative to traditional manual diagnostic methods. The custom CNN model outperformeds pre-trained deep CNN models and achieved a remarkable accuracy of 97.47%, highlighting its potential to support and benefit global agriculture. Rahman et al. [5] contributed to the agricultural sector by utilizing deep learning, specifically convolutional neural networks, for accurate disease and pest detection in rice plants. This study adopts and fine tunes state-of-the-art model architectures such as InceptionV3 and VGG16, demonstrates their effectiveness using real-world datasets. In addition, CNN architecture optimized for mobile devices was proposed, achieving a remarkable accuracy of 93.3% with a reduced model size, which addresses the practical constraints of deployment on resource-limited platforms.

Bari et al. [6] applied a faster R-CNN algorithm for the real-time detection of rice leaf diseases, which overcomes challenges such as image background, ambiguous symptoms, and different image acquisition conditions. The suggested deep learning methodology attains elevated precision in identifying three rice leaf ailments (rice blast, brown spot illness, and hispa disease). It identifies healthy rice leaves, demonstrating its effectiveness in accurately detecting diseases in rice crops in real-time. Poorni et al. [7] contributed to agricultural disease detection by proposing a convolutional neural network-based model for the accurate detection of rice leaf diseases. By using a collection of different images of rice leaf diseases and applying transfer learning with the Inception-V3 model, the developed system achieved a high accuracy of 94.48%. This study addresses the urgent need for early diagnosis of rice plants, enabling farmers to contain diseases and ultimately improving global food security. The crucial role of agriculture in economic development was recognized in the paper [8] published by Shaikh et al. They also highlighted the importance of rice as a staple food for over 65% of the Indians. With the increasing prevalence of diseases affecting rice plants, including bacterial, fungal, and viral pathogens, this study presents a novel approach for the early detection of rice leaf diseases using a Faster-RCNN. The proposed model aims to reduce farmers' problems, increase production, and prevent financial losses in the agricultural industry owing to plant diseases by incorporating a distinctive dataset of field data and utilizing images from the Kaggle dataset.

A significant contribution to the detection of rice diseases is seen was reported by Sharma et al. [9]. They used convolutional neural networks and deep learning to create a method for detecting rice leaf blight (RLB). Using a large real-time image dataset, the study achieved high accuracies of 94.33% for binary classification and 95.3% for multi-classification based on disease severity. Kongara et al. [10] made an important contribution to agriculture by developing a convolutional neural network model that focuses on the identification and classification of paddy diseases. The study used deep learning, transfer learning, and image processing to identify more than 30 disease species, focusing on four common diseases (leaf blast, brown spot, hispa, and bacterial leaf blight). With a high accuracy of 91.23% for Inception-V3, this study provides a valuable tool for early disease detection and subsequent crop yield improvement for farmers in India.

This study by Chaudhari et al. is an essential contribution to agriculture because it suggests a CNN (Inception-ResNet) SVM model for identifying and treating diseases on rice leaves. The model achieved high accuracy (0.97) and precision (0.93) using image processing techniques and a refined dataset collected from agricultural fields. The ability to accurately classify and treat various diseases responds to farmers' challenges. It provides an improved and reliable solution for effective pest management of rice crops, thereby increasing agricultural productivity.

Another robust image-based framework [11] for rice disease detection was presented by Pal et al., specifically targeting BrownSpot disease using a Convolutional Neural Network(CNN). A user-defined dataset with 2223 images under complex conditions enables effective binary classification. The CNN model exhibited high accuracy (96.8%), outperforming state-of-the-art models and demonstrating its effectiveness in automatically detecting rice diseases. Chen et al. in their paper [12], presented a novel lightweight network architecture, MobInc-Net, developed for plant disease detection and recognition, which overcomes challenges such as computational complexity and memory overhead. By improving the inception module and using two-stage transfer learning, the proposed method achieved an impressive average detection accuracy of 99.21% in a public dataset and 97.89% in a local dataset. This contributes to an efficient solution for plant disease detection by emphasizing the performance of authentic experimental materials. For the diagnosis of rice, Shah et al. developed an automated method [13] using deep learning, image processing, and transfer learning with pre-trained models (Inception-V3, VGG16, VGG19, and ResNet50). The modified ResNet50 with connection skipping exhibited the highest accuracy of 99.75%, demonstrating superior performance in disease prediction.

A lightweight convolutional neural network architecture for automatically detecting rice leaf spots and rice leaf blight was proposed [14] by Hussain et al., with the aim of improving quality control in rice cultivation. The challenge of data scarcity is addressed through practical variance modeling and custom filter development, incorporating a reference protocol for filter suppression to improve the result. Thakur et al. presented a CNN model [15], "VGG-ICNN", which was developed to detect plant diseases based on leaves images. The model, which uses only 6 million parameters, outperforms current deep-learning approaches. It gained an accuracy of 99.16% in the PlantVillage dataset and consistently showed strong performance in different datasets for plant species, including apple, corn, and rice. Liang et al. [16] presented a novel approach for rice disease detection using a computer vision method that combines CNN object detection with an image tiling method. Including a leaf width prediction model and using tiled sub-images improves the efficiency of rice disease detection and overcomes the challenges posed by varying object sizes in real-world conditions. The proposed technique shows improved performance in terms of average accuracy on the YOLOv4 architecture compared to conventional methods. An automated approach for accurately detecting and classifying rice plant diseases using computer vision, image processing, machine learning, and deep learning is proposed by Haridasan et al. in their paper [17]. It addresses the most common diseases affecting Indian rice fields and achieved a high validation accuracy of 91.45% by integrating a SVM and CNNs classifer.

3 Methodology

Our study aims to improve the accuracy of machine learning tasks with multiple labels using an ensemble method that

Table 1 Summary of the dataset showing the types of rice leaf diseases and healthy leaves included

Leaf classification	Samples
Healthy	314
Mild Bacterial Blight	314
Mild Blast	314
Mild Brownspot	314
Mild Tungro	314
Severe Bacterial Blight	314
Severe Blast	314
Severe Brownspot	314
Severe Tungro	314

combines the best features of two advanced CNN architectures, Xception and Inception V3.

3.1 Dataset

The dataset was collected from Kaggle [18] for the experiments. The dataset contains healthy and diseased rice leaves images. Among the diseased rice leaves, four diseases were sub-classified as mild or severe. Table 1 presents the distribution of the datasets. Additionally, that table displays the quantity of photos in each class. Figure 1 shows the infected rice leaves, namely, Bacterial Blight, Blast, Brownspot, and Tungro, with a healthy leaf image.

The dataset includes images of four rice leaf diseases (bacterial blight, tuber leaf blight, brown spot, and tungro) with different instances for each disease. The authors combined images from Mendeley, data collected by Ethy Barpanda et al. [19], and additional images of healthy leaves from Kaggle and the UCI Machine Learning Repository. The dataset now comprises 3006 images, with 314 images in each training class and 20 in each test class, categorized by disease severity.

Figure 1 illustrates some sample images of the dataset used along with the many diseases they each have.

3.2 Image Pre-processing

Each image in our dataset was pre-processed to ensure uniformity and optimal model performance. The steps include:

3.2.1 Resizing

All images were resized to a uniform dimension of 120×120 pixels, making them suitable for the input layer of the CNNs.

$$\text{Resized Image} = \text{Resize}(\text{Original Image}, 120, 120) \quad (1)$$

3.2.2 Normalization

Pixel values are normalized to a [0, 1] to facilitate faster and more stable convergence during training.

$$\text{Normalized Image} = \frac{\text{Resized Image}}{255} \quad (2)$$

4 Model Architectures

Model architecture refers to the design or the structure of machine learning and deep learning models. In this section, the architecture of the proposed model is described.

4.1 Xception Model

The Xception architecture, derived from Inception, incorporates depthwise separable convolutions to enhance its performance.

Depthwise Separable Convolution: The process consists of two steps. A depthwise convolution was implemented, wherein a solitary convolutional filter was employed for each input channel. Subsequently, it is pointwise convolution, essentially a straightforward 1×1 convolution, is employed to generate a linear amalgamation of the output from the depthwise layer. This architectural design reduces the parameter count and computational costs while maintaining or improving the model's performance.

Additional Components: Similar to the majority of Convolutional Neural Networks (CNNs), incorporate activation functions, often Rectified Linear Units (ReLU), pooling layers to reduce the size of features, and fully linked layers that contribute to the ultimate output.

Equation: The output of a depthwise separable convolution is given by:

$$\text{Output} = \text{DepthwiseConv}(\text{Input}) + \text{PointwiseConv}(\text{Depthwise Output}) \quad (3)$$

Figure 2 demonstrates the architecture of Xception. It shows a simplified block diagram of the architecture of the Xception model, a type of convolutional neural network. The process starts with an input that passes through several parallel streams. Each stream consists of a sequence of operations, starting with a 1×1 convolution and then a 3×3 convolution for two streams. The third layer contains a 1×1 convolutional layer followed by a 3×3 max-pooling layer. The results of these parallel streams are then combined into a single result. The model's capacity to recognize intricate patterns in the input space is improved by this architecture, which allows it to learn from a variety of representations of the input data across several spatial hierarchies. The Xception architecture efficiently decouples cross-channel and spatial correlations by

Fig. 1 Dataset sample images

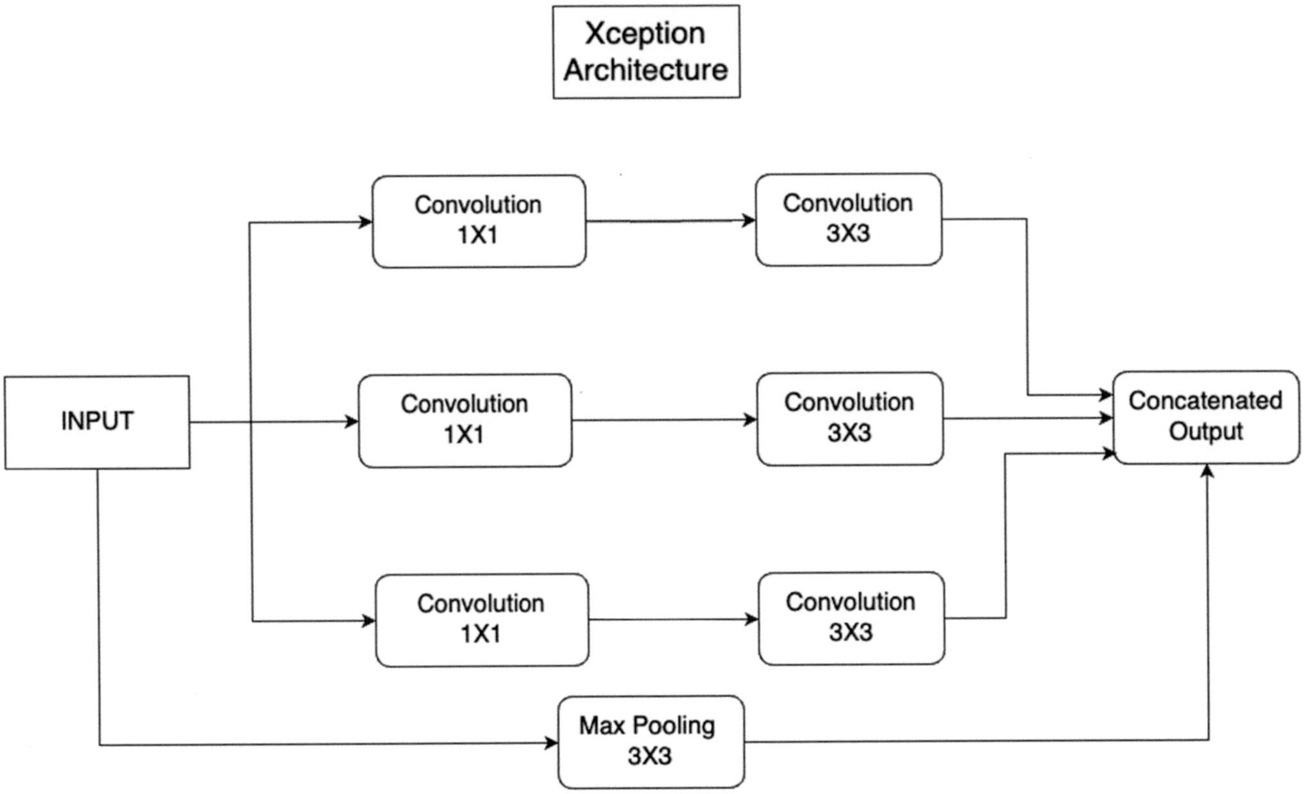

Fig. 2 Model architecture of Xception

utilizing depthwise separable convolutions, leading to more accurate and efficient feature extraction.

4.2 Inception V3 Model

Inception V3, A deep CNN model, has demonstrated its efficacy in feature extraction and classification by achieving state-of-the-art performance in various image recognition tests. It is a member of the inception family of networks, initially developed by Google. To enhance its efficacy and performance, Inception V3 implements numerous critical concepts.

Factored Convolutions: Inception V3 reduces the computational complexity while maintaining high accuracy by factorizing convolutions into simpler convolutions.

Auxiliary Classifiers: These incorporated into the training process to enhance the convergence rate and propagate the gradient.

Grid Size Reduction: It various methods have been implemented to reduce the grid size, thereby enhancing the network's efficiency and manageability.

Multiple-Sized Filters: A notable characteristic of this system is the incorporation of convolutional filters of different sizes (such as 1×1, 3×3, 5×5) inside the same layer which enabled the model to capture information at different scales. Inception V3 employs a factorization technique to break downlarger convolutions into smaller and more efficient ones, hence decreasing the computational workload.

Mixed Convolution Layers: Combines 1×1, 3×3, and 5×5 convolutions, as illustrated in Fig. 3.

Equation: The output of a mixed layer is calculated as:

$$\text{Output} = \sum_i \text{Conv}_{\text{Size } i}(\text{Input}) \tag{4}$$

Figure 3 shows a component of the Inception v3 model architecture, namely an Inception module. This module is characterized by its parallel paths of convolutions with different kernel sizes and a pooling operation. The "base" indicates an input from a previous layer, which is then processed in parallel by convolutions of different sizes: 1×1, 3×3, and combinations of 1×3 followed by 3×1 to capture patterns at different scales. A 1×1 convolution also follows a path with a pooling operation. These parallel paths are designed to capture information at different resolutions and contexts. The outputs of all paths are combined in the "Filter Concat" node, creating a more comprehensive representation of the input. This concatenated output is then usually fed into the next network layer. With the help of introduction modules

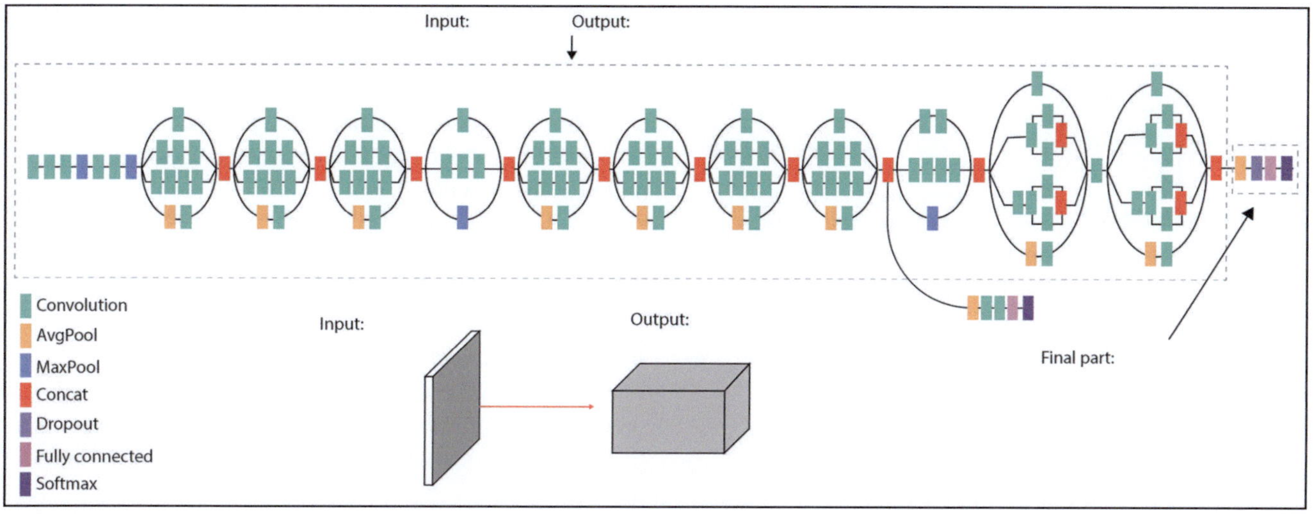

Fig. 3 Model architecture of Inception-V3

like this, the network can be deeper and broader while controlling computational costs.

4.3 Ensemble Method

An ensemble method in machine learning integrates the predictions of numerous models to get a final prediction, typically leading to enhanced accuracy and resilience compared to individual models. This is the operational process, specifically in the context of merging the Xception and InceptionV3 models.

4.3.1 Ensemble Learning Methodology

Training Individual Models: The same dataset is used to train the Xception and Inception V3 models were trained separately at first. Every model learns and makes predictions by utilizing its own design and strengths.

Diversity in Learning: The key to a successful ensemble is the diversity among the models. Xception and Inception V3 have different architectural designs, which means that they learn and generalize the data differently. Xception uses depthwise separable convolutions, focusing on channel-wise relationships, whereas Inception V3 uses multiple-sized filters to capture various features at different scales. This diversity ensures that each model captures multiple aspects of the data.

Combining Predictions: Once models are trained, their predictions are combined to make a final decision. There are several methods to combine these predictions:

- **Averaging**: The simplest method is to take the average of the predictions from both the models. If the models output probabilities, this would mean averaging of these probabilities.
- **Weighted Averaging**: If one model performs better, more weight can be assigned to its predictions.
- **Voting**: To improve accuracy and resilience in classification tasks, each model's prediction can be seen as a "vote," with the final result decided by a majority voting process.
- **Stacking**: A more complex method is training a new model-often called an ensemble model or meta-model-to efficiently integrate the results of the separate models, utilizing their advantages to produce better overall performance.

The final prediction is the average of both models' outputs.

$$\text{Ensemble Output} = \frac{\text{Xception Output} + \text{Inception V3 Output}}{2} \tag{5}$$

4.3.2 Error Reduction

The ensemble method reduced the likelihood of erroneous prediction. If one model makes an incorrect prediction, the other model's correct prediction can offset this mistake, leading to a proper final prediction. A basic and simplified ensemble model is shown on Fig. 4.

Diverse Errors: If base estimators make errors in different parts of the input space (i.e., their errors are uncorrelated), they can cancel out when averaged. Mathematically, for two models with errors e_1 and e_2, if the errors are uncorrelated, the variance of the combined model (assuming equal weights) is:

$$\text{Var}\left(\frac{e_1 + e_2}{2}\right) = \frac{\text{Var}(e_1) + \text{Var}(e_2)}{4} \tag{6}$$

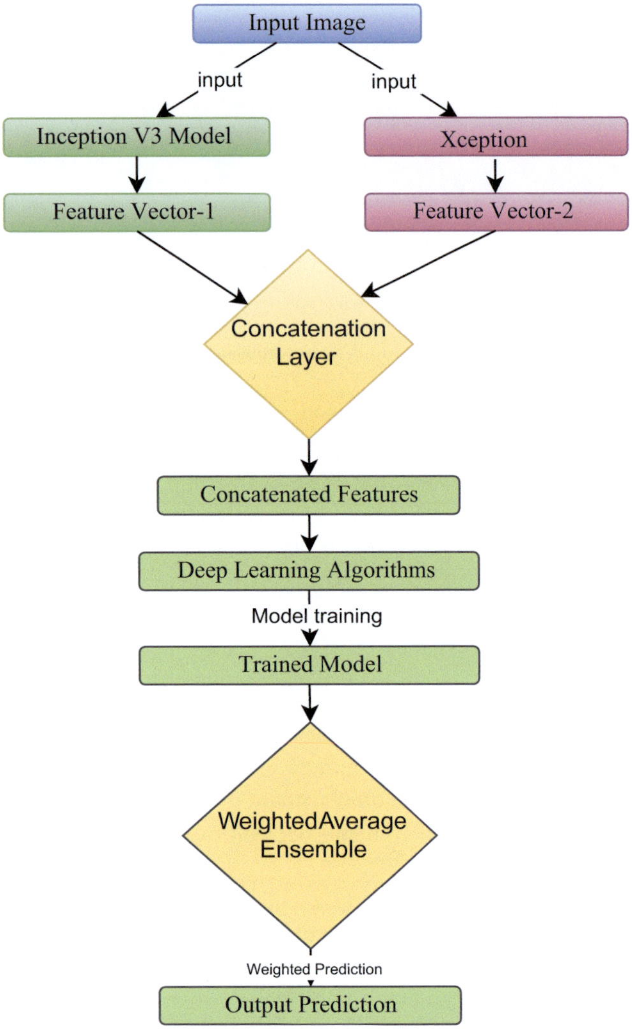

Fig. 4 Architecture of the ensemble model combining Xception and Inception-V3

If $\mathrm{Var}(e_1)$ and $\mathrm{Var}(e_2)$ have less than perfect correlation, this combined variance is less than the variance of the individual errors.

Error Decomposition: Three distinct components may be used to classify a learning method's predicted generalization error: bias, variance, and fundamental error. Mathematically, for a squared loss function, this is often represented as:

$$\text{Error} = \text{Bias}^2 + \text{Variance} + \text{Irreducible Error} \tag{7}$$

- **Bias**: Errors resulting from erroneous assumptions in the learning method. Excessive bias may cause the algorithm to neglect significant correlations between input characteristics and target outputs, leading to underfitting.
- **Variance**: Error brought on by the training set's vulnerability to minute variations. Large variance may lead to overfitting, which is defined by imitating the random noise in the training data.
- **Irreducible Error**: The error introduced by the noise in the actual data.

4.3.3 Improved Generalization

Ensembles are less likely to overfit because they combine predictions from models that might have different overfitting characteristics. This leads to better generalization of the unseen data.

4.3.4 Handling Uncertainty

Ensembles can be particularly effective in cases where the data are noisy or when dealing with complex problems where it's difficult for a single model to capture all the patterns.

The ensemble method exploits these advantages and minimizes the drawbacks of individual models, resulting in more precise and dependable forecasts. This strategy is particularly potent in intricate tasks such as picture recognition, where various models can effectively capture the distinct elements and properties of the input.

5 Evaluation

In order to evaluate the effectiveness of our model, we took three important measures out of the confusion matrix: accuracy, precision, and F1-Score. True-positive (TP), true-negative (TN), false-positive (FP), and false-negative (FN) are the four components that make up the confusion matrix, a key tool in machine learning and deep learning for classification problems. These elements make it possible to accurately assess the predictive power of the suggested design.

5.1 Metrices

We used evaluation criteria to test our models; these are:

Accuracy: The percentage of true findings (including true positives and true negatives) in the total number of instances analyzed is known as accuracy in the context of our model. Mathematically, it is represented as:

$$\text{Accuracy} = \frac{\text{No. of positive Predictions}}{\text{Total No. of Predictions}} \tag{8}$$

Precision: Precision quantifies how well optimistic predictions work. It is calculated as the ratio of genuine positives to the total of false positives and true positives.

$$\text{Precision} = \frac{\text{True Positives}}{\text{True Positives} + \text{False Positives}} \quad (9)$$

F1-Score: A useful statistic for striking a compromise between accuracy and recall is the F1-score, which is the harmonic mean of these two metrics. The formula for the F1-Score is expressed as:

$$\text{F1-Score} = 2 \times \frac{\text{Precision} \times \text{Recall}}{\text{Precision} + \text{Recall}} \quad (10)$$

Recall (Sensitivity):

The percentage of true positives that the model properly detected is measured by a statistic called recall. It is especially important in situations when reducing false negatives is essential. The mathematical formula for recall is given by:

$$\text{Recall} = \frac{\text{True Positives (TP)}}{\text{True Positives (TP)} + \text{False Negatives (FN)}} \quad (11)$$

where:

- **True Positives (TP)** are the correctly identified positive cases.
- **False Negatives (FN)** are positive cases but were negatively predicted by the model.

5.2 Experimental Setup

Python is the primary language for used data collection, preprocessing, conducting experiments, and evaluating model performance [16]. The use of the Keras library facilitates the construction of neural network architectures [17]. Numpy [18] and Pandas were employed for fundamental mathematical computations. Keras has also been utilized to develop deep-learning models. TensorFlow [19] enables the execution of neural networks on GPUs, thereby enhancing computational efficiency. Pre-training with ImageNet weights was adopted for each model to leverage the previously learned features. The dimensionality of the leaf input images was specified as 224 × 224 × 3. The dataset was partitioned into training, validation, and testing subsets, constituting 87.27%, 6.46%, and 6.46% of the data, respectively. The image data were manually shuffled before each model was run to prevent any bias. We also used the following data augmentation techniques: rescaling = 1./255, shear range = 0.2, zoom range = 0.2, horizontal flip, shear range = 0.2, zoom range = 0.2. The training set was the subset utilized for the learning process of the DL model, as commonly referenced in the existing literature. The role of the validation set's is to monitor and tune the performances of the DL models during training. However, the test set is regarded as the crucial final subset to evaluate the model's efficacy on the data it has never encountered before. The Adam optimization algorithm of the Keras framework was employed to optimize the training process across all datasets with a learning rate set to 0.0001. The models were trained in 32 batches to ensure efficient learning and convergence. We used 50 epochs with early stopping and a patience of 10. The ROC curve for the ensemble model is shown in Fig. 5.

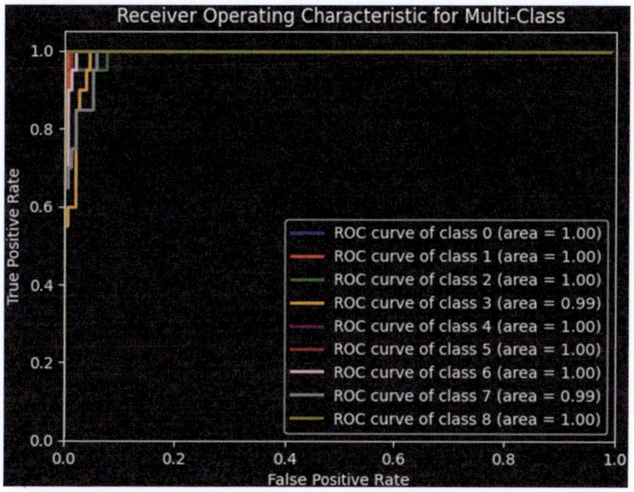

Fig. 5 ROC curve of the ensemble model combining Xception and Inception-V3

5.3 Result Analysis

The proposed ensemble model achieved an accuracy of 93%. Upon analyzing the provided evaluation metrics, accuracy, precision, F1 score, and recall, for various deep learning models, we observed the following:

1. **Performance Metrics Analysis**:

   - In terms of accuracy, precision, F1-score, and recall, the **Ensemble (Inception + Xception)** model outperforms the individual models. This illustrates its improved capacity to correctly classify the data, attaining a significant alignment between predicted and actual values and a well-balanced trade-off between sensitivity and specificity.
   - Models such as **Inception** and **Xception** also show strong performance, which could be attributed to their advanced architectures capable of capturing complex

Table 2 The table shows the result achieved by the experiment

Model	Accuracy	Precision	F1-score	Recall
Inception	0.922222	0.922612	0.922181	0.922222
Xception	0.916667	0.917718	0.916580	0.916667
VGG19	0.911111	0.916604	0.910305	0.911111
InceptionResNet	0.888889	0.889724	0.888819	0.888889
DenseNet	0.888889	0.889724	0.888819	0.888889
NasnetMobile	0.883333	0.887760	0.882499	0.883333
MobileNetV2	0.911111	0.914675	0.910579	0.911111
Proposed ensemble model	0.933333	0.933333	0.933333	0.933333

patterns in the data. However, their limitations are likely to be mitigated when they are combined in the ensemble.

- Models at the lower end of performance, such as **Resnet50**, might be struggling because of factors such as not being adequately tuned, not fitting the specific data characteristics well, or requiring more training data.

From the information in Table 2, the metrics-wise analysis of this experiment is as follows:

- **Accuracy**: The *Ensemble (Inception + Xception)* model showed the highest accuracy (0.933333), indicating its superior overall correctness. This is followed by the *Inception* model (0.922222) and the *Xception* model (0.916667).
- **Precision**: In terms of precision, the ensemble model again yielded a score of 0.933333, demonstrating its effectiveness in correctly identifying positive cases.
- **F1-Score**: The ensemble model achieves the highest F1-score (0.933333), indicating a well-balanced performance between precision and recall.
- **Recall**: The ensemble model correctly identified the most positive samples with the highest recall of 0.933333.

Best Performing Model and Rationale

The **Ensemble (Inception + Xception)** model emerged as the best performer across all the metrics. Its superior performance can be attributed to the following factors:

1. The effective combination of the strengths of both the Inception V3 and Xception models leads to a more comprehensive understanding of the data.
2. Reduction in individual model biases and variances results in robust and generalized performance.
3. The diversity in the features and patterns learned by the Inception V3 and Xception models enhances the predictive capability of the ensemble.

Ensemble Advantage

- The superior performance of the ensemble model supports the theory that combining models can lead to improved generalization. This was likely due to the ensemble's ability to capture diverse patterns and reduce the impact of individual model biases or errors.
- The ensemble benefits from the strengths of both Inception V3 (efficient at multi-level feature extraction) and Xception (optimized for learning cross-channel correlations), resulting in a more robust model.

Underlying Model Architectures

- Individual architectures such as textbfInception, textbfXception, and textbfVGG-19 have unique strengths. For example, Inception networks are known for their efficient and compact representation of images, whereas Xception networks leverage depthwise separable convolutions for enhanced feature extraction with fewer parameters.
- The varied performance of these models indicates the importance of architecture choice relative to the specific characteristics of the dataset.

In conclusion, the ensemble method results in better prediction performance for complex tasks such as disease detection because it effectively utilizes the advantages offered by the ensembled models.

6 Conclusion

Our research has shown that using an ensemble strategy that combines the Inception and Xception models may outperform each model alone in detecting and classifying rice leaf illnesses. Despite the positive results, the dataset size and diversity limitations may restrict the model's applicability to other agricultural contexts. The need for improved model optimization methods is also emphasized because of the computational resources required for the training and inference stages.

By expanding the dataset to include a wider variety of disease kinds and environmental situations, further research will improve the model's robustness and usefulness. Investigating how to include other deep learning architectures into the ensemble framework might increase accuracy and efficiency. Moreover, an essential advancement would include using the model in natural agricultural environments to assess its use and efficiency under various conditions. This research establishes the foundation for advancements in precision agriculture, which enhances sustainable farming practices and boosts food security.

References

1. Dey, B., Haque, M.M.U., Khatun, R., Ahmed, R.: Comparative performance of four CNN-based deep learning variants in detecting Hispa pest, two fungal diseases, and NPK deficiency symptoms of rice (Oryza sativa). Comput. Electron. Agric. **202**, 107340 (2022)
2. Singh, S.P., Pritamdas, K., Devi, K.J., Devi, S.D.: Custom convolutional neural network for detection and classification of rice plant diseases. Procedia Comput. Sci. **218**, 2026–2040 (2023)
3. Liu, H., Cui, Y., Wang, J., Yu, H.: Analysis and research on rice disease identification method based on deep learning. Sustainability **15**(12), 9321 (2023)
4. Mohapatra, S., Marandi, C., Sahoo, A., Mohanty, S., Tudu, K.: Rice leaf disease detection and classification using a deep neural network. In: International Conference on Computing, Communication and Learning, pp. 231–243. Springer (2022)
5. Rahman, C.R., Arko, P.S., Ali, M.E., Khan, M.A.I., Apon, S.H., Nowrin, F., Wasif, A.: Identification and recognition of rice diseases and pests using convolutional neural networks. Biosyst. Eng. **194**, 112–120 (2020)
6. Bari, B.S., Islam, M.N., Rashid, M., Hasan, M.J., Razman, M.A.M., Musa, R.M., Ab Nasir, A.F., Majeed, A.P.A.: A real-time approach of diagnosing rice leaf disease using deep learning-based faster R-CNN framework, PeerJ Comput. Sci. **7**, e432 (2021)
7. Poorni, R., Kalaiselvan, P., Thomas, N., Srinivasan, T.: Detection of rice leaf diseases using convolutional neural network. ECS Trans. **107**(1), 5069 (2022)
8. Shaikh, J., Paradeshi, K.: Faster R-CNN based rice leaf disease detection method. Telematique 5890–5901 (2022)
9. Sharma, R., Kukreja, V., Kaushal, R.K., Bansal, A., Kaur, A.: Rice leaf blight disease detection using multi-classification deep learning model. In: 2022 10th International Conference on Reliability, Infocom Technologies and Optimization (Trends and Future Directions) (ICRITO), pp. 1–5. IEEE (2022)
10. Kongara, R.K.V., Somasila, V.S.C., Revanth, N., Polagani, R.D.: Classification and comparison study of rice plant diseases using pretrained CNN models. In: 2022 International Conference on Inventive Computation Technologies (ICICT), pp. 140–145. IEEE (2022)
11. Pal, C., Pratihar, S., Mukherjee, I.: A custom CNN model for detection of rice disease under complex environment. In: Proceedings of the First Workshop on NLP in Agriculture and Livestock Management, pp. 5–8 (2022)
12. Chen, J., Chen, W., Zeb, A., Yang, S., Zhang, D.: Lightweight inception networks for the recognition and detection of rice plant diseases. IEEE Sens. J. **22**(14), 14628–14638 (2022)
13. Shah, S.R., Qadri, S., Bibi, H., Shah, S.M.W., Sharif, M.I., Marinello, F.: Comparing inception V3, VGG 16, VGG 19, CNN, and ResNet 50: a case study on early detection of a rice disease. Agronomy **13**(6), 1633 (2023)
14. Hussain, M., Al-Aqrabi, H., Munawar, M., Hill, R.: Feature mapping for rice leaf defect detection based on a custom convolutional architecture. Foods **11**(23), 3914 (2022)
15. Thakur, P.S., Sheorey, T., Ojha, A.: VGG-ICNN: a lightweight CNN model for crop disease identification. Multimedia Tools Appl. **82**(1), 497–520 (2023)
16. Liang, K., Wang, Y., Sun, L., Xin, D., Chang, Z.: A lightweight-improved CNN based on VGG16 for identification and classification of rice diseases and pests, in: The international Conference on Image, Vision and Intelligent Systems (ICIVIS 2021), pp. 195–207. Springer
17. Haridasan, A., Thomas, J., Raj, E.D.: Deep learning system for paddy plant disease detection and classification. Environ. Monitor. Assess. **195**(1), 120 (2023)
18. Severity-Based Rice Disease Classification—kaggle.com. https://www.kaggle.com/datasets/isaacritharson/severity-based-rice-leaf-diseases-dataset/data (2023)
19. Sethy, P.K.: Rice leaf disease image samples. Mendeley Data (2020). https://doi.org/10.17632/fwcj7stb8r.1

A Hybrid CNN Architecture for Efficient Detection of Maize Plant Diseases

Md. Asaduzzaman, Md. Johir Uddin Bhuiyan, Habiba Akter Rimi, Hashibul Ahsan Shoaib, and Md. Raisul Islam

Abstract

This research presents a novel deep-learning architecture intended to classify corn leaf diseases with high accuracy and reliability. Our model integrates advanced techniques such as Depthwise Separable Convolutions (DSC), Squeeze-and-Excitation (SE) Blocks, and Residual Connections (RC) to capture intricate patterns and enhance feature extraction effectively. The dataset used comprises 4 distinct categories: Common Rust, Blight, Gray Leaf Spot, and Healthy, with a total of 4,188 images sourced from the PlantVillage and PlantDoc datasets. The obtained results present the accuracy of the proposed method. The model reached a validation accuracy of 88.31% and a test accuracy of 87.35%. The proposed architecture shows superior performance metrics results by achieving a precision (0.87), recall (0.87), and an F1-score (0.87). Specifically, the model exhibited high precision and recall for the Common Rust and Healthy categories, achieving F1-scores of 0.94 and 0.97. The results emphasize the potential of our approach in precisely diagnosing leaf diseases which is crucial for timely and effective agricultural management.

Keywords

Deep learning • Convolutional neural networks • Maize plant diseases • AI in agriculture • Plant diseases

Md. Asaduzzaman · Md. J. Uddin Bhuiyan
Bangladesh University of Business and Technology, Dhaka, Bangladesh

H. A. Rimi
Shanto-Mariam University of Creative Technology, Dhaka, Bangladesh

H. A. Shoaib (✉)
Information Technology, St. Francis College, New York City, USA
e-mail: aahsan.cs@gmail.com

Md. R. Islam
School of Engineering and Technology, Western Illinois University, Macomb, USA

1 Introduction

The agriculture sector is a cornerstone of worldwide food security, and the health of crop plants is paramount to ensuring high yields and quality produce. Among the various crops cultivated worldwide, maize is important, serving as an essential food, animal feed, and raw material for diverse industrial products. However, exposure to various diseases can severely affect the crop productivity and quality of maize plants. Identifying these diseases on time and accurately is necessary for efficient management of disease and minimizing crop losses.

Conventional approaches to disease detection in maize depend on manual evaluation by farmers or experts, a process that is slow, demanding, and susceptible to mistakes. Moreover, the subtle visual differences between different types of leaf diseases can make an accurate diagnosis challenging. To address these limitations, automated disease detection using advanced machine learning (ML) techniques has appeared as a promising solution. In particular, Convolutional Neural Networks (CNNs) have demonstrated remarkable success in various image classification tasks due to their ability to learn and extract hierarchical features from images [1]. Lu et al. [2] proposed a study where they examined recent CNN architectures relevant to the classification of plant foliage diseases. They outlined the principles of DL that are concerned with the classification of plant diseases. In addition, he outlined the primary problems and similar solutions of CNN used for the plant diseases classification. AI is driving the creation of expert systems for the automatic detection of corn disease. These systems analyze digital images to diagnose problems quickly, empowering farmers to protect their crops and maximize yields, such as Zhang et al. [3] and Wiesner et al. [4]. Previous research in corn disease detection has explored deep-learning techniques. While these methods achieve high accuracy, their large model sizes can be complicated to implement due to significant storage requirements. Conversely, studies utilizing conventional machine learning approaches often involve complex feature extraction processes [5, 6], with

M. F. Mridha and N. Dey (eds.), *Machine Vision in Plant Leaf Disease Detection for Sustainable Agriculture*, Studies in Computational Intelligence 1202, https://doi.org/10.1007/978-981-96-4520-6_6

images from expensive cameras [7] adding a layer of difficulty to real-world applications.

This study presents a new CNN-based architecture for classifying maize leaf diseases. Our model leverages the strengths of multiple advanced techniques, including DSC, SE Blocks, and RC. These components are integrated to form a hybrid model that enhances feature extraction and improves classification performance while maintaining computational efficiency.

DSC [8], introduced in the MobileNet architecture [9], decomposes the standard convolution operation into depthwise and pointwise convolutions. This separation noticeably degrades the number of parameters and computational costs, making the model more efficient without sacrificing performance. SE blocks [10], on the other hand, adaptively recalibrate channel-wise feature responses by explicitly modeling interdependencies between channels, thus enhancing the representational power of the network. Inspired by the ResNet architecture [11], Residual Connections address the vanishing gradient problem [12] and enable the training of deeper networks by allowing gradients to flow through shortcut connections.

- Developed a novel hybrid CNN architecture that incorporates depth-wise separable convolutions, squeeze-and-excitation blocks, and residual connections.
- Achieved high accuracy in maize leaf disease classification, attaining a validation accuracy of 88.31% and a test accuracy of 87.35%, demonstrating the effectiveness of the model across different disease categories.
- Provided detailed performance metrics and analysis, showcasing the strength and potential of the model for practical use in the detection and management of automated maize diseases.

2 Related Works

Deep learning (DL) applications, particularly CNNs, for plant disease detection have gained significant interest in recent years. Researchers have created models to decrease crop losses, with machine-learning-based techniques promising future research. Professionals in computer vision [13–17] have published several significant papers in this field. Various ML techniques have been suggested to diagnose plant diseases using picture data. These strategies are customized for many plant species. wheat [18], cucumbers [19], banana [20], tomato [21].

This section reviews the relevant literature on using CNNs and other machine-learning techniques for identifying plant diseases, highlighting key advancements and their limitations. Panigrahi et al. [22] explored using ML for maize disease detection via image analysis. It compares several algorithms (Naive Bayes (NB) [23], Decision Tree (DT) [24], K-Nearest Neighbor (KNN) [25], Support Vector Machine (SVM) [26], and Random Forest (RF) [27]) and finds Random Forest most accurate (79.23%). However, the research lacks details on dataset size and source, training methods, and accuracy, which might be insufficient for real-world use depending on the disease and misclassification cost.

Deep learning algorithms were initially developed for plant image recognition using leaf vein patterns[28]. Mohanty et al. [29] introduced a DL model to learn the species of 14 crops and diseases of 26 crops. Their proposed model demonstrated an accuracy of 99.35%. Ma et al. [30] used CNN to distinguish symptoms of cucumber diseases. The recognition accuracy was 93.4%. To pinpoint wheat disease areas in images, Qiu et al. [31] investigated using Mask-RCNN, a deep learning model built on ResNet50 or ResNet101 backbones. Their model attained a remarkable accuracy of 92.01%. Bi et al. [32] compared different deep learning models for recognizing apple leaf diseases. They found that ResNet152 achieved the highest accuracy (77.65%) for identifying apple leaf spots and rust, followed by Inception V3 (75.59%) and MobileNet (73.50%). Interestingly, these accuracies were obtained using datasets collected by agricultural experts, suggesting potential for real-world application.

Ozguven et al. [33] improved leaf spot detection on sugar beet by modifying Faster R-CNN. They increased the input image size from a tiny 32 × 32 pixels to a much larger 600 × 600 pixels. This significant adjustment led to a more accurate system, achieving 95.48% accuracy compared to the original Faster R-CNN's 92.89%. Their approach effectively identified three types of disease rigidity: mild, moderate, and severe.

Wu et al. [34] introduced a two-channel CNN for maize leaf disease detection. This CNN, built upon VGG and ResNet architectures, achieved 98.33% accuracy on a dataset of three diseases downloaded from Kaggle. While promising, the research lacks details on dataset size and doesn't compare its performance to other CNN architectures besides AlexNet. Ouppaphan et al. [35] introduced a lightweight CNN for identifying corn leaf diseases (rust, blight, spot, healthy) with an accuracy of 97.09% on a dataset of 8,506 corn leaf images from PlantVillage.org. While the accuracy is good, it sacrifices some performance compared to more complex models for the benefit of faster processing on devices like smartphones. Tian et al. [36] explored using a five-layer CNN for maize disease recognition. Their CNN achieved an accuracy of 96.8%, highlighting the potential of CNNs for practical applications in this area.

Singh et al. [37] investigated using the AlexNet model for maize leaf disease detection (spot and rust) on the PlantVillage dataset. While attaining a high accuracy of 99.16%, the research doesn't explore how many training iterations were optimal or if this accuracy translates well to real-world scenarios with more diverse disease variations. Malliga et al. [38] proposed using DL for automatic maize disease detection. They compare a regular CNN (87% accuracy) to the AlexNet architecture (98.5% accuracy) on an unspecified dataset. While the AlexNet achieves high accuracy, the lack of information about the dataset and the significant performance gap between the two models raise questions about the generalizability of the results.

Conventional disease detection techniques frequently rely on visual inspection and can be inaccurate and slow. This demands more complex strategies. Enter the world of advanced hybrid approaches, these strategies intentionally combine the advantages of multiple machine learning models. Madhavi et al. [39] presented a novel hybrid model joining CNNs and MLPs for maize disease classification. The model achieves high accuracy (98.98% training, 96.19% validation, 95.76% test) on an unspecified dataset. While the results are promising, the lack of dataset details and the potential challenges of MLP interpretability in complex tasks like image recognition require further investigation.

While considerable advancements have been achieved in plant disease detection using DL, there are still some gaps that remain. Most existing studies focus on general plant disease datasets, with limited attention to maize leaf diseases. Furthermore, there is a need for more efficient frameworks capable of operating on low-resource devices to enable instant detection in agricultural settings.

Our study addresses these gaps by proposing a novel hybrid CNN architecture. This model is specifically designed for maize leaf disease classification, aiming to achieve high accuracy while maintaining computational efficiency. By utilizing these advanced methods, our goal is to enhance state-of-the-art automated maize disease detection and support the overarching aim of boosting agricultural productivity through technological innovation.

3 Methodology

This section details the approach used to design and assess a model for classifying maize leaf diseases using high-resolution images. The methodology is structured into several main components: dataset overview, data preprocessing steps, model design, training procedure, and evaluation metrics.

Table 1 Number of images in each category of our used datasets

Category	No. of images
Blight (Class-I)	1,146
Common Rust (Class-II)	1,306
Gray leaf spot (Class-III)	574
Healthy (Class-IV)	1,162

3.1 Dataset Description

The PlantVillage and PlantDoc datasets are utilized for this research. They contain high-quality images of maize leaves, categorized into four classes, as summarized in Table 1.

The total dataset consists of 4,188 images and it is split into 80% training, 10% validation, and 10% test sets (Fig. 1).

3.2 Data Preprocessing

Each image is resized to 256×256 pixels to ensure uniform input dimensions for the model. Image augmentation methods, including rotation, flipping, zooming, and shifting are utilized to enhance the diversity of the training dataset and mitigate overfitting.

3.3 Proposed Model Architecture

The proposed model architecture leverages DSC, SE Blocks, and RC to enhance feature extraction and classification performance. The detailed components are described as follows (Fig. 2).

3.3.1 Depthwise Separable Convolutions (DSC)

DSC decomposes the standard convolution operation into two separate layers: depthwise convolution and pointwise convolution. This reduces the number of parameters and cost of computation.

Let $\mathbf{A}$ represent the input tensor of shape $X \times Y \times Z$, where X and Y are the height and width, and Z is the number of channels. A depthwise convolution applies a single convolutional filter per input channel, defined as:

$$\mathbf{B}_d = \mathbf{A}_d * \mathbf{F}_d \quad (1)$$

Fig. 1 Example of PlantVillage and PlantDoc datasets used in the research on maize plant disease detection

where $\mathbf{A}_d$ is the d-th channel of the input, $\mathbf{F}_d$ is the corresponding filter, and $*$ represents the convolution operation. The resulting $\mathbf{B}_d$ is the d-th channel of the output tensor $\mathbf{B}$.

Following the depthwise convolution, a pointwise convolution (1×1 convolution) is applied to combine the depthwise convolution output channels:

$$\mathbf{Z} = \sum_{c=1}^{C} \mathbf{Y}_c * \mathbf{W}_{1\times1,c} \tag{2}$$

where $\mathbf{W}_{1\times1,c}$ are the pointwise convolution filters. The final output $\mathbf{Z}$ has the same height and width as $\mathbf{Y}$ but can have a different number of channels.

3.3.2 Squeeze-and-Excitation (SE) Blocks

SE Blocks adjust channel-specific feature responses in real-time by modeling the relationships between channels. This block is made up of two key operations: squeezing and excitation.

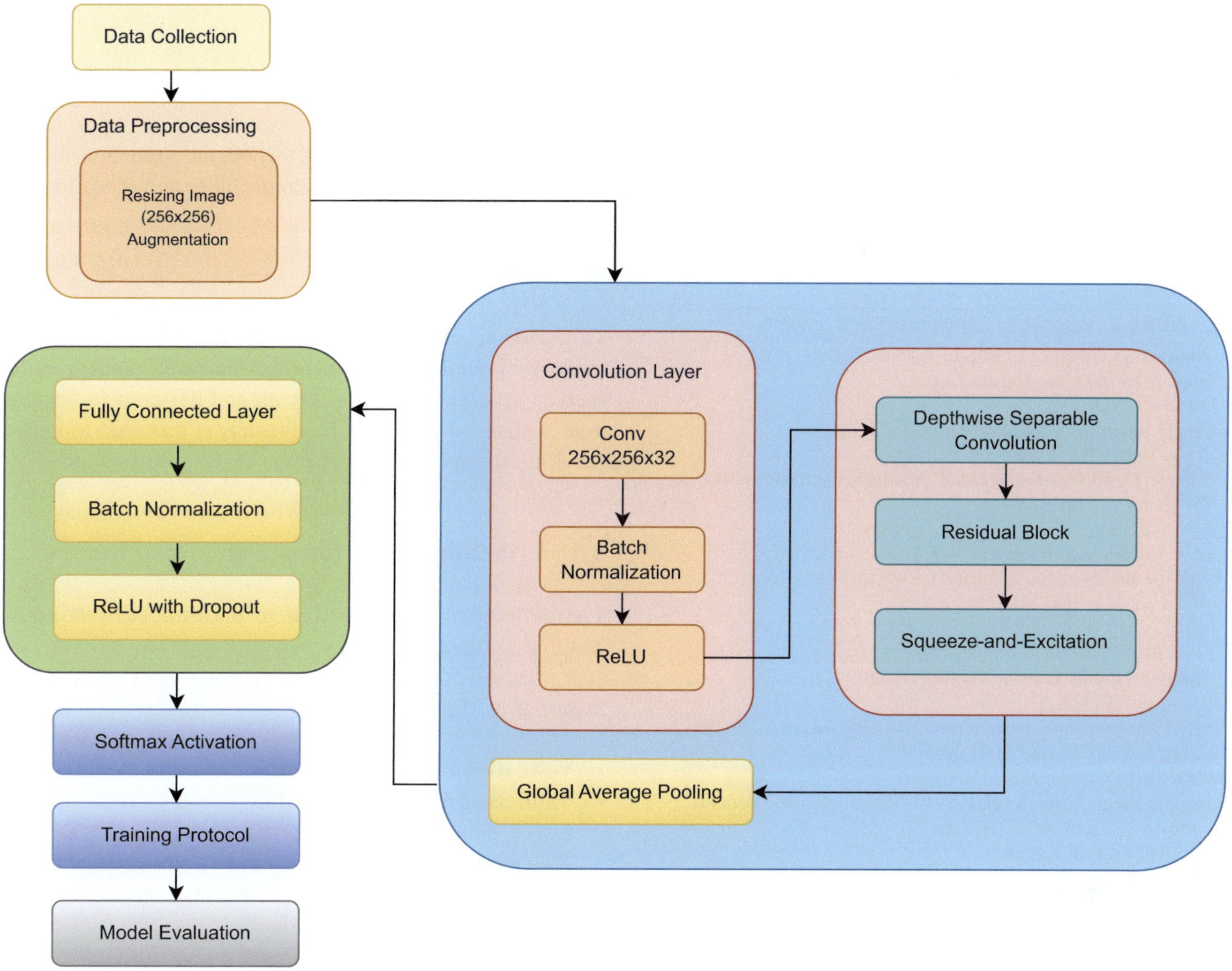

Fig. 2 Proposed hybrid CNN architecture with data preprocessing and model evaluation for efficient detection of maize plant diseases

Given an input tensor **X** with dimensions $M \times N \times P$, the squeeze operation performs global average pooling to compute statistics for each channel:

$$\mathbf{y}_p = \frac{1}{M \times N} \sum_{m=1}^{M} \sum_{n=1}^{N} \mathbf{X}_{mnp}$$

where **y** is a vector of length P.

The excitation operation then passes **z** through two fully connected (FC) layers with non-linear activations:

$$\mathbf{s} = \sigma(\mathbf{N}_2 \delta(\mathbf{N}_1 \mathbf{z})) \tag{3}$$

where $\mathbf{N}_1$ and $\mathbf{N}_2$ are the weights of the FC layers, δ is the ReLU activation function, and σ is the sigmoid activation function.

The final output is obtained by scaling the input tensor **X** using the activations **t**:

$$\mathbf{V}_{ijc} = \mathbf{X}_{mnp} \cdot \mathbf{s}_c \tag{4}$$

3.3.3 Residual Connections

Residual connections help mitigate the problem of vanishing gradients and facilitate the training of deeper neural networks.

Let $\mathbf{F}(\mathbf{X})$ be the output of a series of transformations applied to the input **X**. In a residual block, the output is computed as:

$$\mathbf{Y} = \mathbf{X} + \mathbf{F}(\mathbf{X}) \tag{5}$$

If the dimensions of **X** and $\mathbf{F}(\mathbf{X})$ is unmatched, a 1x1 convolution is applied to **X** to match the dimensions:

$$\mathbf{X}' = \mathbf{W}_{1\times1}\mathbf{X} \tag{6}$$

Then, the output becomes:

$$\mathbf{Y} = \mathbf{X}' + \mathbf{F}(\mathbf{X}) \tag{7}$$

3.3.4 Complete Architecture

The complete architecture combines these components to form a powerful maize leaf disease classification model. The model is structured as follows, as outlined in Algorithm 1:

Algorithm 1 Algorithm of our proposed system.

Input: Input image I of size $H \times W \times C$
Step 1: Initial Convolution Layer
$X_1 = \text{Conv}(I, 32)$
Apply Batch Normalization and ReLU activation:
$X_2 = \text{ReLU}(\text{BN}(X_1))$
Step 2: Depthwise Separable Convolution + Residual Blocks
for each $i = 1, 2, 3$ **do**
  $X_i = \text{DepthwiseConv}(X_{i-1}, F_i)$
  $X_i = \text{SE}(\text{Residual}(X_{i-1}, X_i))$
  where $F_i \in \{64, 128, 256\}$
end for
Step 3: Global Average Pooling (GAP)
$X_{\text{GAP}} = \text{GAP}(X_3)$
Reduces spatial dimensions to $1 \times 1 \times F_3$, where $F_3 = 256$
Step 4: Fully Connected Layer
$X_{\text{FC}} = \text{FC}(X_{\text{GAP}}, 256)$
Apply Batch Normalization, ReLU activation, and Dropout:
$X_{\text{FC}} = \text{Dropout}(\text{ReLU}(\text{BN}(X_{\text{FC}})))$
Step 5: Output Layer
$Y = \text{Softmax}(X_{\text{FC}})$
where $Y \in \{\text{Category}_1, \text{Category}_2, \text{Category}_3, \text{Category}_4\}$

3.4 Model Training

Adamax optimizer is used for model compilation with 0.001 learning rate. The loss function used is categorical cross-entropy, with accuracy as the evaluation metric. The model is trained for 50 epochs, utilizing a batch size of 32. To avoid overfitting, early stopping is applied with a patience of 5 epochs.

3.5 Evaluation Metrics

The model is evaluated using precision, recall, F1-score, and overall accuracy. These metrics comprehensively assess the model's classification capabilities, particularly in handling imbalanced datasets. The formulas for these metrics are as follows:

- Precision: Precision $= \frac{\text{TP}}{\text{TP+FP}}$
- Recall: Recall $= \frac{\text{TP}}{\text{TP+FN}}$
- F1-score: F1-score $= 2 \times \frac{\text{Precision}\times\text{Recall}}{\text{Precision}+\text{Recall}}$
- Accuracy: Accuracy $= \frac{\text{TP+TN}}{\text{Total}}$.

The evaluation is conducted on both the validation and test sets to ensure the model's generalizability and robustness.

4 Results

This section showcases the performance of the proposed model on both the validation and test datasets, using metrics such as accuracy, precision, recall, and F1-score for evaluation. A detailed analysis of the results is provided to understand the strengths and limitations of the proposed model.

4.1 Validation Performance

The training and validation of our model are done by the maize leaf disease dataset. To prevent overfitting, The training process involved 50 epochs with early stopping. The validation results are as follows:

- **Validation Loss**: 0.5687
- **Validation Accuracy**: 88.31%.

These outcomes highlight the model's strong capability to generalize to the validation data, delivering high accuracy and a relatively low loss.

4.2 Test Performance

The model's effectiveness was also assessed using the test dataset. A summary of the test results is provided below:

- **Test Loss**: 0.7620
- **Test Accuracy**: 87.35%.

The test accuracy is slightly lower than the validation accuracy, indicating a good generalization but also suggesting room for improvement.

4.3 Detailed Performance Metrics

The detailed performance metrics for each class are presented in Table 2.

The model exhibits strong performance across most classes, particularly for Common Rust and Healthy leaves, with F1-scores of 0.94 and 0.97, respectively. However, the performance for Gray Leaf Spot is relatively lower, with an

Table 2 Result of each class

Class	Precision	Recall	F1-score	Support
Class-I	0.78	0.82	0.80	114
Class-II	0.95	0.94	0.94	130
Class-III	0.73	0.60	0.66	58
Class-IV	0.94	0.99	0.97	117
Accuracy	0.87 (419)			
Macro Avg	0.85	0.84	0.84	419
Weighted Avg	0.87	0.87	0.87	419

F1-score of 0.66, indicating a need for further improvement in distinguishing this class.

4.4 Confusion Matrix

Figure 3 show the confusion matrix which evaluates a classification model's performance on a test dataset by identifying four classes. The model correctly identifies 93 Blight cases but misclassifies four as Common Rust, 12 as Gray Leaf Spot, and 5 as Healthy. In Gray Leaf Spot, the model misclassifies 20 cases but correctly identifies 35, while the Healthy class has 116 valid identifications. The matrix shows that while the model performs well, there are areas for improvement, particularly in separating between identical illness classifications.

4.5 ROC and AUC Analysis

The Receiver Operating Characteristic (ROC) curve and Area Under the Curve (AUC) analysis in Fig. 4 and Table 3 evaluates the model's classification performance for all classes. The AUC score measures the overall discriminatory ability of the model. The Healthy class achieves a perfect AUC score of 1, indicating flawless classification. Common Rust also performs well with an AUC score of 0.98. Blight has an AUC score of 0.94, showing good performance, while Gray Leaf Spot has the lowest AUC score of 0.93, indicating room for improvement. Overall, the micro-average ROC curve with an AUC of 0.98 reflects high performance across all classes,

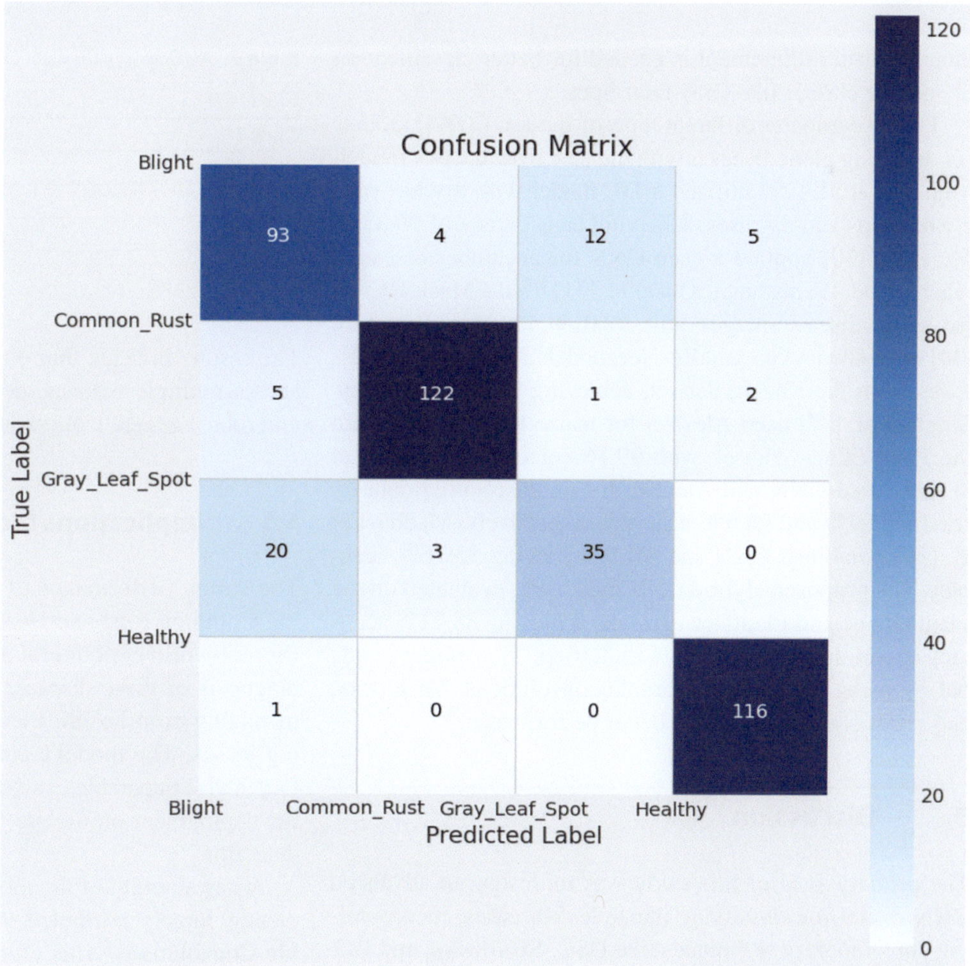

Fig. 3 Confusion matrix for test dataset

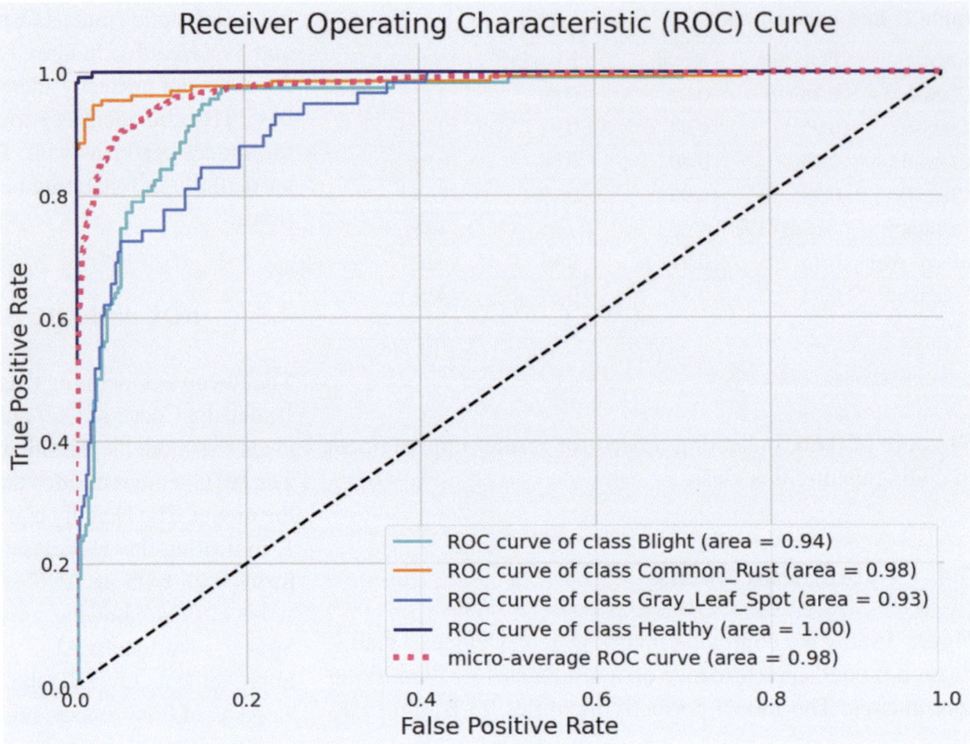

Fig. 4 ROC curves for each class

though further refinement is needed for better classification of specific classes like Gray Leaf Spot.

Table 4 compares different state-of-the-art (SOTA) models for detecting plant diseases with the new hybrid CNN model. Mohanty et al. [29] utilized a DL model with a wide variety of crops and diseases, achieving an accuracy of 99.35%. Ma et al. [30] applied a deep CNN for cucumber diseases, reaching 93.4% accuracy. Qiu et al. [31] used a Mask-RCNN for wheat disease images, with 92.01% accuracy. Wu et al. [34] combined VGG and ResNet models for maize leaf diseases from the Kaggle dataset, achieving 98.33% accuracy. Singh et al. [37] used AlexNet for maize leaf diseases from the PlantVillage dataset, with 99.16% accuracy. Malliga et al. [38] used CNN and AlexNet for an unspecified dataset, reaching 87% and 98.5% accuracy, respectively. Madhavi et al. [39] combined CNN and MLP, achieving 95.76% accuracy. The proposed Hybrid CNN model was evaluated on the PlantVillage and PlantDoc datasets, achieving 88.31% accuracy on validation and 87.35% on testing. This model stands out by using depthwise separable convolutions, SE blocks, and residual connections for better performance.

5 Discussion

The primary goal of this study was to design an advanced CNN model for classifying maize leaf diseases, incorporating contemporary techniques like DSC, SE Blocks, and RC.

Table 3 AUC scores for each class

Class	AUC score
Class-I	0.94
Class-II	0.98
Class-III	0.93
Class-IV	1

The results indicate that our proposed model performs well across multiple metrics, demonstrating both high accuracy and robust generalization capabilities.

5.1 Implications for Real-World Applications

The strong performance of the proposed model in classifying Common Rust and Healthy leaves indicates its potential for real-world agricultural applications. Accurate and timely diagnosis of these diseases can significantly aid farmers in managing crop health, thereby improving yield and reducing losses. The model's computational efficiency, owing to Depthwise Separable Convolutions, also makes it appropriate for deployment on mobile devices, enabling on-site disease detection.

A key strength of the model lies in its computational efficiency, largely attributed to the use of Depthwise Separable Convolutions. This efficiency makes the model suitable

Table 4 Comparison of SOTA models with the hybrid CNN architecture

Study	Model/method	Dataset description	Accuracy	Key features
Mohanty et al. [29]	Deep learning model	[29]	99.35%	Comprehensive dataset, high accuracy
Ma et al. [30]	Deep CNN	[30]	93.4%	High recognition accuracy
Qiu et al. [31]	Mask-RCNN (ResNet50/101)	[31]	92.01%	Pinpoints disease areas accurately
Wu et al. [34]	Two-channel CNN (VGG, ResNet)	[34]	98.33%	High accuracy, robust architectures
Singh et al. [37]	AlexNet	[37]	99.16%	High accuracy, simple architecture
Malliga et al. [40]	CNN, AlexNet	[40]	87% (CNN), 98.5% (AlexNet)	High accuracy with AlexNet
Madhavi et al. [39]	Hybrid CNN-MLP	[39]	95.76%	Combines CNN and MLP for high accuracy
Proposed model	Hybrid CNN	PlantVillage and PlantDoc datasets	88.31% (Val), 87.35% (Test)	Combines depthwise separable convolutions, SE blocks, residual connections

for resources-constrained devices such as smartphones or portable edge devices. The ability to process images and classify diseases directly on-site empowers farmers with real-time decision-making tools, even in regions with limited access to advanced infrastructure or computing facilities. Furthermore, integrating such technology into smart farming ecosystems can lead to more sustainable agricultural practices by optimizing resource usage and reducing wastage caused by disease outbreaks.

In addition, the generalizability of the proposed architecture suggests potential applicability to other crops. By fine-tuning the model on datasets of different plant species and diseases, it can serve as a scalable solution for broader agricultural disease management.

The adoption of AI-driven solutions in agriculture has broader implications beyond yield improvement. By enabling precise disease management, these technologies contribute to reducing the overuse of chemical pesticides, which can harm the environment and human health.

However, deploying such technologies also raises ethical concerns. Ensuring equitable access to these tools, particularly for small-scale farmers in developing regions, is critical. Collaboration with local stakeholders to provide affordable and user-friendly implementations of the technology would ensure its widespread adoption and benefits.

5.2 Limitations and Future Work

Regardless of its promising results, the study has certain limitations. The relatively lower performance for Gray Leaf Spot suggests that the model may benefit from further refinement. Future work could focus on:

- Enhancing the dataset with more representative images of underperforming classes.
- Incorporating advanced data augmentation techniques to improve generalization.
- Exploring the integration of attention mechanisms to focus on relevant features.
- Applying ensemble learning techniques to combine multiple models for improved accuracy.

By integrating this model into Internet of Things (IoT)-based systems, continuous monitoring and disease prediction can be achieved through automated image capturing and analysis. Additionally, combining this approach with weather and soil condition data could further enhance predictive capabilities, providing a holistic solution for precision agriculture.

Future work could also explore extending this architecture to detect multiple diseases in a single image or diagnosing early-stage symptoms that are difficult to identify with the naked eye. Incorporating multi-modal data, such as hyperspectral or thermal imaging, may further refine the model's diagnostic accuracy.

Additionally, real-world deployment would require extensive testing in diverse agricultural environments to ensure the model's robustness across different conditions and variations in image quality.

6 Conclusions

This study shows the development of a novel CNN architecture incorporating DSC, SE Blocks, and RC to classify maize leaf diseases. This architecture achieved high accuracy and demonstrated robust generalization capabilities, particularly excelling in identifying Common Rust and Healthy leaves. While the model demonstrated slightly decreased performance for Gray Leaf Spot, this highlights opportunities for further enhancement. Our results underscore the potential of advanced CNN techniques in agricultural disease management, offering an efficient and accurate tool for early disease detection. Future work will concentrate on refining the performance of the model, particularly for challenging disease classes, and exploring its Implementation in agricultural settings in real world to support farmers in improving crop health and productivity.

References

1. Mazumder, M.K.A., Kabir, M.M., Rahman, A., Abdullah-Al-Jubair, M., Mridha, M.: Densenet201plus: cost-effective transfer-learning architecture for rapid leaf disease identification with attention mechanisms. Heliyon **10**(15) (2024)
2. Lu, J., Tan, L., Jiang, H.: Review on convolutional neural network (CNN) applied to plant leaf disease classification. Agriculture **11**(8), 707 (2021)
3. Zhang, X., Qiao, Y., Meng, F., Fan, C., Zhang, M.: Identification of maize leaf diseases using improved deep convolutional neural networks. IEEE Access **6**, 30370–30377 (2018)
4. Wiesner-Hanks, T., Stewart, E.L., Kaczmar, N., DeChant, C., Wu, H., Nelson, R.J., Lipson, H., Gore, M.A.: Image set for deep learning: field images of maize annotated with disease symptoms. BMC Res. Notes **11**, 1–3 (2018)
5. Sankaran, S., Mishra, A., Ehsani, R., Davis, C.: A review of advanced techniques for detecting plant diseases. Comput. Electron. Agric. **72**(1), 1–13 (2010)
6. Khaled, A.Y., Abd Aziz, S., Bejo, S.K., Nawi, N.M., Seman, I.A., Onwude, D.I.: Early detection of diseases in plant tissue using spectroscopy-applications and limitations. Appl. Spectrosc. Rev. **53**(1), 36–64 (2018)
7. Rumpf, T., Mahlein, A.-K., Steiner, U., Oerke, E.-C., Dehne, H.-W., Plümer, L.: Early detection and classification of plant diseases with support vector machines based on hyperspectral reflectance. Comput. Electron. Agric. **74**(1), 91–99 (2010)
8. Kamal, K., Yin, Z., Wu, M., Wu, Z.: Depthwise separable convolution architectures for plant disease classification. Comput. Electron. Agric. **165**, 104948 (2019)
9. Chen, H.-Y., Su, C.-Y.: An enhanced hybrid mobilenet. In: 2018 9th International Conference on Awareness Science and Technology (iCAST), pp. 308–312. IEEE (2018)
10. Hu, J., Shen, L., Sun, G.: Squeeze-and-excitation networks. In: Proceedings of the IEEE Conference on Computer Vision and Pattern Recognition, pp. 7132–7141 (2018)
11. Wu, Z., Shen, C., Van Den Hengel, A.: Wider or deeper: revisiting the Resnet model for visual recognition. Pattern Recogn. **90**, 119–133 (2019)
12. Borawar, L., Kaur, R.: Resnet: Solving vanishing gradient in deep networks. In: Proceedings of International Conference on Recent Trends in Computing: ICRTC 2022, pp. 235–247. Springer (2023)
13. Too, E.C., Yujian, L., Njuki, S., Yingchun, L.: A comparative study of fine-tuning deep learning models for plant disease identification. Comput. Electron. Agric. **161**, 272–279 (2019)
14. Chen, J., Chen, J., Zhang, D., Sun, Y., Nanehkaran, Y.A.: Using deep transfer learning for image-based plant disease identification. Comput. Electron. Agric. **173**, 105393 (2020)
15. Khan, R.U., Khan, K., Albattah, W., Qamar, A.M.: Image-based detection of plant diseases: from classical machine learning to deep learning journey. Wirel. Commun. Mob. Comput. **2021**(1), 5541859 (2021)
16. Rao, D.S., Ch, R.B., Kiran, V.S., Rajasekhar, N., Srinivas, K., Akshay, P.S., Mohan, G.S., Bharadwaj, B.L.: Plant disease classification using deep bilinear CNN. Intell. Autom. Soft Comput **31**(1), 161–176 (2022)
17. Barbedo, J.G.A.: Plant disease identification from individual lesions and spots using deep learning. Biosyst. Eng. **180**, 96–107 (2019)
18. Johannes, A., Picon, A., Alvarez-Gila, A., Echazarra, J., Rodriguez-Vaamonde, S., Navajas, A.D., Ortiz-Barredo, A.: Automatic plant disease diagnosis using mobile capture devices, applied on a wheat use case. Comput. Electron. Agric. **138**, 200–209 (2017)
19. Fujita, E., Kawasaki, Y., Uga, H., Kagiwada, S., Iyatomi, H.: Basic investigation on a robust and practical plant diagnostic system. In: 2016 15th IEEE International Conference on Machine Learning and Applications (ICMLA), pp. 989–992. IEEE (2016)
20. Amara, J., Bouaziz, B., Algergawy, A.: A deep learning-based approach for banana leaf diseases classification (2017)
21. Fuentes, A., Yoon, S., Kim, S.C., Park, D.S.: A robust deep-learning-based detector for real-time tomato plant diseases and pests recognition. Sensors **17**(9), 2022 (2017)
22. Panigrahi, K.P., Das, H., Sahoo, A.K., Moharana, S.C.: Maize leaf disease detection and classification using machine learning algorithms. In: Progress in Computing, Analytics and Networking: proceedings of ICCAN 2019, pp. 659–669. Springer (2020)
23. Rish, I., et al.: An empirical study of the Naive Bayes classifier. In: IJCAI 2001 Workshop on Empirical Methods in Artificial Intelligence, vol. 3, pp. 41–46. Citeseer (2001)
24. Quinlan, J.R.: Induction of decision trees. Mach. Learn. **1**, 81–106 (1986)
25. Cover, T., Hart, P.: Nearest neighbor pattern classification. IEEE Trans. Inf. Theory **13**(1), 21–27 (1967)
26. Cortes, C., Vapnik, V.: Support-vector networks. Mach. Learn. **20**, 273–297 (1995)
27. Ho, T.K.: The random subspace method for constructing decision forests. IEEE Trans. Pattern Anal. Mach. Intell. **20**(8), 832–844 (1998)
28. Grinblat, G.L., Uzal, L.C., Larese, M.G., Granitto, P.M.: Deep learning for plant identification using vein morphological patterns. Comput. Electron. Agric. **127**, 418–424 (2016)
29. Mohanty, S.P., Hughes, D.P., Salathé, M.: Using deep learning for image-based plant disease detection. Front. Plant Sci. **7**, 215232 (2016)
30. Ma, J., Du, K., Zheng, F., Zhang, L., Gong, Z., Sun, Z.: A recognition method for cucumber diseases using leaf symptom images based on deep convolutional neural network. Comput. Electron. Agric. **154**, 18–24 (2018)
31. Qiu, R., Yang, C., Moghimi, A., Zhang, M., Steffenson, B.J., Hirsch, C.D.: Detection of fusarium head blight in wheat using a deep neural network and color imaging. Remote Sens. **11**(22), 2658 (2019)
32. Bi, C., Wang, J., Duan, Y., Fu, B., Kang, J.-R., Shi, Y.: Mobilenet based apple leaf diseases identification. Mob. Netw. Appl. 1–9 (2022)
33. Ozguven, M.M., Adem, K.: Automatic detection and classification of leaf spot disease in sugar beet using deep learning algorithms. Phys. A: Stat. Mech. Appl. **535**, 122537 (2019)
34. Wu, Y.: Identification of maize leaf diseases based on convolutional neural network. J. Phys.: Conf. Ser. **1748**, 032004 (IOP Publishing) (2021)
35. Ouppaphan, P.: Corn disease identification from leaf images using convolutional neural networks. In: 2017 21st International Computer Science and Engineering Conference (ICSEC), pp. 1–5. IEEE (2017)
36. Tian, J., Zhang, Y., Wang, Y., Wang, C., Zhang, S., Ren, T.: A method of corn disease identification based on convolutional neural network. In: 2019 12th International Symposium on Computational Intelligence and Design (ISCID), vol. 1, pp. 245–248. IEEE (2019)

37. Singh, R.K., Tiwari, A., Gupta, R.K.: Deep transfer modeling for classification of maize plant leaf disease. Multimedia Tools Appl. **81**(5), 6051–6067 (2022)
38. Malliga, S., Nandhini, P., Kogilavani, S., Harini, R.J., Shree, S.J., Jeeva, G.: Maize leaf disease classification using convolutional neural network. AIP Conf. Proc. **2387** (AIP Publishing) (2021)
39. Madhavi, P.B., Subaranjani, T., Sigtia, S., Mehta, V.: A hybrid deep learning approach for accurate and transparent maize plant disease classification. In: 2024 11th International Conference on Computing for Sustainable Global Development (INDIACom), pp. 229–234. IEEE (2024)
40. Indira, K., Mallika, H.: Classification of plant leaf disease using deep learning. J. Inst. Eng. (India): Ser. B 1–12 (2024)

Detection of Corn Leaf Diseases Using Deep CNN with SHAP-Based Explainability for Enhanced Agriculture

Maharin Afroj, Israt Jahan, Md. Raisul Islam, and Md. Polash Islam

Abstract

Corn, also known as maize, is a vital crop globally, with significant economic and nutritional value. However, maize production faces major challenges owing to various diseases that cause substantial crop losses and threaten food security. Advancements in technology, particularly the widespread availability of smart devices offers a promising solution for mitigating these losses through automatic disease detection. This study introduces a method for recognizing corn leaf diseases using a DCNN. The proposed model incorporates SHAP (Shapley Additive Explanations) for enhanced explainability, allowing for better understanding and interpretation of the diagnostic process.The experimental results demonstrate that the custom Deep CNN model achieves an accuracy of 97.53%, highlighting its potential for effective and reliable disease detection in maize crops. This approach is designed to assist farmers and agricultural stakeholders in the early detection and management of diseases, thereby fostering enhanced crop productivity and contributing to the overall improvement of food security at a broader scale.

Keywords

Deep learning • CNN • Image classification • Corn leaf • Leaf disease

1 Introduction

One of the key staple crops globally, corn is essential for the food and, animal feed industry. However, many illnesses that can negatively affect both yield and quality present serious obstacles to corn production. Timely intervention and efficient management methods are contingent upon the early and precise detection of illnesses affecting corn leaves. Using deep neural network techniques over vector machine classifiers has become an advanced practical way to overcome the limitations of visual inspection and laboratory tests [1].

DCNNs can repeatedly Acquire and extract pertinent features from raw image data and can achieve high accuracy in challenging visual recognition tasks. DCNNs provide a potent tool for the classification and diagnosis of different diseases from images of maize leaves thus eliminating the necessity for extensive preprocessing. This method offers a scalable and effective way to monitor diseases by using massive datasets of annotated photos to train models that can generalize Effectively handles new, unseen data [2]. This, in turn, aims to support more efficient and sustainable agricultural practices. Our approach consists of obtaining and preprocessing image data, creating and training a suitable DCNN architecture, and evaluating the model's performance against preset benchmarks. In this work, we demonstrate how DCNNs might transform agricultural diagnostics and contribute to the global effort to protect food supplies from crop disease risk [3].

Moreover, combining DCNNs with the current farm management systems may result in more comprehensive and data-driven agricultural practices. Farmers can improve crop health and productivity by utilizing real-time disease detection in conjunction with other data, such as weather patterns,

M. Afroj · Md. P. Islam (✉)
Department of CSE, Bangladesh University of Business and Technology, Mirpur, Dhaka, Bangladesh
e-mail: polashprince2@gmail.com

I. Jahan
Department of Information Technology, Stockholm University, Stockholm, Sweden

Md. R. Islam
School of Engineering and Technology, Western Illinois University, Macomb, Illinois, USA
e-mail: raisulislamjony@gmail.com

M. F. Mridha and N. Dey (eds.), *Machine Vision in Plant Leaf Disease Detection for Sustainable Agriculture*, Studies in Computational Intelligence 1202, https://doi.org/10.1007/978-981-96-4520-6_7

soil conditions, and crop growth stages, to enable more educated decision-making. By lowering the need for unnecessary chemical treatments, this all-encompassing strategy not only promotes sustainable farming methods but also increases production and quality [4].

The creation and application of a DCNN-based system for the detection and categorization of corn leaf diseases were examined in this paper. Our goal was to improve illness diagnostic accuracy and reliability by utilizing deep learning skills, which will facilitate more productive and sustainable farming methods. We accomplish this by first collecting and preprocessing picture data, then building and training an appropriate DCNN architecture, and finally comparing the model's output to predetermined benchmarks. We hope to illustrate how DCNNs can revolutionize agricultural diagnostics and support international efforts to safeguard food supplies from crop disease risk by showcasing their effectiveness in this setting.

To sum up, the application of DCNNs to maize leaf disease detection and classification marks a substantial development in agricultural technology. This strategy not only addresses the key concerns of traditional disease detection methods but also offers a scalable, accurate, and efficient solution to a prevalent challenge in maize production. We can use deep learning to support more resilient and productive agricultural techniques, thereby ensuring a more stable food supply for the future, by conducting more research and development. Our work offers several vital contributions:

- **Development of a High-Accuracy DCNN Model**: We propose a custom DCNN model specifically designed for recognition of leaf diseases, achieving an impressive accuracy of 97.53%.
- **Integration of Explainability with SHAP**: To enhance the interpretability of the diagnostic process, we integrate SHAP into our model, providing clear insights into feature contributions and allowing for a better understanding of the model's decision-making.

The paper follows the following structure: In Sect. 2, an exploration of recent research findings on the topic is provided, offering a comprehensive overview. Section 3 outlines the detailed methodology proposed for model development. The results of the proposed models and the experimental findings of our research are presented in Sect. 4. Further, several critical points have been discussed in Sects. 5 and 6 abbreviates the significant conclusions.

2 Related Work

Due to the critical impacts of plants on the industrial production of raw materials, global food security, livestock feed, etc, plant-related disease classification and detection have been extensively studied. As maize or corn is one of the essential food supplies worldwide, leaf diseases such as blight [5], rust [6], and spot [7] optimization is a crucial factor for the agriculture industry. Traditional methods such as laboratory tests, and chemical control for detecting diseases are always time-consuming, acute labor, and induce delusion in humans as most of the cases are handled by visual inspection. As a consequence, systematic and automated approaches are demanded in disease detection and are still in the process of finding models that are convenient as well as generate accurate results with no error [8, 9]. The field of image analysis has proven its impact on plant-based disease detection and it affects fast-forwarding changes by deep learning, specifically DCNNs with various integrations [10, 11]. Particularly in detecting leaf diseases, Neural Networks perform better and explore impactful feature extraction methods with efficient results [12, 13].

The survey [14] reviewed the application of DCNNs for predicting plant leaf diseases, focusing on challenges like dataset size, preprocessing techniques, and model architecture selection. Various DCNN architectures, such as SqueezeNet, and DenseNet, ResNet are compared using datasets that primarily include images captured from the upper side of leaves. The study marked DCNNs' advantages in feature extraction and can be improved disease localization, and automatic estimation of disease severity to enhance practical agricultural applications. Several methods with various feature (color, texture, shape, etc. [15]) extracting techniques such as local binary patterns (LBP) [16], gradient explainability [17], as well as multiple classifier fusion [18–21], emerge drastic improvement. The paper [22] evaluated the PlantVillage dataset employing a CNN called SqueezeNet for classification and Grad-CAM for visual explanation, achieving performance measures of 95.2% accuracy. Advanced visual explanation method Grad-CAM++ can compete with the challenges related to class imbalance and optimizing epoch values for better performance. Authors [23] addressed the challenge of classifying maize small leaf spot disease severity using a deep learning network, DISE-Net, which enhances multi-scale feature extraction and incorporates attention mechanisms to focus on key features. The dataset included 1268 images of maize leaves with different disease grades and healthy leaves. The performance result achieved 97.12% test accuracy, outperforming other models like VGG16, ResNet50, and DenseNet121.

Table 1 provides an overview of the challenges and gaps identified across various related studies, highlighting key areas of concern and limitations.

3 Methods and Materials

The efficacy of the proposed framework for detecting corn leaf diseases is projected through the rigorous implementation of systematically crafted experimental protocols. This section delineates the critical facets of the adopted methodology Fig. 1, encompassing a comprehensive exposition of the dataset, intricate preprocessing strategies, and an in-depth examination of the model architectures employed.

3.1 Dataset Specifications

The dataset utilized in this study is a comprehensive collection derived from the renowned PlantVillage [29] and PlantDoc [30] datasets, encompassing a total of 4188 images categorized into four distinct classes. Specifically, it includes 1306 images of Common Rust, 1146 images of Blight, 574 images of Gray Leaf Spot, and 1162 images of Healthy plants. This diverse and well-balanced dataset facilitates effective training and testing for plant disease classification models, ensuring conductive performance across different disease conditions and healthy states. The integration of images from these two prominent sources provides a rich variety of visual representations, enhancing the model's ability (Figs. 2 and 3, Table 2).

3.2 Data Pre-processing

We used several advanced techniques in our preprocessing pipeline to raise the quality of dataset like normalization, data augmentation etc. We also used the Min–Max Scaler to modify the numbers to fit inside a specific range while maintaining the original distribution. This was particularly helpful in light of the different magnitudes and anomalies present in our dataset. The Min–Max formula is represented by the following equation, in where m is the new value, x is the original cell value, $x_{\min}$ is the column's minimum value, and $x_{\max}$ is the column's highest value.

$$m = \frac{(x - x_{\min})}{(x_{\max} - x_{\min})} \tag{1}$$

The raw dataset is partitioned into 80% training and 20% validation, with augmentation techniques. In contrast, for the validation and test datasets, only pixel normalization is performed.

Table 1 Overview of existing studies and their limitations

References	Method	Dataset	Performance	Highlighting key concerns
Zhang et al. [24]	GA-SVM	Images collected by digital camera under sunlight condition	Average 92.82%	The paper addressed the challenge of determining parameters manually in classic SVMs, thereby genetic algorithm is used to select optimal parameters automatically
Zhang et al. [19]	K-means	Cucumber leaf image database	85.7%	It needs more efficiency for sparsity in large leaf image databases
Lin et al. [25]	Multichannel Convolutional Neural Network (MCNN)	10,820 RGB maize images	92.31%	Image noise remains in the research and comprehensive data is needed for method analysis
Sibiya et al. [13]	CNN using the Neuroph Studio framework	Images clicked by smartphone cameras	92.85%	The research included northern corn common rust, leaf blight, and gray leaf spot . The model can be refined for gray-scale image evaluation to implement in large-scale settings
Shoaib et al. [17]	Explainable Gradient-based CNN (EG-CNN)	A diverse dataset of gene expression, metabolite, and hyperspectral images	95%	Collected data affected by four common diseases: powdery mildew, leaf spot, rust , and blight
Zhang et al. [26]	MAF with ResNet50	4428 images	97.41%	The method can replace linear activation functions with nonlinear ones and may increase the number of network parameters to solve issues
Bansal et al. [27]	An ensemble model	dataset of 3642 apple leaf images	90% for leaf diseases	The research can be expanded by using more disease categories with a large number of dataset with enhanced model
Zeng et al. [20]	SKPSNet-50	Natural scene images	92.9%	The paper identified early maize leaf diseases with small spots and irregular shapes
Padilla et al. [28]	CNN	Verified images of a barangay farm	93%LB, 89% for LR, and 89% for LS	The process combined two models to improve execution time

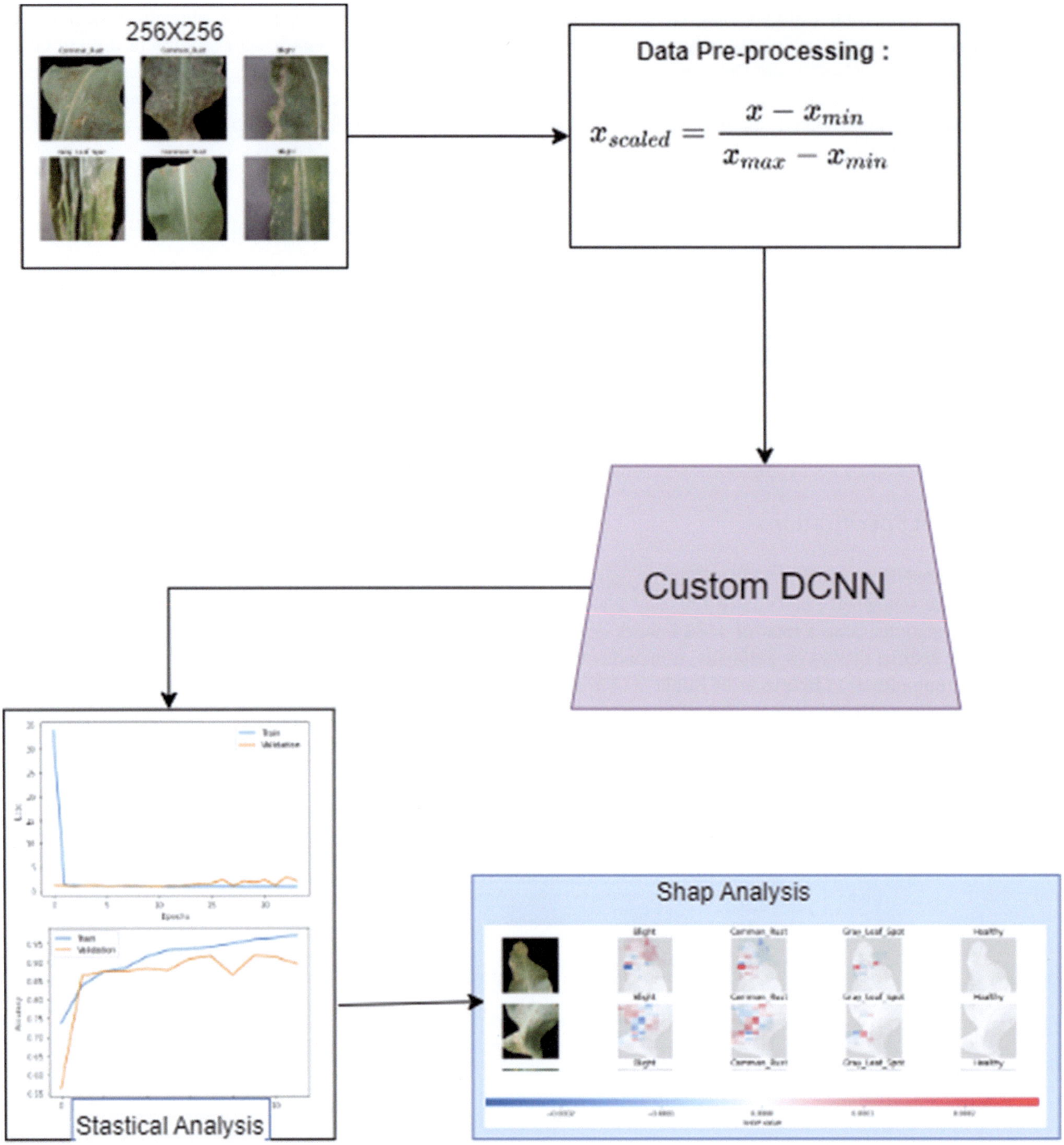

Fig. 1 Unveiling the power of our methodology: a step-by-step workflow

3.3 Architecture of Proposed Model

We propose a CNN model for image classification tasks, with a focus on plant disease detection. The architecture consists of five Conv2D layers, each followed by MaxPooling2D layers to capture and downsample spatial features effectively. A Dropout layer is included to prevent overfitting, thereby including effective generalization. The feature maps are then flattened and passed through two Dense layers, leading to a final output layer with four neurons for class classification. Our model comprises a total of 183,812 trainable parameters. A summarization of the model's details has been given in Table 3 (Fig. 4).

Fig. 2 Sample images of dataset

3.4 Conv2D

The Conv2D layer conducts 2-dimensional convolution on input data, extracting crucial features for image processing. Mathematically, for an input X and kernel K, the convolution is defined as:

$$(X * K)(i, j) = \sum_m \sum_n X(i + m, j + n)K(m, n)$$

3.5 MaxPooling2D

MaxPooling2D reduces dimensionality, boosting computational efficiency. The layer partitions input into regions and outputs maximum values. The operation is defined as:

$$\text{MaxPooling2D}(X)(i, j, k) = \max_{m,n} X(2i + m, 2j + n, k)$$

3.6 Dense

The Dense layer conducts a linear transformation on input data, followed by the Softmax activation. It connects every neuron, learning weights (W) and biases (b). The output (O) is calculated as:

$$O = \text{Softmax}(\text{WX} + b)$$

3.7 Dropout

Dropout is a regularization technique that stochastically drops out neurons during training, preventing overfitting. For an input X, the dropout operation is defined as:

$$\text{Dropout}(X)_i = 0 \,\text{with probability, else}\, \frac{X_i}{1 - p}$$

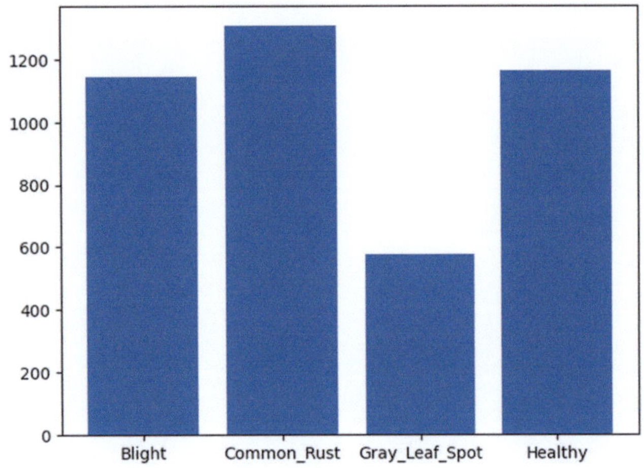

Fig. 3 Training set distribution odata

Table 2 Dataset description

Class	Image count
Common rust	1306
Gray leaf spot	574
Blight	1146
Healthy	1162

4 Experimental Analysis

This section analyzes the results of extensive testing and evaluation of deep learning models for classifying crop or maize leaf diseases.

The Table 4 presents a comparative analysis of existing works on corn leaf disease detection with our proposed DCNN model. Bansal et al. [27] utilized a pre-trained combined model of classic networks, achieving 90% accuracy. Zhang et al. [24] employed a GA-SVM model, resulting in 92.82% accuracy, while Sibiya et al. [13] used a CNN model, attaining 92.85% accuracy. Another work by Zhang et al. [19] implemented K-means clustering, with an accuracy of 85.7%. In contrast, our proposed DCNN model significantly outperforms these methods, achieving an accuracy of 97.53%, highlighting its superior performance in disease detection and classification.

The Table 5 provides a comparative analysis of various models with our proposed custom DCNN. The Inception model achieves an accuracy of 72.87%, while SqueezeNet performs better with an accuracy of 81.81%. VGG16 shows significant improvement, reaching an accuracy of 91.55%. ResNet101 further enhances performance with 96.15% accuracy. Our proposed custom DCNN surpasses all these models, achieving the highest accuracy of 97.53%, demonstrating its superior effectiveness in the task.

4.1 Performance Metrics

Table 6 summarizes the model's performance in classifying leaf diseases across four categories, achieving 97% overall accuracy. For Blight, Common Rust, Gray Leaf Spot, and Healthy, the Precision, Recall, and F1-Scores range between

Table 3 Model summary

Layer (type)	Output shape	Param #
Conv2D	None, 254, 254, 32)	896
MaxPooling2D	None, 127, 127, 32)	0
Conv2D	(None, 125, 125, 64)	18496
MaxPooling2D	(None, 62, 62, 64)	0
Conv2D	(None, 60, 60, 64)	36928
MaxPooling2D	(None, 30, 30, 64)	0
Conv2D	(None, 28, 28, 64)	36928
MaxPooling2D	(None, 14, 14, 64)	0
Conv2D	(None, 12, 12, 64)	36928
MaxPooling2D	(None, 6, 6, 64)	0
Conv2D	(None, 4, 4, 64)	36928
MaxPooling2D	(None, 2, 2, 64)	0
Dropout	(None, 2, 2, 64)	0
Flatten	(None, 256)	0
Dense	(None, 64)	16448
Dense	(None, 4)	260
Total parameters		183812 (718.02 KB)
Trainable parameters		183812 (718.02 KB)

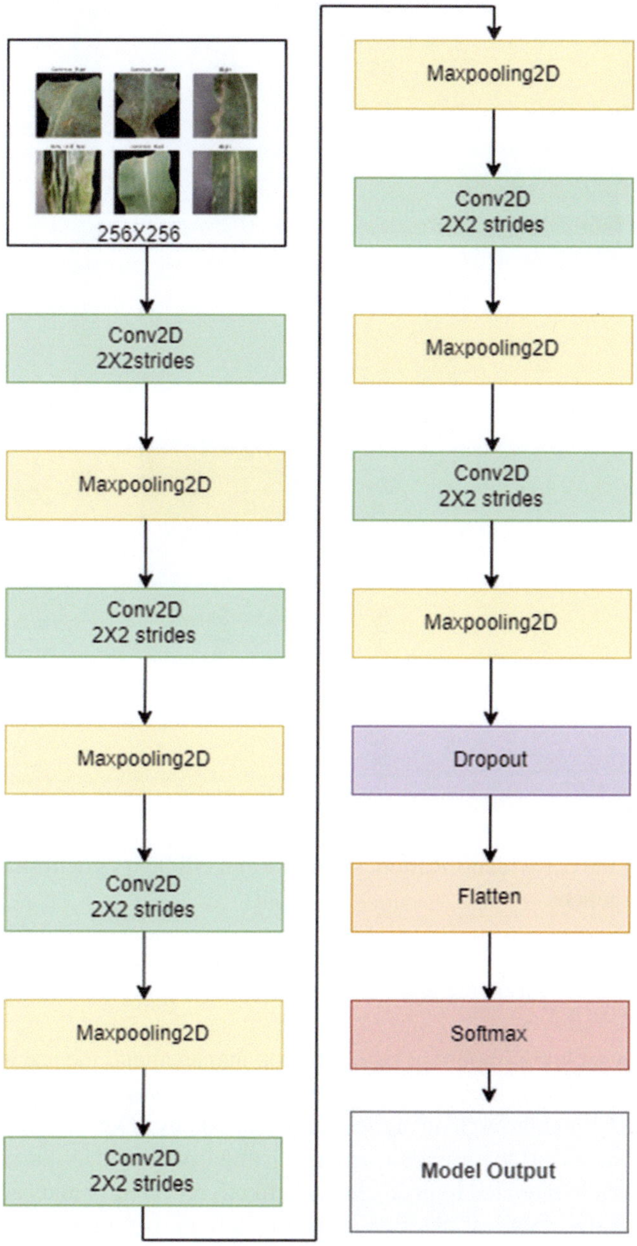

Fig. 4 Revealing the cutting-edge architecture of our proposed model

Table 4 Comparison of existing works with our proposed DCNN

Existing work	Model	Accuracy (%)
Bansal et al. [27]	A pre-trained combined model of classic networks	90%
Zhang et al. [24]	GA-SVM	92.82
Sibiya et al. [13]	CNN	92.85
Zhang et al. [19]	K-means clustering	85.7
Our proposed	DCNN	97.53

Table 5 Comparison between various models and our custom DCNN

Model	Accuracy (%)
Inception	72.87
SqueezeNet	81.81
VGG16	91.55
ResNet101	96.15
Our proposed custom DCNN	97.53

Table 6 Classification report

	Precision	Recall	F1-score	Support
0	0.96	0.93	0.95	1146
1	0.98	0.99	0.99	1306
2	0.88	0.93	0.91	574
3	0.99	0.98	0.99	1162
Accuracy	0.97			

88 and 99%, demonstrating the model's high efficacy and reliability across all classes.

Accuracy refers to the ratio of accurately predicted data points to the total number of data points. The accuracy is calculated according to the method shown in Eq. 2.

$$\text{Accuracy} = \frac{TP + TN}{TP + TN + FP + FN} \tag{2}$$

FP refers to the instances where attacks are incorrectly classified as normal, while TP denotes the instances that are accurately identified as attacks. The formula that is shown in Eq. 3, is a probability measure that evaluates the probability of a positive prediction being accurate.

$$\text{Precision} = \frac{TP}{TP + FP} \tag{3}$$

The term TPR refers to the proportion of correctly predicted positive occurrences inside the positive class. The algorithm for calculating this rate is provided in Eq. 4.

$$\text{Recall} = \frac{TP}{TP + FN} \tag{4}$$

The F1 score, often known as the F-measure, is a quantitative measure used to assess the precision of a test (Fig. 5).

$$\text{f1 score} = 2 * \frac{\text{Precision} * \text{Recall}}{\text{Precision} + \text{Recall}} \tag{5}$$

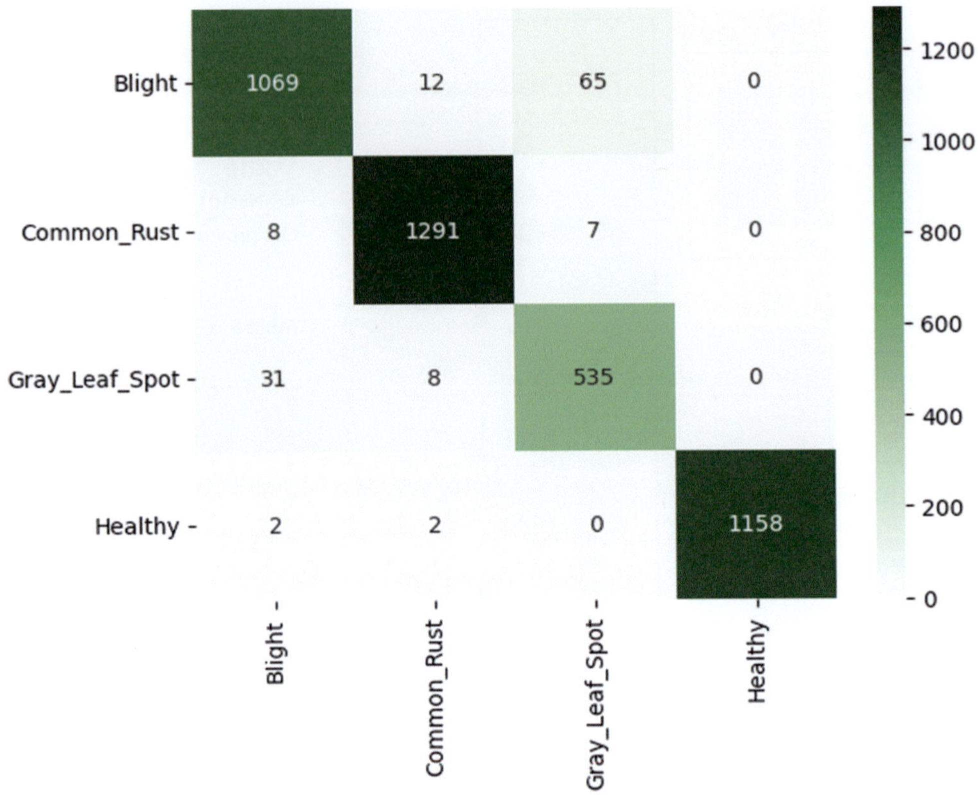

Fig. 5 Comprehensive graphical representation showcasing the dynamic performance metrics

Figure 6a, b presents two graphs that illustrate the performance of the deep learning model during training and validation.

4.2 Explainability with SHAP

The SHAP approach has been included in our deep CNN model for leaf disease detection Fig. 6 in order to depict interpretability. By measuring each pixel or region in the input image, SHAP offers important insights into how the model produces predictions. This helps us to comprehend which characteristics have the greatest bearing on whether a leaf is infected or healthy. By providing localized explanations for each prediction, SHAP contributes to the visualization and interpretation of CNN model's decision-making, providing farmers and other agricultural stakeholders with transparency and useful information. This strategy not only increases confidence in the model's accuracy but also facilitates educated crop management decision-making, assisting in the early identification of disease and efficient agricultural practices.

5 Discussion

By tackling important issues in crop management, the use of DCNNs for maize leaf disease detection marks a substantial breakthrough in agricultural technology. Our research focuses on the creation and application of a DCNN-based system intended to precisely and effectively identify and categorize corn leaf diseases. Rather than requiring a lot of preprocessing, this method makes use of DCNNs' capacity to learn and extract important features straight from raw picture data, improving scalability across huge datasets.

Across a range of visual identification tasks, including the categorization of diseases from images of maize leaves, DCNNs have demonstrated efficacy. Our customized DCNN model successfully learns and improves feature representations over time, yielding an astounding accuracy of 97.53%.

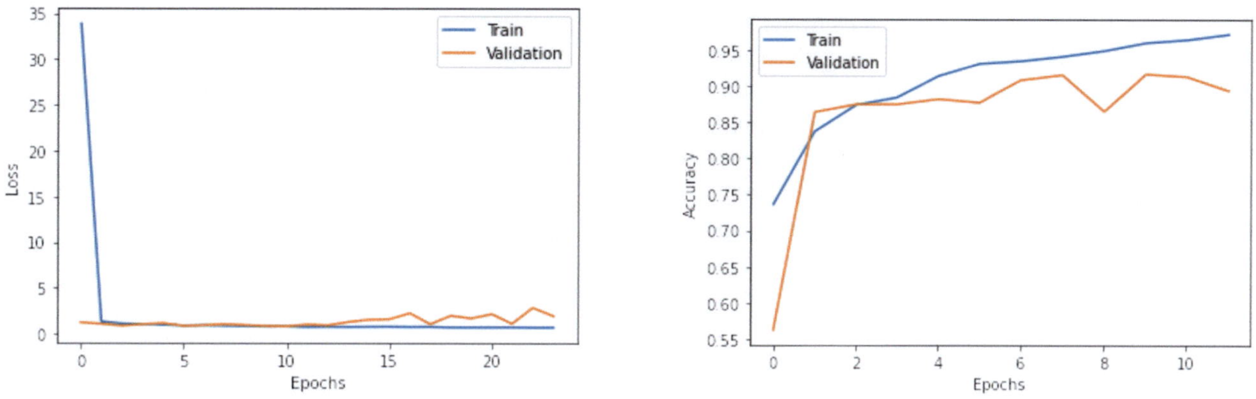

(a) Train and validation loss.

(b) Train and validation accuracy.

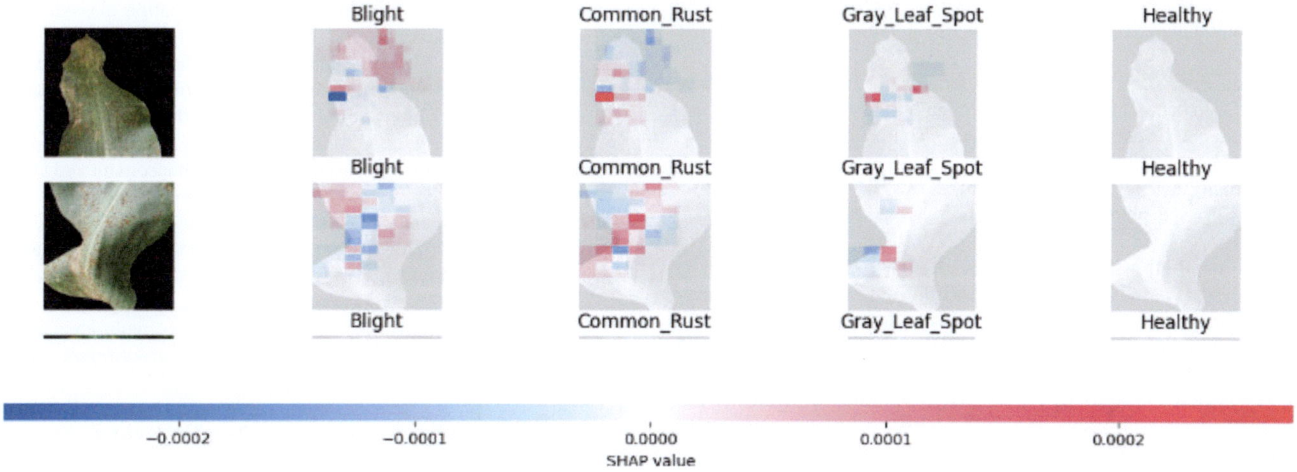

Fig. 6 The heatmap illustrates the feature contributions to the model's prediction, helping to visualize the impact of various symptoms like leaf color, texture, and spots on diagnosis

This high accuracy is essential for early disease diagnosis, allowing farmers to efficiently minimize crop losses and put in place appropriate management plans. One limitation of our work is the potential challenge of ensuring consistent model performance across diverse environmental conditions and varying image quality in real-world agricultural settings.

The incorporation of SHAP improves our model's interpretability even more. A better understanding of how the DCNN makes judgments is made possible by SHAP. Building confidence among agricultural stakeholders and enabling well-informed decision-making in disease control depends on this transparency.

Our research also emphasizes the possibility of integrating DCNNs with current agricultural management systems. Farmers may embrace more data-driven and sustainable farming practices by fusing disease finding capabilities with other agricultural data, such as weather and soil health. By reducing needless chemical treatments, this all-encompassing strategy not only increases crop health and yield but also fosters environmental sustainability.

6 Conclusions

Our study underscores the critical importance of applying technological advancements in combating diseases that afflict maize, a vital global crop crucial for economic stability and food security. It has established a reliable solution with 97.53% accuracy by presenting an approach for maize disease recognition based on a DCNN, supplemented by SHAP for interpretability. This approach has great potential for managing diseases early on and providing farmers and other stakeholders with useful information to reduce crop losses and increase yields. The integration of multi-modal data sources should constitute a focal point for subsequent research endeavors, as it holds the potential to substantially augment the precision and robustness of algorithms designed for disease detection. Adding more data streams, such as infrared or hyperspectral imaging, may yield supplemental information that is particularly helpful in identifying disease signs that are not immediately apparent. Furthermore, there is potential for success when our disease detection system is integrated with precision agriculture technologies. This integration uses automated systems for pest control, fertilization, and irrigation that are based on real-time disease assessments to maximize crop health management techniques.

References

1. Ward, J.M., Stromberg, E.L., Nowell, D.C., Nutter, F.W., Jr.: Gray leaf spot: a disease of global importance in maize production. Plant Dis. **83**(10), 884–895 (1999)
2. Dillard, H., Seem, R., et al.: Incidence-severity relationships for common maize rust on sweet corn. Phytopathology **80**(9), 842–849 (1990)
3. Sladojevic, S., Arsenovic, M., Anderla, A., Culibrk, D., Stefanovic, D.: Deep neural networks based recognition of plant diseases by leaf image classification. Comput. Intell. Neurosci. **2016**(1), 3289801 (2016)
4. Mohanty, S.P., Hughes, D.P., Salathé, M.: Using deep learning for image-based plant disease detection. Front. Plant Sci. **7**, 215232 (2016)
5. Sharma, R., Singh, A., Dutta, M.K., Riha, K., Kriz, P., et al.: Image processing based automated identification of late blight disease from leaf images of potato crops. In: 2017 40th International Conference on Telecommunications and Signal Processing (TSP), pp. 758–762. IEEE (2017)
6. Ashourloo, D., Aghighi, H., Matkan, A.A., Mobasheri, M.R., Rad, A.M.: An investigation into machine learning regression techniques for the leaf rust disease detection using hyperspectral measurement. IEEE J. Sel. Top. Appl. Earth Observ. Remote Sens. **9**(9), 4344–4351 (2016)
7. Aravind, K., Raja, P., Mukesh, K., Aniirudh, R., Ashiwin, R., Szczepanski, C.: Disease classification in maize crop using bag of features and multiclass support vector machine. In: 2018 2nd International Conference on Inventive Systems and Control (ICISC), pp. 1191–1196. IEEE (2018)
8. Khakimov, A., Salakhutdinov, I., Omolikov, A., Utaganov, S.: Traditional and current-prospective methods of agricultural plant diseases detection: A review. In: IOP Conference Series: earth and Environmental Science, vol. 951, p. 012002. IOP Publishing (2022)
9. Kotwal, J., Kashyap, R., Pathan, S.: Agricultural plant diseases identification: from traditional approach to deep learning. Mater. Today: Proc. **80**, 344–356 (2023)
10. LeCun, Y., Bengio, Y., Hinton, G.: Deep learning. Nature **521**(7553), 436–444 (2015)
11. Khan, A., Sohail, A., Zahoora, U., Qureshi, A.S.: A survey of the recent architectures of deep convolutional neural networks. Artif. Intell. Rev. **53**, 5455–5516 (2020)
12. Jeon, W.-S., Rhee, S.-Y.: Plant leaf recognition using a convolution neural network. Int. J. Fuzzy Log. Intell. Syst. **17**(1), 26–34 (2017)
13. Sibiya, M., Sumbwanyambe, M.: A computational procedure for the recognition and classification of maize leaf diseases out of healthy leaves using convolutional neural networks. AgriEngineering **1**(1), 119–131 (2019)
14. Dhaka, V.S., Meena, S.V., Rani, G., Sinwar, D., Ijaz, M.F., Woźniak, M.: A survey of deep convolutional neural networks applied for prediction of plant leaf diseases. Sensors **21**(14), 4749 (2021)
15. Li, Y., Chen, Y., Wang, Y.: Disease recognition of maize leaf based on KNN and feature extraction. Int. J. Pattern Recogn. Artif. Intell. **36**(11), 2257004 (2022)
16. Rachmad, A., Syarief, M., Rifka, S., Sonata, F., Setiawan, W., Rochman, E.M.S.: Corn leaf disease classification using local binary patterns (LBP) feature extraction. J. Phys.: Conf. Ser. **2406**, 012020 (IOP Publishing) (2022)
17. Shoaib, M., Shah, B., Sayed, N., Ali, F., Ullah, R., Hussain, I.: Deep learning for plant bioinformatics: an explainable gradient-based approach for disease detection. Front. Plant Sci. **14**, 1283235 (2023)
18. Xu, L., Xu, X., Hu, M., Wang, R., Xie, C., Chen, H.: Corn leaf disease identification based on multiple classifiers fusion. Trans. Chinese Soc. Agric. Eng. **31**(14), 194–201 (2015)
19. Zhang, S., Wu, X., You, Z., Zhang, L.: Leaf image based cucumber disease recognition using sparse representation classification. Comput. Electron. Agric. **134**, 135–141 (2017)
20. Zeng, W., Li, H., Hu, G., Liang, D.: Identification of maize leaf diseases by using the SKPSNet-50 convolutional neural network model. Sustain. Comput.: Inform. Syst. **35**, 100695 (2022)
21. Lv, M., Zhou, G., He, M., Chen, A., Zhang, W., Hu, Y.: Maize leaf disease identification based on feature enhancement and DMS-robust alexnet. IEEE Access **8**, 57952–57966 (2020)
22. Setiawan, W., Rulaningtyas, R.: Visual explanation of maize leaf diaseases classification using squeezenet and gradient-weighted class activation map. AIP Conf. Proc. **2679** (AIP Publishing) (2023)
23. Yin, C., Zeng, T., Zhang, H., Fu, W., Wang, L., Yao, S.: Maize small leaf spot classification based on improved deep convolutional neural networks with a multi-scale attention mechanism. Agronomy **12**(4), 906 (2022)
24. Zhang, Z., He, X., Sun, X., Guo, L., Wang, J., Wang, F.: Image recognition of maize leaf disease based on GA-SVM. Chem. Eng. Trans. **46**, 199–204 (2015)
25. Lin, Z., Mu, S., Shi, A., Pang, C., Sun, X., et al.: A novel method of maize leaf disease image identification based on a multichannel convolutional neural network. Trans. ASABE **61**(5), 1461–1474 (2018)
26. Zhang, Y., Wa, S., Liu, Y., Zhou, X., Sun, P., Ma, Q.: High-accuracy detection of maize leaf diseases CNN based on multi-pathway activation function module. Remote Sens. **13**(21), 4218 (2021)
27. Bansal, P., Kumar, R., Kumar, S.: Disease detection in apple leaves using deep convolutional neural network. Agriculture **11**(7), 617 (2021)
28. Padilla, D.A., Pajes, R.A.I., De Guzman, J.T.: Detection of corn leaf diseases using convolutional neural network with openmp implementation. In: 2020 IEEE 12th International Conference on Humanoid, Nanotechnology, Information Technology, Communication and Control, Environment, and Management (HNICEM), pp. 1–6. IEEE (2020)
29. Geetharamani, G., Pandian, A.: Identification of plant leaf diseases using a nine-layer deep convolutional neural network. Comput. Electr. Eng. **76**, 323–338 (2019)
30. Singh, D., Jain, N., Jain, P., Kayal, P., Kumawat, S., Batra, N.: Plantdoc: a dataset for visual plant disease detection. In: Proceedings of the 7th ACM IKDD CoDS and 25th COMAD, pp. 249–253 (2020)

Harnessing Convolutional Neural Networks for Potato Leaf Disease Detection: A Proposed Model

Mir Maruf Ahmed, Rakin Sad Aftab, Sultanul Arifeen Hamim, Md. Abdullah-Al-Jubair, and Dip Nandi

Abstract

This study introduces a novel approach to potato leaf disease detection using a DenseNet121-based convolutional neural network (CNN) model that achieves remarkable accuracy and efficiency. By utilizing a carefully curated dataset comprising six categories: early blight, late blight, healthy, virus, fungi, and pest and employing advanced preprocessing and augmentation techniques, the model demonstrated exceptional robustness and generalization. The proposed system achieved an impressive testing accuracy of 97.40%, precision of 97.10%, recall of 97.19%, and F1-score of 97.12%. Additionally, the model showed efficient performance, with an inference time of 0.1847 s per batch and an average inference time of 0.0029 s per image, highlighting its suitability for real-time applications. The optimized training pipeline effectively extracts complex patterns and differentiates subtle variations between visually similar categories, thereby ensuring high-confidence predictions. This scalable, real-time solution bridges the gap between traditional manual diagnostics and modern agricultural needs, and offers a practical tool for timely and accurate disease management. By addressing critical challenges in potato farming, such as yield losses and quality deterioration, this study establishes a new standard for automated plant disease diagnostics. It promises to revolutionize agricultural practices by fostering efficiency, sustainability, and equitable access to advanced technological solutions.

Keywords

Potato • Conv2D • NumPy • Convolutional neural networks (CNN) • DenseNet

1 Introduction

Plant diseases have emerged as a major threat to global agriculture, with significant impacts on crop yield, food security, and economic stability. Of the many crops, potatoes are one of the most important staple foods for millions [1]. Potato plants are susceptible to several diseases caused by fungal, bacterial, and viral infections. Such diseases, if not attended to properly, could ultimately lead to great losses in both the quality and quantity of produce. Early and timely diagnosis of such diseases is very important to reduce their adverse effects [2]. Traditional methods for identifying plant diseases mostly rely on manual inspection by experts, which is time-consuming, laborious, and prone to human error. With the development of machine learning and image processing, there is a promising alternative to traditional approaches, such as automated disease detection systems [3, 4].

In recent years, there has been great interest in the application of image processing techniques in agriculture. Building on computational power and deep learning algorithms, current research has focused on the development of systems that can aid in the identification of plant diseases in real time with high accuracy. Such systems extract visual information (color, shape, and texture) from images of plants to classify the health status of crops. Among the many advanced architectures, deep learning models, including convolutional neural networks (CNNs), have been outstanding in performing the tasks involved in feature extraction and classification [5, 6].

M. M. Ahmed · R. S. Aftab (✉) · S. A. Hamim ·
Md. Abdullah-Al-Jubair · D. Nandi
Department of Computer Science, American International University-Bangladesh, Dhaka, Bangladesh
e-mail: rakinsadaftab@gmail.com

Md. Abdullah-Al-Jubair
e-mail: abdullah@aiub.edu

D. Nandi
e-mail: dip.nandi@aiub.edu

M. F. Mridha and N. Dey (eds.), *Machine Vision in Plant Leaf Disease Detection for Sustainable Agriculture*,
Studies in Computational Intelligence 1202, https://doi.org/10.1007/978-981-96-4520-6_8

DenseNet121 has been found to be promising since this network can use dense connections between layers, which provides efficient feature propagation with less redundancy [7, 8].

Potatoes, which are more susceptible to diseases such as early blight and late blight, present a very strong case for the development of more advanced detection methods. Recent research has recognized the potential of using image-processing techniques to accurately diagnose these diseases [2, 9]. Although some progress has been made, many current models are still constrained by their limited generalizability, lower accuracy in real-world applications, and computational inefficiencies. These gaps highlight the need for innovative approaches that balance accuracy with computational efficiency and robustness to diverse environmental conditions.

The aim of this study is to design a robust and efficient system for the detection of potato plant diseases using an enhanced CNN model. The proposed approach leverages recent advancements in image processing along with CNN architecture to extract and classify features from segmented images of potato leaves. Aligned with this, the system focuses on six major categories: early blight, late blight, virus, fungi, pest and healthy potato plant leaf, providing precise and reliable classifications to enable timely interventions in agricultural practices. This is to improve existing approaches and address the challenges of scalability and real-time deployment for these applications in agricultural settings. The contributions of this study to this field are multifaceted.

1. Introduction of a novel adaptation of the CNN model tailored for potato plant disease detection, including optimization techniques that enhance feature extraction and classification accuracy.
2. A comprehensive dataset encompassing diverse conditions and disease variations was developed and utilized to ensure the robustness of the model.
3. The model achieved an impressive overall classification accuracy of 97.40%, thereby setting a new benchmark for similar applications. Additionally, it demonstrated strong performance metrics with a precision of 97.10%, recall of 97.19%, and F1-score of 97.12%. The efficiency of the model is notable, with an inference time of 0.1847 s for one batch and an average inference time of 0.0029 s per image. These results highlight the robustness and suitability of the model for real-time applications, combining its high accuracy with exceptional computational efficiency.
4. Provision of valuable insights into the potential of deep learning models in transforming traditional agricultural practices and paving the way for future innovations in plant disease detection.

The structure of this paper is organized as follows: Sect. 2 offers a detailed examination of relevant literature, highlighting prior studies in the field. Section 3 outlines the methodology adopted in this research, along with an in-depth description of the proposed model. The results and their analysis are presented in Sect. 4, while Sect. 5 provides an in-depth discussion of the findings. Lastly, Sect. 6 concludes the paper by summarizing the key outcomes and proposing directions for future research.

2 Related Work

2.1 Hybrid and Explainable Deep Learning Models

In recent years, the classification of potato leaf diseases has evolved through the integration of hybrid and explainable deep learning models. Advanced frameworks such as PLDP-Net, a novel hybrid deep learning model, incorporating U-Net for precise segmentation alongside VGG19, Inception-V3, and vision transformer for robust feature extraction, have enabled the isolation and categorization of disease-affected regions with high accuracy [1]. In addition, explainable deep classification models, including variations using variational autoencoders (VAE) combined with you only look once version 3 (YOLOv3) and faster region-based convolutional neural networks (R-CNN), offer clarity on classification decisions and enhance robustness under varying field conditions [3]. DenseNet, combined with 1D CNNs, has shown efficacy in segmenting and classifying diseased areas at the pixel level, thereby allowing more granular disease identification [10]. Furthermore, hierarchical models such as hierarchical deep learning convolutional neural network (HDLCNN) utilize preprocessing techniques such as median filtering to extract textural features that support decision systems for disease notification [11]. The generative approach using generative adversarial network (GANs) combined with CNNs has introduced synthetic image generation, augmenting datasets to improve model generalization across disease categories [12]. Techniques such as Fisher Vector, extracted through scale-invariant feature transform (SIFT) features, coupled with classifiers such as multilayer perceptron (MLP) and support vector machine (SVM), enhance disease classification by leveraging handcrafted features for refined predictions [13]. Finally, the integration of transfer learning with deep convolutional neural networks (DCNNs), as observed in models such as single-stage detection (SSD) Inception V2 and recurrent fully-convolutional network (RFCN) residual neural network (ResNet101), underscores the power of fine-tuning pre-trained networks to detect subtle disease characteristics, pushing the boundaries of potato disease classification [14].

To synthesize recent advancements in potato disease classification, this study leverages the strengths observed in hybrid and explainable deep learning models. CNN was chosen for its densely connected layers and aligned with techniques such as those seen in PLDPNet and HDLCNN, which excel in extracting complex features and improving segmentation accuracy. Models integrating VAE with YOLOv3 and GANs underscored the value of interpretability and data augmentation, contributing to reliable disease detection. By applying CNN, this study builds upon these established methods, seeking to further enhance the classification accuracy and resilience across diverse environmental conditions, thus advancing potato disease diagnostic capabilities.

2.2 Segmentation and Deep Convolutional Models

The classification of potato diseases has greatly benefited from integrating segmentation methods with deep convolutional networks, each of which enhances feature extraction and classification accuracy. Initial techniques using image localization and color transformation, along with logistic regression, MLP, and SVM classifiers, laid the groundwork for accurate disease-boundary identification [15]. The graph-cut algorithm further refines segmentation, providing a solid foundation for subsequent classification [5]. Techniques such as K-means clustering, paired with VGG16, VGG19, and ResNet50, extract high-quality feature maps to improve classification robustness [6]. The application of CNN-especially models such as ResNet, VGG, and MobileNet-allowed deep-layer feature extraction, building on pretrained layers for generalizing across various disease types [16]. Enhanced models, such as an adapted AlexNet with SVM and MobileNetV2 with ResNet, have shown versatility, enabling both rapid processing and high accuracy [17, 18]. DenseNet121, widely utilized for its densely connected architecture, captures intricate leaf patterns, proving effective for precise disease categorization [7, 19].

Building on these segmentation and deep learning advancements, this study employs DenseNet121, noted for its densely connected structure, to enhance the potato disease classification accuracy. Previous models, such as CNNs adaptations using ResNet and MobileNet, and segmentation techniques, such as K-means and graph-cut-highlighted the importance of robust feature extraction and precise boundary identification in disease classification. DenseNet121 leverages these strengths by drawing on layered architectures and data preprocessing techniques to offer a more refined and resilient solution for detecting potato diseases across varied conditions, thereby advancing the diagnostic landscape in agriculture.

2.3 Transfer Learning and Optimization-Enhanced CNN Models

The field of potato disease classification has advanced significantly with the adoption of transfer learning and optimization techniques that leverage pretrained models for efficient and accurate disease identification. Studies applying VGG19, NASNetMobile, and DenseNet169 have demonstrated the effectiveness of transfer learning in reducing the training time while enhancing model performance, thereby achieving precise disease differentiation [20]. Optimization approaches, including the greylag goose and grey wolf optimizers paired with CNNs, have further improved feature extraction layers and refined hyperparameters to increase classification accuracy [21]. Deep CNN architectures, such as VGG19, DenseNet121, and ResNet50 have consistently proven their robustness in capturing complex leaf disease patterns and strengthening their classification capabilities [22]. Dual-stream deep fusion models that incorporate CNN-SVM and CNN-softmax techniques combine complementary strategies to improve disease detection accuracy, which is a powerful approach for multi class classification [23]. DenseNet121's dense connectivity facilitates the capture of intricate image details, an asset in handling complex disease presentations across diverse environments [24].

This study builds on transfer learning and optimization strategies that have demonstrated notable success in potato disease classification. By implementing DenseNet121, known for its dense layers and effective feature mapping, this study seeks to capitalize on the strengths of established models, such as VGG and ResNet, which distinguish complex disease patterns. Enhanced by transfer learning, DenseNet121 refines feature extraction, achieves accurate disease classification even under challenging conditions, and continues the trajectory set by optimized CNN models to improve agricultural diagnostics.

2.4 Spectral Analysis and Lightweight Models

Integrating spectral analysis with lightweight CNNs has opened new avenues for potato disease classification, particularly by harnessing spatial and spectral dimensions. Dual architectures, such as 2D-CNN and 3D-CNN combined with SE-ResNet, enable detailed spatial-spectral feature extraction and enhance model sensitivity to disease markers [2]. Hyperspectral remote sensing, paired with multivariate regression, has been effective in analyzing spectral reflectance, offering precise, non-invasive diagnostic potential [25]. Key spectral indices, such as normalised difference vegetation index (NDVI705) and dual-way seeds indexing (DWSI), provide additional clarity on disease signatures and bolster classification reliability [26]. Lightweight CNN models, such as MobileNet-V2 with Atrous Spatial Pyramid Pooling, efficiently capture multi-scale features, making these models ideal for real-time applications requiring high processing speed [27]. Innovations, such as octave convolution in MobileNet-V2 further improve high-frequency feature extraction, reduce redundancy and adapt well to complex field conditions [28].

This study builds upon spectral analysis and CNN techniques by applying DenseNet121, a model known for its robust feature mapping and dense layer connections. Previous studies integrating spectral indices and optimized lightweight models such as MobileNet-V2 has shown the importance of efficient feature capture in real-time disease detection. DenseNet121 leverages these spectral and CNN advancements, refining feature extraction to accurately classify potato diseases across variable conditions, thus advancing both the precision and applicability of agricultural diagnostics.

2.5 Spectral Reflectance Analysis with Artificial Neural Networks (ANN) and Chemometric Methods

Spectral reflectance analysis plays a critical role in potato disease classification, leveraging artificial neural networks (ANNs) and chemometric techniques. By employing densely connected multilayer perceptrons (MLPs), ANNs effectively capture unique spectral signatures associated with disease states, enabling reliable differentiation between healthy and diseased potato plants [29]. Visible/near-infrared (Vis/NIR) spectroscopy combined with partial least squares (PLS) regression has been instrumental in modeling disease severity through spectral measurements. This method integrates physicochemical indicators such as chlorophyll content and enzyme activity, refined further through machine learning classification techniques [30].

Building on these approaches, this study applies DenseNet121, an advanced model optimized for intricate feature extraction. Dense connectivity within DenseNet121 enhances the detection of subtle disease features, improving classification performance under diverse disease manifestations and environmental conditions. Prior work with Vis/NIR spectroscopy and PLS regression highlights the importance of capturing specific spectral responses, which DenseNet121 utilizes to achieve robust disease classification in potato agriculture.

A study leveraging the publicly accessible PlantVillage database used 450 images of healthy and diseased potato leaves to classify disease states. Feature extraction employed global descriptors such as Hu moments, Haralick textures, and color histograms. Seven classifiers-random forest, logistic regression, k-nearest neighbors (KNN), decision trees, naive Bayes, linear discriminant analysis (LDA), and support vector machines (SVM)-were evaluated, with the random forest classifier achieving the highest performance, effectively detecting and classifying potato leaf diseases [31]. This study demonstrates the efficacy of combining image processing with machine learning for autonomous potato disease diagnosis.

Another study explored deep learning techniques to classify potato diseases, analyzing healthy leaves alongside black scurf, common scab, black leg, and pink rot. Using a dataset of 5,000 potato images, the study compared CNN-based approaches, including AlexNet, GoogLeNet, VGG, R-CNN, and transfer learning. Results revealed that the proposed deep learning approach outperformed existing methods, achieving high classification accuracy and precision rates across all disease categories [32]. A separate study proposed a CNN architecture to distinguish healthy potato leaves from those affected by early and late blight. The dataset size was augmented from 1,722 to 9,822 images, significantly enhancing the model's performance during testing [33].

A recent study developed an alternative approach utilizing RGB imaging combined with machine learning for disease classification in potato leaves. This method, which achieved strong classification performance, avoids the need for costly hyperspectral imaging or computationally intensive deep learning algorithms, highlighting the advantages of simpler and cost-effective approaches [34].

The effectiveness of deep learning methods extends beyond potatoes, with studies applying these techniques to a wide range of crops. For instance, deep learning models have been used for the segmentation and classification of banana leaf diseases [35]. In another study, CNNs were applied to diagnose potato diseases, achieving robust classification results. The findings emphasized the importance of larger training datasets and included a confusion matrix analysis showcasing the reliability of CNNs in agricultural disease classification [36].

3 Methodology

Potato crops are vital for ensuring global food security and sustaining the agricultural economy. However, the prevalence of diseases poses a significant challenge, often resulting in reduced yields and compromised crop quality. Consequently, the ability to detect and diagnose diseases accurately and promptly is critical for effective management strategies. Recent advancements in computer vision and deep learning, including models like CNN and DenseNet, have demonstrated great potential in automating the identification and classification of potato diseases. The proposed methodology of this research is illustrated in Fig. 1.

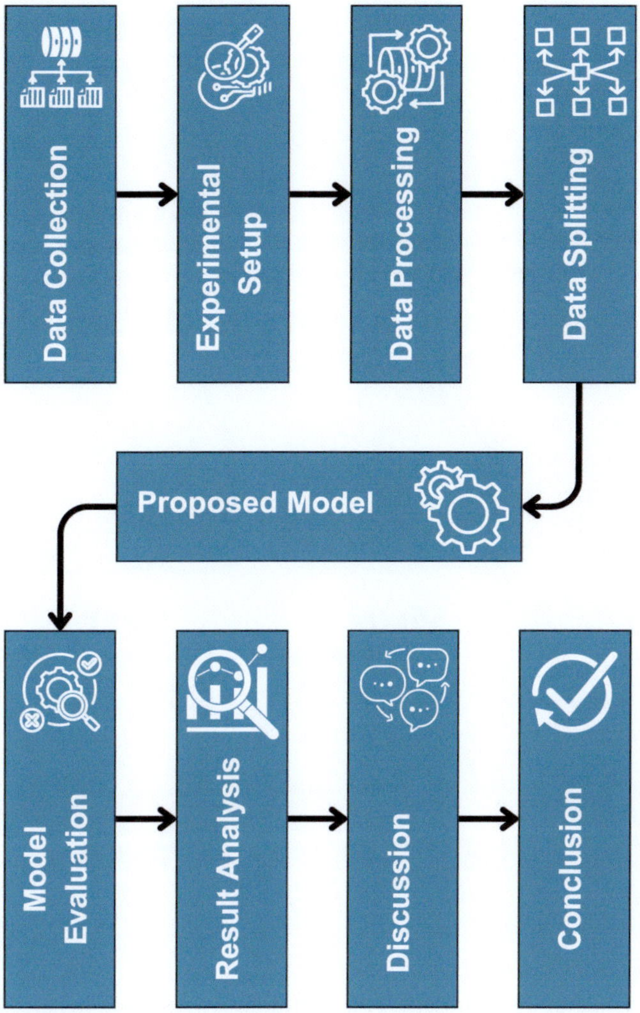

Fig. 1 Proposed research methodology

3.1 Experimental Setup

The experimental setup for this research involved configuring the necessary tools and resources to process and analyze the dataset efficiently. The dataset was stored on Google Drive, which is mounted in a working environment to enable seamless access. The dataset was compressed in a .zip format was extracted to the local directory for further use. Essential libraries were imported, including TensorFlow for deep learning, matplotlib for visualization, seaborn for enhanced data visualization, and PIL for image processing. To load and preprocess the dataset, the TensorFlow *image_dataset_from_directory* function was utilized, which enabled the automated loading, batching, and resizing of images from the dataset directory.

Table 1 Label list of the classified images

Label name	Number of images
Early blight	1000
Fungi	748
Healthy	1000
Late blight	1000
Pest	1000
Virus	1000

3.2 Data Collection

The availability of high-quality datasets is crucial for the success of CNN-based models in disease detection. Publicly available datasets from open-source platforms, such as Kaggle, were explored to assemble a comprehensive dataset for potato leaf disease detection. Several challenges were encountered during the dataset selection process, as many datasets were either highly imbalanced or contained a small number of image files, often fewer than 1000 images per class. Some datasets featured a large number of images but were limited in class diversity, restricting their applicability. To address these limitations, a representative dataset [37] was curated by focusing on six classes relevant to potato leaf health, including five diseased leaf classes-early blight, fungi, late blight, pest, and virus-and one healthy class. Images were carefully selected to ensure adequate representation, with each class containing 1000 images, except for the "fungi" class, which comprises 748 images owing to constraints in data availability. This curated dataset provides a robust foundation for training and evaluating the proposed CNN model, ensuring a balance between class diversity and data sufficiency, while enabling effective generalization across various disease types. Table 1 lists the class-wise image counts.

3.3 Image Pre-processing

Data preprocessing was crucial for improving the performance of the proposed CNN model. In this stage, the input images were prepared to improve the performance of the model and its generalizability. All the images are first resized to a unified size of 244 × 244. The dataset was then shuffled with a shuffle size of 2000 to eliminate any potential biases in the order of data during training. Data augmentation techniques are further applied to increase variability: random rotations up to 0.3 radians and random flips in both vertical and horizontal directions. These augmentations create a more

Fig. 2 Label-wise leaves samples

diverse training dataset to help the model learn the robust features. The batch size was set to 64 to balance the computational efficiency and model performance during the training. Adam was used as the optimizer, and its adaptive learning rate helped the model converge efficiently and stably. The model was trained for 80 epochs to allow sufficient iterations to learn complicated patterns. The collective implementation of these preprocessing steps serves to standardize and diversify the dataset, thereby facilitating the capacity of the CNN model to accurately learn and classify the six distinct categories of potato leaf conditions. Figure 2 shows the label-wise leaf samples.

3.4 Dataset Partitioning

The dataset was partitioned into training, validation, and test sets in an 80:10:10 ratio. A training set, consisting of 80% of the data, was used to train the proposed model. The validation set, comprising 10% of the data, was used to evaluate the performance of the model during training, and the remaining 10% was allocated to the test set to assess the model's performance on unseen data. This partitioning ensures a balanced distribution of data for training, validation, and testing.

3.5 Proposed Model

The proposed model shown in Fig. 3 for potato leaf disease detection is a CNN designed to achieve high accuracy in multi-class classification. The model begins with an image preprocessing pipeline that resizes all images to a resolution of 244 × 244 pixels and rescales the pixel values to a range of 0–1. Data augmentation techniques, including random rotations of up to 30% and random flips (horizontal and vertical), were applied to enhance the generalization capabilities of the model.

The CNN architecture consisted of five convolutional layers, each configured with 64 filters, a kernel size of 3 × 3, and rectified linear unit (ReLU) activation functions to extract spatial features. Each convolutional layer was followed by a max-pooling layer with a pooling window of 2 × 2 to reduce the spatial dimensions while retaining the critical features. The feature maps are then flattened and passed through two dense layers, both with 64 units, where the first dense layer

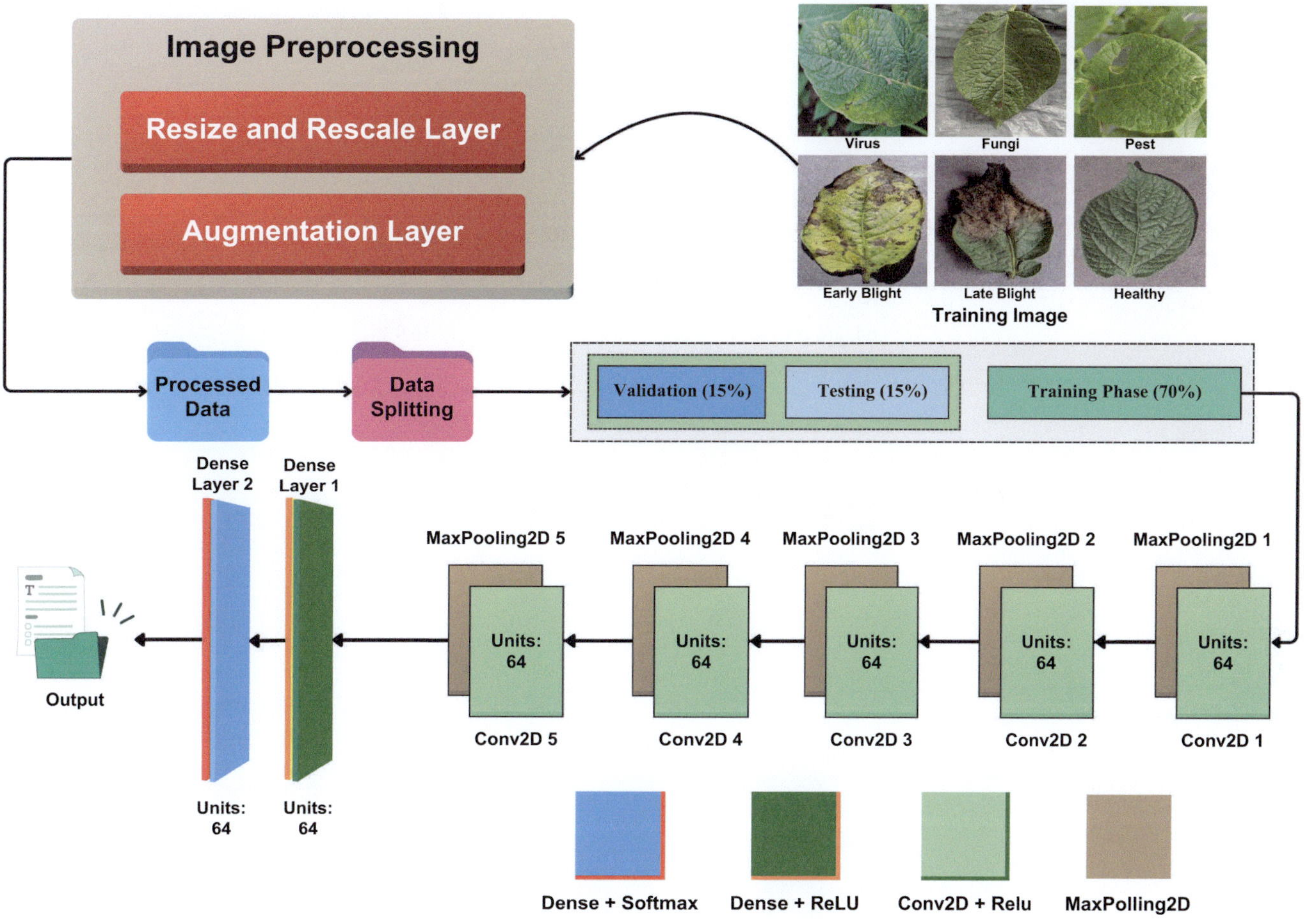

Fig. 3 Proposed model

uses ReLU activation, and the output layer employs a softmax activation function to classify images into six classes: virus, fungi, pest, early blight, late blight, and healthy.

The training process used a batch size of 64, Adam optimizer for efficient parameter updates, and shuffle buffer size of 10,000 to ensure randomness during training. The model was trained over 80 epochs, and caching and prefetching mechanisms were applied to optimize the data-loading efficiency. These hyperparameter settings and preprocessing techniques ensure a reliable and well-optimized model for disease detection.

Table 2 summarizes the architecture of the proposed model for potato leaf disease detection. The model began with an input layer of shape 244 × 244 × 3, followed by five convolutional layers, each with 64 filters and a kernel size of 3 × 3, and max-pooling layers that progressively reduced the spatial dimensions of the feature maps. After the final max-pooling layer, the feature maps are flattened into a one-dimensional vector with a size of 1600. This flattened vector was passed through a dense layer with 64 units, followed by the final dense layer with six units corresponding to the six classes. The model had a total of 252,358 trainable parameters and no non-trainable parameters. This table outlines the layer-wise output shapes and the number of parameters contributing to the model.

4 Result Analysis

Infectious diseases of crops pose a global threat to agricultural production; however, advancements in AI for detection and classification offer effective solutions to manage and mitigate these risks at early stages. The proposed model, which was extensively evaluated for classifying six different classes of potato leaves, demonstrated exceptional performance. It achieved an impressive accuracy of 97.40%, precision of 97.10%, recall of 97.19%, and F1-score of 97.12%. Additionally, the model exhibited remarkable efficiency, with an inference time of 0.1847 s per batch and an average inference time of 0.0029 s per image. These results highlight the reliability and practicality of the proposed model for real-world agricultural applications.

The confusion matrix in Fig. 4 presents a detailed breakdown of the classification outcomes. The model showed

Table 2 Summary of the proposed model

Layer (type)	Output shape	Param
Sequential	64, 244, 244, 3	0
conv2d	64, 242, 242, 64	1,792
max_pooling2d	64, 121, 121, 64	0
conv2d_1	64, 119, 119, 64	36,928
max_pooling2d_1	64, 59, 59, 64	0
conv2d_2	64, 57, 57, 64	36,928
max_pooling2d_2	64, 28, 28, 64	0
conv2d_3	64, 26, 26, 64	36,928
max_pooling2d_3	64, 13, 13, 64	0
conv2d_4	64, 11, 11, 64	36,928
max_pooling2d_4	64, 5, 5, 64	0
Flatten	64, 1600	0
Dense	64, 64	102,464
Dense_1	64, 6	390
Total params: 252,358 Trainable params: 252,358 Non-trainable params: 0		

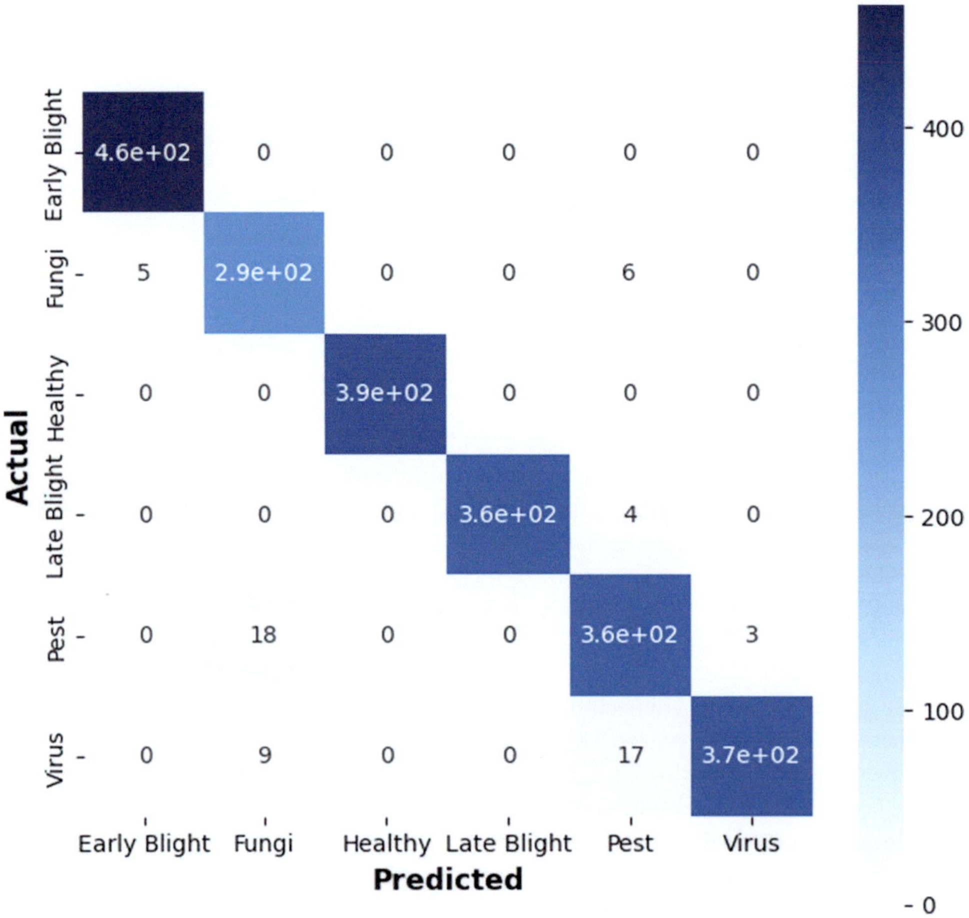

Fig. 4 Confusion matrix of the proposed model

excellent accuracy in most classes, accurately classifying 460 early blight samples and 390 healthy samples. Good performance of the model is also evident in the late blight, pest, and virus classes, for which accurate predictions were recorded

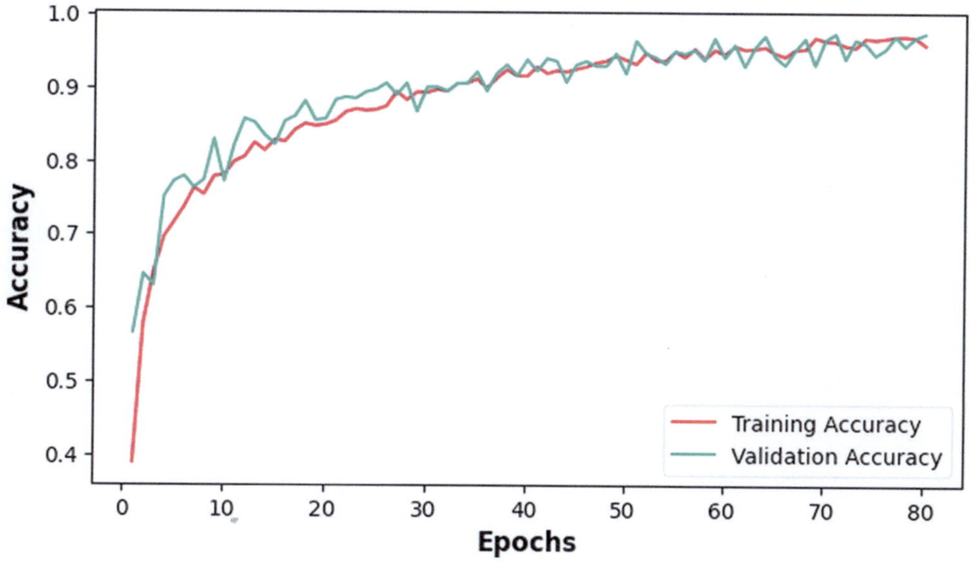

Fig. 5 Training and validation accuracy curve

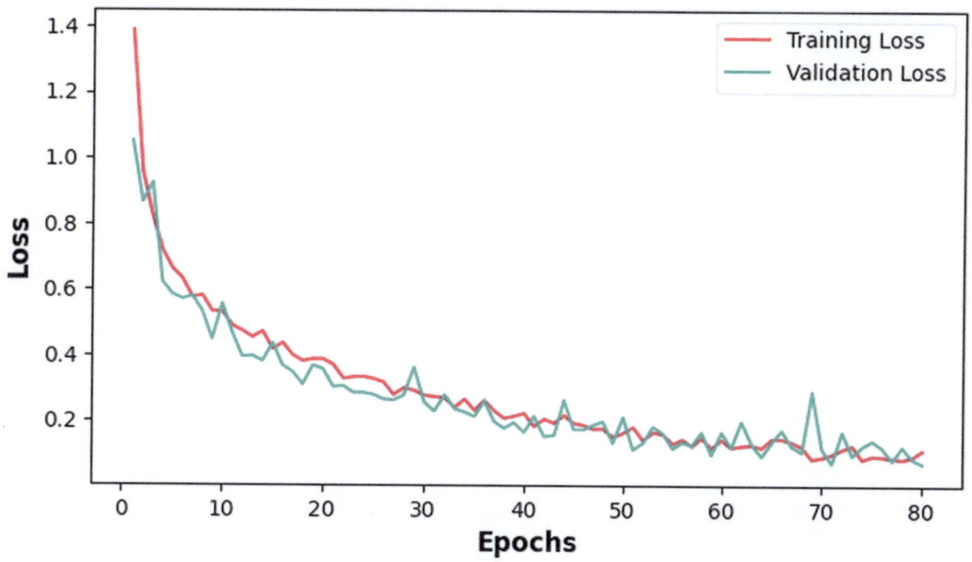

Fig. 6 Training and validation loss curve

on 360, 360, and 370 samples, respectively. Some minor misclassifications also occurred, like 18 cases where Fungi were misclassified into Pest, and 9 Virus samples that were mistakenly labeled as Healthy. The misclassifications that are evident can be attributed to overlapping visual characteristics and texture similarities between these classes. However, despite these small inaccuracies, the confusion matrix highlights the overall accuracy and robustness of the model in dealing with various leaf conditions.

Figures 5 and 6 illustrate the effectiveness of the model during the training and validation processes. The loss for both the training and validation datasets was continuously reduced and reached almost an optimum by the 80th epoch. This ensures that the model learns suitably without major overfitting because the trend of validation loss follows the training loss almost throughout the entire training course. The accuracy trajectories rose quickly and consistently to ultimately provide a final training accuracy of 97% and a validation accuracy of 98%. The findings confirm to the robustness of the architecture and effectiveness of the training methodology.

For qualitative analysis, Fig. 7 shows the representative images for each class, along with their predicted labels and corresponding confidence scores. The model proved remarkably reliable, as all predictions were made with a confidence level exceeding 98%. Notably, some of the Early Blight and Pest samples were classified with 100% confidence, indicating that the model can exactly extract complex patterns and discern subtle differences. For instance, it correctly identified the diseases, which demonstrates good generalization across diverse manifestations of diseases.

The ablation study presented in Table 3 evaluates the performance of the model under varying configurations of

Fig. 7 Actual and predicted classes

key parameters, including the resolution, batch size, number of epochs, random rotation, shuffle size, and augmentation strategies. The proposed configuration, comprising a batch size of 64, ReLU and softMax activation functions, an ADAM optimizer, an input resolution of 244 × 244, a random rotation of 0.3, a shuffle size of 2000, and vertical and horizontal flipping, achieved the highest accuracy of 97.40%, establishing it as the optimal setup. Deviations in parameters, such as smaller or larger resolutions (e.g., 232 × 232 or 252 × 252), alternative augmentation strategies (e.g., horizontal or vertical flipping only), or changes in batch size (e.g., 32 or 128), result in reduced accuracy, highlighting the sensitivity of the model to these factors. Additionally, using alternative optimizers such as SGD significantly reduced the accuracy to

Table 3 Ablation study with different parameters

S/N	Units	Activation	Batch size	Optimizer	EPOCH	Resolution	Random rotation	Shuffle size	Random flip	Accuracy (%)
1	64	ReLU + SoftMax	64	ADAM	80	248 × 248	0.3	2000	vertical_and_ horizontal	95.66
2	64	ReLU + SoftMax	64	ADAM	80	240 × 240	0.3	2000	vertical_and_ horizontal	95.83
3	64	ReLU + SoftMax	64	ADAM	80	244 × 244	0.1	2000	vertical_and_ horizontal	96.53
4	64	ReLU + SoftMax	64	ADAM	80	244 × 244	0.2	2000	vertical_and_ horizontal	94.27
5	64	ReLU + SoftMax	64	ADAM	80	244 × 244	0.4	2000	vertical_and_ horizontal	94.62
6	64	ReLU + SoftMax	64	ADAM	80	244 × 244	0.3	1000	vertical_and_ horizontal	94.79
7	64	ReLU + SoftMax	64	ADAM	80	244 × 244	0.3	1500	vertical_and_ horizontal	95.49
8	64	ReLU + SoftMax	64	ADAM	80	244 × 244	0.3	3000	vertical_and_ horizontal	96.22
9	64	ReLU + SoftMax	64	ADAM	80	244 × 244	0.3	2000	horizontal_and_ vertical	94.44
10	64	ReLU + SoftMax	64	ADAM	80	244 × 244	0.3	2000	horizontal	95.49
11	64	ReLU + SoftMax	64	ADAM	80	244 × 244	0.3	2000	vertical	95.31
12	64	ReLU + SoftMax	64	ADAM	70	244 × 244	0.3	2000	vertical_and_ horizontal	94.27
13	64	ReLU + SoftMax	64	ADAM	60	244 × 244	0.3	2000	vertical_and_ horizontal	95.49
14	64	ReLU + SoftMax	64	ADAM	50	244 × 244	0.3	2000	vertical_and_ horizontal	94.97
15	64	ReLU + SoftMax	64	SGD	80	244 × 244	0.3	2000	vertical_and_ horizontal	83.16
16	64	ReLU + SoftMax	32	ADAM	80	244 × 244	0.3	2000	vertical_and_ horizontal	93.92
17	64	ReLU + SoftMax	128	ADAM	80	244 × 244	0.3	2000	vertical_and_ horizontal	93.44
18	64	ReLU + SoftMax	64	ADAM	80	232 × 232	0.3	2000	vertical_and_ horizontal	91.15
19	64	ReLU + SoftMax	64	ADAM	80	252 × 252	0.3	2000	vertical_and_ horizontal	95.83
Proposed	64	ReLU + SoftMax	64	ADAM	80	244 × 244	0.3	2000	vertical_and_ **horizontal**	97.40

83.16%, reaffirming the superiority of the ADAM optimizer in this context. This systematic evaluation underscores the importance of fine-tuning hyperparameters and preprocessing techniques to optimize the model performance, with the proposed configuration serving as a benchmark for achieving state-of-the-art results in potato leaf disease detection.

5 Discussion

This study proposes a novel CNN model based on DenseNet121 for classifying six categories of potato leaf conditions, achieving a remarkable testing accuracy of

97.40%. The model proved to be a reliable and robust tool for plant disease detection, delivering high-confidence predictions with an accuracy of 98% for most samples. It also demonstrated strong evaluation metrics, including a precision of 97.10%, recall of 97.19%, and F1-score of 97.12%. Furthermore, the model exhibited exceptional computational efficiency, with an inference time of 0.1847 s per batch and an average inference time of 0.0029 s per image. Compared with earlier approaches, which are often hampered by computational inefficiencies or limited generalizability, this model effectively addresses these challenges. Although some misclassifications occur, they are minimal and generally attributable to the inherent visual similarities between certain disease categories, such as fungi and pests.

These errors, while significant, do not by any means deviate from the overall performance of the model and are, therefore, areas possibly warranting refinement in future work. This study has several real-world implications. Agriculture remains a backbone of global food security, and disease susceptibility is especially high in potato crops, which can cause huge losses in yield and quality. It would also provide farmers with a usable, efficient, and accurate tool for real-time disease diagnostics, reducing their dependence on visual inspection or expert intervention. This saves time and labor and allows timely interventions to reduce losses and improve productivity and sustainability within the agricultural sector. Moreover, with a deep learning-based solution, the gap in resources for farmers in regions with limited access to expert diagnostic services can be bridged, thus contributing to equitable advancements in agricultural technology. Analytically, the results were consistent line with the overall project objectives. The high accuracy, robustness against overfitting, and effective handling of complex data show that the proposed methodology is quite successful. Moreover, the ability to generalize across such diverse conditions brings a special kind of practical applicability to the model and meets the expectations of this type of research. The results justify the strong effectiveness of this approach for one of the major concerns of present-day agriculture, making concrete impacts on technology and society.

6 Conclusion

This study demonstrates a critical advancement in the realm of potato leaf disease detection and provides a strong and scalable solution with the proposed architecture for the task of highly accurate classification into six classes. A system achieving 97.40% testing accuracy was developed using advanced preprocessing, augmentation techniques, and a well-designed training pipeline, demonstrating the robustness of the approach and setting a new standard for similar studies. It is a holistic integration of deep learning principles and practical agricultural applications, thereby bridging a critical gap in the existing ways of doing things. The initial goal was to develop a model capable of accurately and efficiently identifying potato disease under diverse conditions. Although the introduced system is well-adjusted with this aim, the results reveal tiny gaps between expectations and achievements. For example, high confidence scores affirm the system's reliability, but minor misclassifications in the visually similar disease categories of fungi and pests point out areas in which further refinement is needed. These gaps simultaneously suggest that more advanced techniques should be adopted, such as hybrid models or explainable AI, to further improve the interpretability of classification accuracy for such edge cases. Although there is still some margin for improvement, especially in dealing with subtle interclass ambiguities, the overall achievements of this research are undeniable. This study provides a solid foundation for real-world agricultural disease detection, which will facilitate further development. This study not only verifies the potential of deep learning models in solving agricultural challenges but also opens an opportunity for iterative enhancement, which can bring us closer to a fully optimized solution. Therefore, this study represents an important step forward, laying the foundation for future innovation in plant disease diagnostics.

References

1. Arshad, F., Mateen, M., Hayat, S., Wardah, M., Al-Huda, Z., Gu, Y.H., Al-Antari, M.A.: PLDPNet: end-to-end hybrid deep learning framework for potato leaf disease prediction. Alex. Eng. J. **78**, 406–418 (2023). https://doi.org/10.1016/j.aej.2023.07.076
2. Qi, C., Sandroni, M., Westergaard, J.C., Sundmark, E.H.R., Bagge, M., Alexandersson, E., Gao, J.: In-field classification of the asymptomatic biotrophic phase of potato late blight based on deep learning and proximal hyperspectral imaging. Comput. Electron. Agric. **205**, 107585 (2023). https://doi.org/10.1016/j.compag.2022.107585
3. Oishi, Y., Habaragamuwa, H., Zhang, Y., Sugiura, R., Asano, K., Akai, K., et al.: Automated abnormal potato plant detection system using deep learning models and portable video cameras. Int. J. Appl. Earth Observ. Geoinf. **104**, 102509 (2021). https://doi.org/10.1016/j.jag.2021.102509
4. Hamim, S.A., Jony, A.I.: Enhanced deep learning model architecture for plant disease detection in Chilli plants. J. Edge Comput. **3**(2), 136–146 (2024). https://doi.org/10.55056/jec.758
5. Hou, C., Zhuang, J., Tang, Y., He, Y., Miao, A., Huang, H., Luo, S.: Recognition of early blight and late blight diseases on potato leaves based on graph cut segmentation. J. Agric. Food Res. **5**, 100154 (2021). https://doi.org/10.1016/j.jafr.2021.100154
6. Nishad, M.A.R., Mitu, M.A., Jahan, N.: Predicting and classifying potato leaf disease using K-means segmentation techniques and deep learning networks. Procedia Comput. Sci. **212**, 220–229 (2022). https://doi.org/10.1016/j.procs.2022.11.006
7. Billah, M.M., Sultana, A., Sad Aftab, R., Ahmed, M.M., Shorif Uddin, M.: Leaf disease detection using convolutional neural networks: a proposed model using tomato plant leaves. Neural Comput. Appl. **36**(32), 20043–20053 (2024). https://doi.org/10.1007/s00521-024-10283-2
8. Kabir, M.M., Ohi, A.Q., Mridha, M. F.: A multi-plant disease diagnosis method using convolutional neural network. In: Computer

Vision and Machine Learning in Agriculture, pp. 99–111 (2021). https://doi.org/10.1007/978-981-33-6424-0_7
9. Siddique, A.Z.B., Das, S., Tabassum, P., Tasir, A.M., Roy, S., Rahman, M.S., et al.: Late and early blight diseases identification of potatoes with a light weight hybrid transfer learning model. In: 2023 IEEE World AI IoT Congress (AIIoT), pp. 0105–0111. IEEE (2023). https://doi.org/10.1109/AIIoT58121.2023.10174429
10. Chen, J., Deng, X., Wen, Y., Chen, W., Zeb, A., Zhang, D.: Weakly-supervised learning method for the recognition of potato leaf diseases. Artif. Intell. Rev. **56**(8), 7985–8002 (2023). https://doi.org/10.1007/s11042-022-12147-0
11. Kumar, A., Patel, V.K.: Classification and identification of disease in potato leaf using hierarchical based deep learning convolutional neural network. Multimedia Tools Appl. **82**(20), 31101–31127 (2023). https://doi.org/10.1007/s11042-023-14663-z
12. Al-Kateb, G., Mijwil, M.M., Aljanabi, M., Abotaleb, M., Priya, S.K., Mishra, P.: AI-PotatoGuard: leveraging generative models for early detection of potato diseases. Potato Res. 1–15 (2024). https://doi.org/10.1007/s11540-024-09751-y
13. Kurmi, Y., Gangwar, S., Agrawal, D., Kumar, S., Srivastava, H.S.: Leaf image analysis-based crop diseases classification. Signal Image Video Process. **15**(3), 589–597 (2021). https://doi.org/10.1007/s11760-020-01780-7
14. Wang, C., Xiao, Z.: Potato surface defect detection based on deep transfer learning. Agriculture **11**(9), 863 (2021). https://doi.org/10.3390/agriculture11090863
15. Kurmi, Y., Gangwar, S.: A leaf image localization based algorithm for different crops disease classification. Inf. Process. Agric. **9**(3), 456–474 (2022). https://doi.org/10.1016/j.inpa.2021.03.001
16. Gülmez, B.: A comprehensive review of convolutional neural networks based disease detection strategies in potato agriculture. Potato Res. 1–35 (2024). https://doi.org/10.1007/s11540-024-09786-1
17. Suarez Baron, M.J., Gomez, A.L., Diaz, J.E.E.: Supervised learning-based image classification for the detection of late blight in potato crops. Appl. Sci. **12**(18), 9371 (2022). https://doi.org/10.3390/app12189371
18. Kang, F., Li, J., Wang, C., Wang, F.: A lightweight neural network-based method for identifying early-blight and late-blight leaves of potato. Appl. Sci. **13**(3), 1487 (2023). https://doi.org/10.3390/app13031487
19. Billah, M.M., Aftab, R.S., Ahmed, M.M., Uddin, M.S.: Deep Facial Recognition: unraveling kinship patterns among strangers using CNN. In: 2024 IEEE International Conference on Computing, Applications and Systems (COMPAS), pp. 1–9. IEEE (2024). https://doi.org/10.1109/COMPAS60761.2024.10797096
20. Lanjewar, M.G., Morajkar, P.P.: Modified transfer learning frameworks to identify potato leaf diseases. Multimedia Tools Appl. **83**(17), 50401–50423 (2024). https://doi.org/10.1007/s11042-023-17610-0
21. El-Kenawy, E.S.M., Alhussan, A.A., Khafaga, D.S., Abotaleb, M., Mishra, P., Arnous, R., Eid, M.M.: Optimizing potato disease classification using a metaheuristics algorithm for deep learning: a novel approach for sustainable agriculture. Potato Res. 1–35 (2024). https://doi.org/10.1007/s11540-024-09755-8
22. Kang, F., Li, J., Wang, C., Wang, F.: A lightweight neural network-based method for identifying early-blight and late-blight leaves of potato. Appl. Sci. **13**(3), 1487 (2023). https://doi.org/10.3390/app13031487
23. Haque, M.R., Sohel, F.: Deep network with score level fusion and inference-based transfer learning to recognize leaf blight and fruit rot diseases of eggplant. Agriculture **12**(8), 1160 (2022). https://doi.org/10.3390/agriculture12081160
24. Nakib, M.I., Ghose, S., Ahmed, M.M., Mahomuda, M.A., Mridha, M.F.: Tomato disease classification using convolutional neural networks. In: 2023 4th International Conference on Data Analytics for Business and Industry (ICDABI), pp. 495–498 (2023). IEEE. https://doi.org/10.1109/ICDABI60145.2023.10629433
25. Nishad, M.A.R., Mitu, M.A., Jahan, N.: Predicting and classifying potato leaf disease using k-means segmentation techniques and deep learning networks. Procedia Comput. Sci. **212**, 220–229 (2022). https://doi.org/10.1016/j.atech.2021.100019
26. Anim-Ayeko, A.O., Schillaci, C., Lipani, A.: Automatic blight disease detection in potato (Solanum tuberosum L.) and tomato (Solanum lycopersicum, L. 1753) plants using deep learning. Smart Agric. Technol. **4**, 100178 (2023). https://doi.org/10.1016/j.atech.2023.100178
27. Chen, J., Deng, X., Wen, Y., Chen, W., Zeb, A., Zhang, D.: Weakly-supervised learning method for the recognition of potato leaf diseases. Artif. Intell. Rev. **56**(8), 7985–8002 (2023). https://doi.org/10.1007/s10462-022-10374-3
28. Chen, W., Chen, J., Zeb, A., Yang, S., Zhang, D.: Mobile convolution neural network for the recognition of potato leaf disease images. Multimedia Tools Appl. **81**(15), 20797–20816 (2022). https://doi.org/10.1007/s11042-022-12620-w
29. Griffel, L.M., Delparte, D., Whitworth, J., Bodily, P., Hartley, D.: Evaluation of artificial neural network performance for classification of potato plants infected with potato virus Y using spectral data on multiple varieties and genotypes. Smart Agric. Technol. **3**, 100101 (2023). https://doi.org/10.1016/j.atech.2022.100101
30. Hou, B., Hu, Y., Zhang, P., Hou, L.: Potato late blight severity and epidemic period prediction based on vis/nir spectroscopy. Agriculture **12**(7), 897 (2022). https://doi.org/10.3390/agriculture12070897
31. Iqbal, M.A., Talukder, K.H.: Detection of potato disease using image segmentation and machine learning. In: 2020 International Conference on Wireless Communications Signal Processing and Networking (WiSPNET), pp. 43–47. IEEE (2020). https://api.semanticscholar.org/CorpusID:221846532
32. Arshaghi, A., Ashourian, M., Ghabeli, L.: Potato diseases detection and classification using deep learning methods. Multimedia Tools Appl. **82**(4), 5725–5742 (2023). https://doi.org/10.1007/s11042-022-13390-1
33. Khalifa, N.E.M., Taha, M.H.N., Abou El-Maged, L.M., Hassanien, A.E.: Artificial intelligence in potato leaf disease classification: a deep learning approach. In: Machine Learning and Big Data Analytics Paradigms: analysis, Applications and Challenges, pp. 63–79 (2021). https://doi.org/10.1007/978-3-030-59338-4_4
34. Islam, M., Dinh, A., Wahid, K., Bhowmik, P.: Detection of potato diseases using image segmentation and multiclass support vector machine. In: 2017 IEEE 30th Canadian Conference on Electrical and Computer Engineering (CCECE), pp. 1–4. IEEE (2017). https://doi.org/10.1109/CCECE.2017.7946594
35. Elsharif, A.A., Dheir, I.M., Mettleq, A.S.A., Abu-Naser, S.S.: Potato classification using deep learning (2020). www.ijeais.org/ijapr
36. Oppenheim, D., Shani, G.: Potato disease classification using convolution neural networks. Adv. Anim. Biosci. **8**(2), 244–249 (2017). https://doi.org/10.1017/S2040470017001376
37. Ahmed, M.M., Sad Aftab, R., Hamim, S.A., Mridha, M.F.: Potato disease classification dataset. IEEE Dataport (2024). https://doi.org/10.21227/n3ce-a694

Automated Plant Diseases Analysis Using Lightweight Deep Learning Models

Shafiul Ajam Opee, Arifa Akter Eva, Mustak Hasan Sayem, and Ahmed Taj Noor

Abstract

Plants are important and essential elements for humans. Automatic detection and analysis of plant diseases are vital for conserving resources, minimizing yield losses, and enhancing treatment efficacy. This has resulted in healthier plants and more efficient farming practices. A novel automated system was developed to accurately identify and classify diseases using plant images. Using data from plantvillages, this system demonstrates computer vision techniques, such as image processing, machine learning (ML), and deep learning (DL), to go beyond conventional approaches to disease management for important plant diseases such as bacterial spot of tomato, tomato early blight. Diseases were then identified based on visual characteristics using lightweight models (Xception, EfficientNet, VGG16, ResNet50, NASNet Mobile, and MobileNet). The DL approach, utilizing ReLU and softmax functions, achieved an impressive validation accuracy of 99.39%. After identifying the diseases, the system suggests predictive treatments to help farmers and agricultural organizations take effective measures against these threats.

Keywords

Plant disease • Deep learning • Agriculture • Image classification • Convolutional neural networks (CNNs)

S. A. Opee · A. A. Eva (✉) · M. H. Sayem
Computer Science and Engineering, American International University Bangladesh, Dhaka, Bangladesh
e-mail: arifaeva2019@gmail.com

A. T. Noor
Computer Science and Engineering, Southeast University, Dhaka, Bangladesh

1 Introduction

Agricultural decision-making is crucial from both economic and social perspectives. With agriculture and related sectors contributing around 25% of the GDP and nearly 70% of Indians relying on it for their livelihood, effective agricultural planning is vital for overall national growth, especially in the context of today [1]. Global population continues to rise, and demand for agricultural products has also increased. Careful planning and decision-making are required to ensure food security and agricultural sustainability. This includes addressing key issues, such as assessing soil quality, choosing the appropriate irrigation methods for specific areas, and understanding the risks associated with farming [2]. Diseases significantly reduce crop yields, directly affecting both national and global food manufacturing and leading to substantial economic losses [3]. According to the Food and Agriculture Organization (FAO) and the United Nations International Plant Protection Convention (2017), pests and diseases of plants cause 20–40% of global agricultural losses. Plant diseases reduce worldwide food yields by an estimated 13% [4].

This plant is an important element. In this rapidly evolving environment, early prevention has grown more crucial [5]. Although there are a number of ways to identify plant pathologies, some diseases may not exhibit obvious symptoms, or the response time may be too delayed to take effective action, necessitating a more sophisticated analysis [6]. However, many illnesses produce visible symptoms, making visual examination a primary method for disease detection. A pathologist must have observational skills to diagnose diseases accurately based on these characteristic symptoms [7]. Misinterpretation of symptoms by non-experts may lead to incorrect diagnoses, highlighting the challenges faced by amateurs and hobbyists compared with professional pathologists. Both novice gardeners and seasoned experts can benefit significantly from an automated system that can detect plant conditions by analyzing visual symptoms [8]. Vision

M. F. Mridha and N. Dey (eds.), *Machine Vision in Plant Leaf Disease Detection for Sustainable Agriculture*,
Studies in Computational Intelligence 1202, https://doi.org/10.1007/978-981-96-4520-6_9

improvements have opened up new avenues for improving precise plant protection and growing the market for applications of precision agriculture [9].

For image processing, various algorithms can be applied from machine learning (ML) and deep learning (DL) because they have become increasingly popular in plant disease detection, demonstrating great potential for accurately identifying diseases [10]. Researchers have recently concentrated on ML, especially DL, which has demonstrated remarkable performance because of the availability of large datasets, lower computational costs, and techniques [11, 12]. Additionally, the rise of DL has had a transformative impact on computer vision, proving to be highly effective in tackling many challenges. Transfer learning [13] and its model, convolutional neural networks (CNNs) [14], 1D CNN [15], deep convolutional neural networks (DCNNs) [16], and different lightweight models have demonstrated remarkable success in identifying and categorizing crop diseases. DL transforms evidence-based decision-making across a wide range of sectors. Its strength lies in its ability to apply multiple layers of non-linear feature transformations to data through representation learning, effectively handling the challenges posed by large datasets. Despite its growing influence, DL is still a relatively new and evolving field, and comprehensive surveys covering the full range of DL architectures are rare [17]. In this book chapter, we use various lightweight DL models to train a popular dataset. The offerings and goals of this study are as follows:

- Exploring DL and lightweight architecture for plants and crops by analyzing images of leaves for identifying diseases. These lightweight methods are designed to alleviate the challenges that farmers face in diagnosing plant diseases, eliminating the need to rely on plant scientists.
- Using CNNs models to categorize a collection of leaf photos that were acquired from PlantVillage.
- Training accuracy, validation accuracy, testing accuracy, the number of trainable parameters, and performance in comparison to previous efforts were among the criteria used to assess the performance.

For the experimental phase, this paper's remaining sections are organized as follows: Section 2 provides a literature review of existing methods, Sect. 3 discusses the dataset in detail, and Sect. 4 presents the experimental setup and results, the conclusion is in Sect. 5.

2 Related Work

Understanding prior research in this field is crucial for making meaningful progress in this direction. Plant disease identification has been a major field of research, and good classification requires both image processing and DL approaches. In this paper, we review the most commonly utilized techniques in the literature related to this subject.

Borhani et al. conducted experiments on the Wheat Rust Classification and Rice Leaf Disease datasets. For wheat rust classification, eight models were evaluated at resolutions of 50×50, 100×100, and 200×200, with Model 3 at 200×200 demonstrating the best balance between accuracy and computational cost, outperforming others, including EfficientNet. Rice Leaf Disease dataset, simpler models were selected because of the smaller dataset size, and experiments were conducted at resolutions of 50×50 and 100×100. Both experiments involved training the models for 100 epochs, with model 3 proving to be the most effective for wheat rust classification [18]. Shahi et al. showed that UAVs with RGB, multispectral, and hyperspectral sensors are the key tools in crop disease detection. DL methods, using low-altitude RGB imagery achieved higher performance 85–100% compared to traditional ML approaches, which prefer multispectral sensors and have accuracies ranging from 72 to 98%. Fungal diseases, particularly Fusarium wilt, have dominated the research. This study highlights trends of cost-effective sensors and the need for lightweight DL models and benchmark datasets [19]. Furthermore, Arun et al. presented a Complete Concatenated DL (CCDL) method for multi-crop disease detection. They introduced a Complete Concatenated Block (CCB), which includes reducing parameters and enhancing feature map utilization. Plant Village dataset was later pruned to create the Pruned CCDL (PCCDL) model. Among three variants, the PCCDL with Partial Standard Convolution Technique (PCCDL-PSCT) achieved the highest result of 98.14% [20]. Haridasan et al. used an automated method for identifying and categorizing rice crop diseases to decrease resource waste and increase treatment effectiveness. They used CNN and support vector machine classifiers to detect and classify illnesses. A high validation accuracy of 91.45% was attained using the DL technique [21].

Furthermore, Jung et al. created an automated crop disease detection system that uses five pre-trained models and a CNN algorithm to identify photos of pairings of healthy and infected plants. Their system includes three classification stages: crop classification, disease classify and detection, and it has a performance of 97.09% in classifying crops and disease types [22]. Meena et al. focused on using DL to detect weeds, diseases, and pests in agriculture. By acquiring and augmenting images of plant diseases, pests, and weeds, various DCNN were evaluated. Fine-tuning different models, the InceptionV3 model achieved an accuracy of 87.85%, Mobilenet reached 91.85%, VGG16 had 78.71%, DenseNet201 performed best with 99.62%, and the Hyperparameter Search model had 71.07% accuracy. The results indicate that DenseNet201 is the most effective for classifying weeds, plant diseases, and pests, demonstrating high performance across these tasks [23]. Zanzaney et al. demonstrated

Table 1 The class analysis of plantvillage dataset

Class	Species diseases	Healthy categories	Amount of images
Apple	3	1	33,172
Cherry	1	1	1906
Blueberry	0	1	1502
Grape	3	1	4063
Corn	3	1	3852
Orange	1	0	5507
Potato	2	1	2152
Peach	1	1	2657
Bell pepper	1	1	2475
Raspberry	0	1	371
Squash	2	0	1835
Soybean	0	1	5090
Tomato	9	1	18,162
Strawberry	1	1	1565

how technological advancements, particularly DL models and ML techniques such as ResNet50, were employed to recognize illnesses and insect assaults from photos of plant leaves. Their approach offered comprehensive countermeasures in both English and the local tongue, along with specifics on the pest's scientific name and genus, ideal climate, suitable countermeasures, and pesticide application. The model, which can be customized for other crops, achieved high accuracy with 99.05% for tomatoes and 99.52% for potatoes[24]. Chug et al. proposed a novel framework combining techniques of DL and ML to detect diseases, addressing the limitations of both approaches. The framework included 40 Hybrid DL (HDL) models, integrating eight variants of EfficientNet as feature extractors with ML classifiers. The Indian Agricultural Research Institute's real-time tomato early blight disease picture collection was used to evaluate the models achieving accuracy rates between 87.55 and 100% [25].

Mallick et al. used CNNs have demonstrated encouraging outcomes in picture classification tasks, such as identifying diseases and pests. To get over the problem of limited mung bean crop photos, DL-based method that uses transfer learning. From healthy and afflicted leaves gathered over the seasons, they were able to correctly identify four different pests and six different mung bean diseases. The smartphone-based model detected pests and diseases with an average accuracy of 93.65% [26]. Furthermore, a dataset of photos of both healthy and diseased crops was used to train Jha et al., which was demonstrated for crop disease identification. In order to distinguish between healthy and sick crops, it successfully learned characteristics including crop shape and color. The model correctly recognized diseases in photos, including classes like tomato leaf bacterial spot, potato early blight, and corn-common rust. Using a dataset of 900 photos, the model classified crop diseases with an accuracy of 98.5%, and the results were in line with what was anticipated [27]. Adem et al. investigated the use of DL models, and a hybrid Yolov4 model and image processing to automatically detect and classify the severity of leaf spot disease on sugar beet. The models were trained and tested on 1,040 images, with the hybrid method achieving the highest classification accuracy of 96.47% [28] (Tables 1 and 2).

3 Methodology

A complete description of the methodology for the book chapter is provided in this section regarding what we will implement for our experiment.

3.1 Dataset

Datasets are a key element in AI-based research. Selecting the right dataset is necessary for an experiment because there are different types of datasets available, and for a good experiment, the dataset must be well structured. Therefore, in our experiment on plant disease classification, a popular and publicly available dataset. The plantvillage dataset is a widely used collection of plant images, originally comprising 54,309 images from 2016 [29]. These images represent 14 crop species. Table 3 lists individual crops in the dataset. The dataset consisted of expertly labeled images of both healthy and diseased leaves, accessible through the PlantVillage platform. The dataset categorizes plant diseases into 17 fungal diseases, 4 bacterial diseases, 2 mold (oomycete) diseases, 2 viral diseases, and 1 mite disease, totaling 38 disease classes.

From these 38 disease classes, we used 25 for training and testing. In Fig. 1, we graphically represent every category employed in our experiment and the numbers.

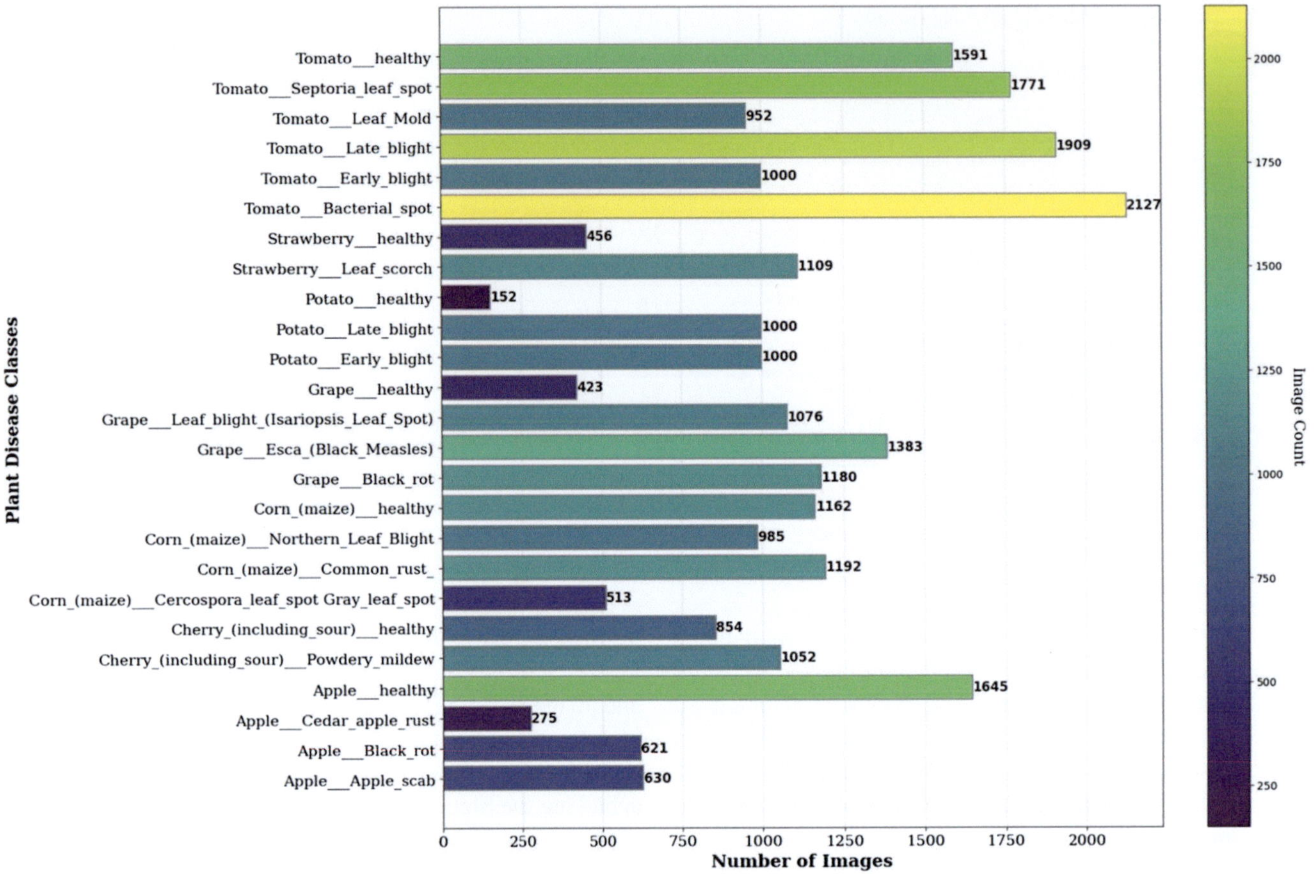

Fig. 1 Number of sub-classes of our dataset are presented in the graph

3.2 Data Preprocessing

Prior to classification, preprocessing images of sick plants is essential since the quality of these photos has a big impact on how accurately the findings are classified. The steps in the procedure are as follows:

3.2.1 Data Augmentation

By boosting the quantity of training images, image augmentation is a technique for artificially enhancing the data. This method improves the model's performance and lessens overfitting problems [30]. The training images can undergo a number of changes, including rotation, flipping, zooming, and shifting. Eight different augmentation kinds were used in this instance, as shown in Fig. 2: movement, shear, zoom, height shift, horizontal flipping, vertical filliping, and fill mode.

3.3 Base Architecture

In the book Chapter on Plant Diseases, we implemented and focused on different classification methods. We used six different types of CNNs to evaluate our work strongly. The base network for our models was the same as that of CNN, and the base architecture of CNN is described below [31]. The base architecture of a CNN typically includes the following components:

- **Input Layer**: It accepts the raw pixel values of an image. The pixel value can differ based on its use.
- **Convolutional Layers**: Convolutional layer is the most common and useful layer for all the CNN models. Apply convolution operations with filters (kernels).
- **Activation Function**: Although other activation function types can be employed for assessment, non-linearity is often introduced using ReLU (Rectified Linear Unit).
- **Pooling Layers**: Use downsampling techniques, such Max Pooling, to minimize spatial dimensions while preserving significant information.
- **Fully Connected Layers**: Flatten the output and connect to dense layers for classification or regression tasks.
- **Output Layer**: For every class, probabilities are generated using the softmax function.

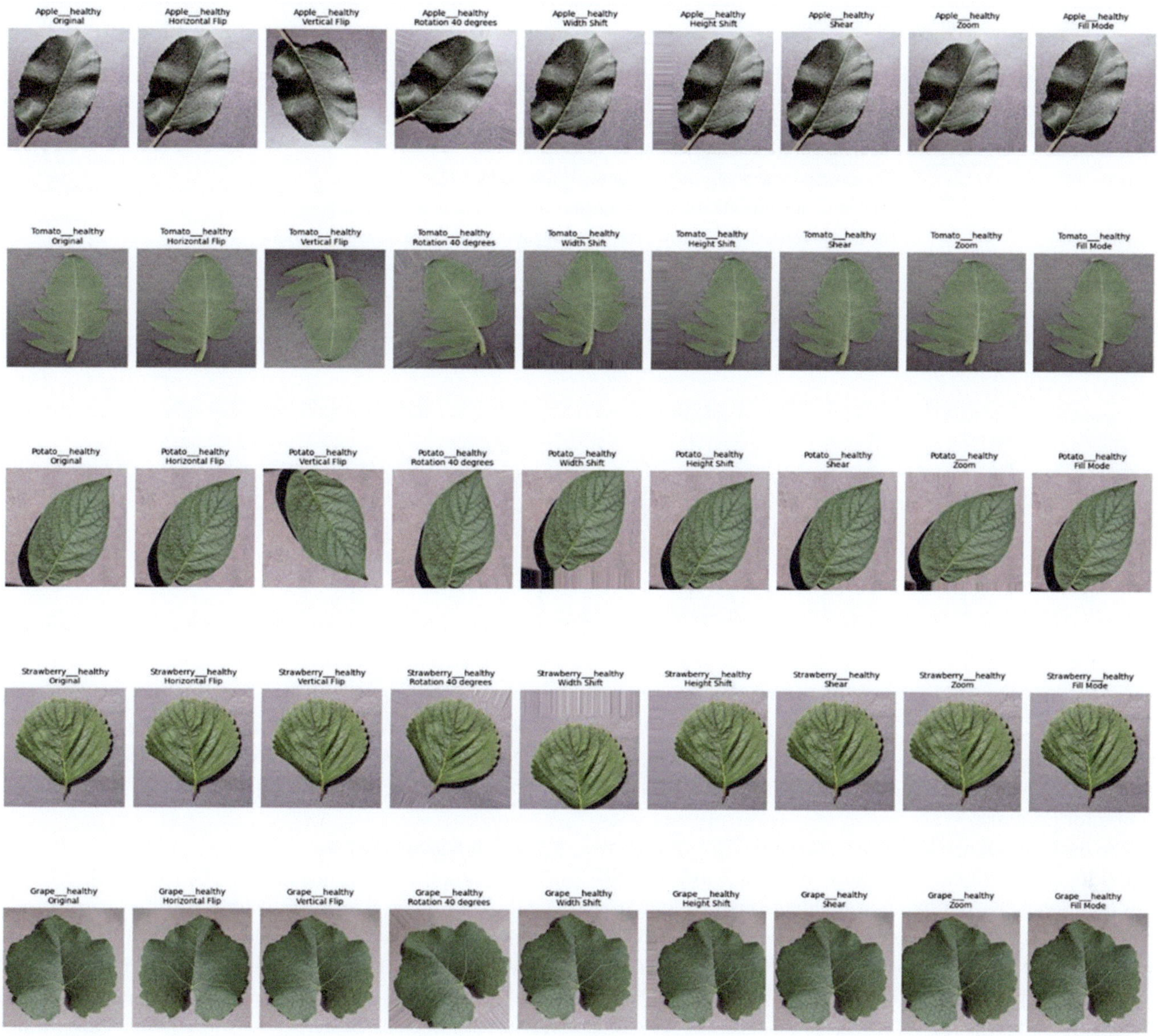

Fig. 2 Augmentation image for the plantvillage dataset

3.3.1 VGG16

Simonyan et al. proposed VGG16 in 2014 [32]. VGG16 was proposed to explore the impact of increasing the depth of convolutional networks on large-scale image recognition tasks. VGG16 demonstrated that deeper networks could significantly improve the performance in image classification while maintaining a straightforward and scalable design. The primary concept of VGG16 is to use a deep network architecture with 16 layers to capture complex features while maintaining a simple and consistent design. This approach enhances image classification performance by increasing the network depth. The architecture of the VGG166 model is shown in Fig. 3.

The basic differences between standard the CNN and VGG16 are their depth and filter size. While a standard CNN may have fewer layers and varied filter sizes, VGG16 uses a much deeper architecture with 16 layers and consistently small 3×3 convolutional filters, enabling it to collect fine features and enhance classification performance through increased depth.

3.3.2 Xception

Chollet et al. proposed the XceptionNet in 2017 [33]. Depthwise separable convolutions, a more effective technique that lowers the amount of parameters and computational cost, were proposed as Xception's replacement for Inception's intricate modules. This design allows the network to capture spatial correlations and channel-wise features separately, leading to better performance while maintaining efficiency. Xception

Table 2 Data augmentation type for our work

Augmentation type	Description
Original	Applied no augmentation, used rescaling the pixel values by 1./255
Horizontal flip	The image horizontally flip, rescale by 1./255
Vertical flip	The image Vertically flip, rescale by 1./255
Rotation 40°	Image rotation randomly angle within 40° rescale by 1./255
Width shift	The image horizontally shifts by a fraction of width (up to 20%).rescale by 1./255
Height shifts	The image vertically shifts by a fraction of width (up to 20%).rescale by 1./255
Shear	Used random shear transformation
Zoom	Used random factor zoom in/out on the image
fill mode	Used fill mode to the new pixels created during transformations

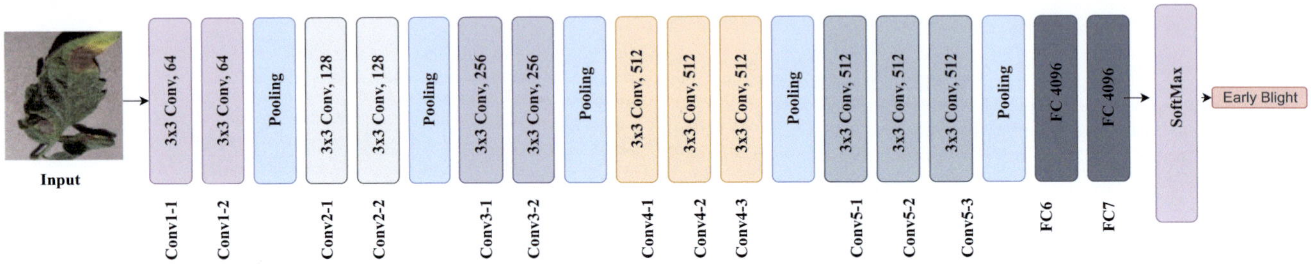

Fig. 3 Architecture of VGG16

enhances network efficiency and performance by using depthwise separable convolutions, which decouple the spatial and channel-wise feature learning streamlined and effective architecture for DL tasks.

The basic difference between a standard CNN and Xception is its convolutional operations. While a standard CNN uses traditional convolutions that simultaneously capture spatial and channel-wise information, Xception employs depthwise separable convolutions, which first process spatial features independently for each channel and then combine them. This separation reduces the computational complexity and improves the model efficiency without compromising performance.

3.3.3 EfficientNet B0

Tan et al. presented EfficientNet in 2019 [34]. The problem of scaling deep neural networks to achieve performance while preserving computational efficiency was addressed by the proposal of EfficientNet. EfficientNet achieves excellent accuracy with fewer parameters and reduced computing costs by optimizing performance and efficiency of network depth, width, and resolution. Figure 4 presents the EfficientNet architecture.

The primary architectural distinction between CNN and EfficientNet is that the latter optimizes the trade-offs between model size and accuracy by proportionally scaling technique. Conventional CNNs, on the other hand, usually scale one dimension at a time, which might result in less than ideal performance and resource consumption. The goal of EfficientNet's architecture is to improve performance while using fewer parameters and processing power.

3.3.4 NASNet Mobile

In 2018, Zoph et al. proposed the NASNet [35]. NASNet Mobile was proposed to create an efficient, high-performance neural network architecture tailored for mobile and edge devices. NASNet Mobile provides a lightweight yet powerful model for resource-constrained environments. NASNet uses Neural Architecture Search (NAS) to automatically design an optimal DL model by searching for the best architectural configurations.

The key difference between a standard CNN and NASNet Mobile is its architectural design approach. While a standard CNN relies on manually designed layers with fixed configurations, NASNet Mobile uses NAS to automatically generate an optimized architecture tailored for mobile devices. Compared to conventional CNNs, this produces a more effective model that preserves high accuracy while lowering processing demands.

3.3.5 MobileNet

In 2017, Howard et al. introduced MobilenNet [36]. MobileNet was proposed to address the need for efficient CNN that can operate on mobile and edge devices with limited computational resources. The difference between a standard CNN and MobileNet lies in their convolution

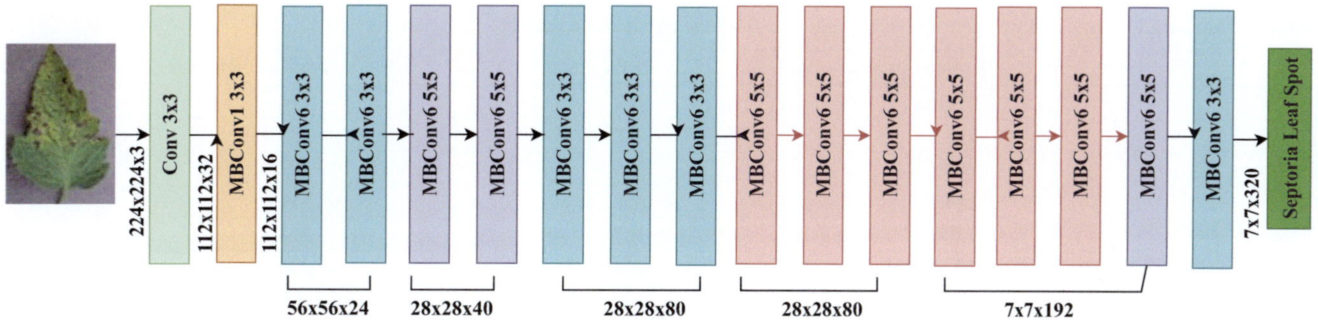

Fig. 4 Architecture of EfficientNet B0

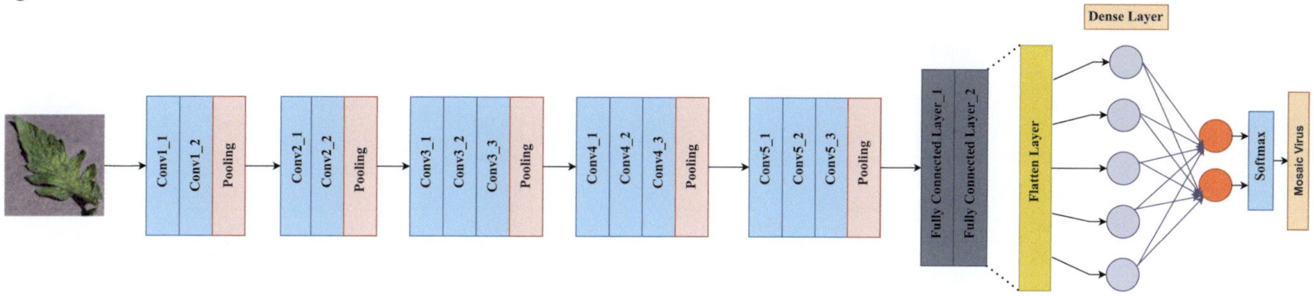

Fig. 5 Architecture of ResNet50

operations. Whereas standard CNNs use regular convolutions with input channels, depthwise separable convolutions, which divide the convolution into depthwise and pointwise layers, are used by MobileNet. MobileNet is more effective for mobile and edge devices because of this method, which drastically lowers the computational cost and model size while maintaining excellent accuracy.

3.3.6 ResNet50

He et al. proposed this method in 2016 [37]. To solve the issue of disappearing gradients and deterioration in deep neural networks, ResNet50 was proposed. Figure 5 shows the VGG16. By adding residual connections, also known as skip connections gradients can move through the network more readily during training, making it easier to train much deeper networks without experiencing performance issues and increasing accuracy overall.

ResNet50 builds on the standard CNN architecture by incorporating residual connections. ResNet50 architecture reduces the issue of the vanishing gradient. It allows for deeper networks by facilitating gradient flow through skip connections, leading to improved performance and easier training compared to traditional CNNs.

4 Result Analysis

4.1 Experiment Setup

Our experiment was conducted on Google Colab pro [38] to utilize its T4 GPU, which has the following specifications: GPU Memory of 16 GB, a Memory Clock of 1.32 GHz, and a Performance capability of 9.3 TFLOPS. We used the Python language to run our programming and also used various libaries such as Keras, Matplotlib, TensorFlow, and pandas. In this experiment, various convolutional layers are added to our model and input sizes of $224 \times 224 \times 3$, and $299 \times 299 \times 3$. The training epoch used for the experiment was 20 for every model. The learning rate is set as 0.001. Adam and softmax were the activation functions and the batch sizes are 16 and 32, respectively. Our dataset plantvillage was split for training, testing, and validation, and the ratio was 70:15:15. Table 3 presents the number of images that we used for testing, training, and validation in our experiment. We used 18232 images for training, 3957 images for testing, and 3960 images for validation.

Table 3 Number of images that we used for our experiments from the plantvillage dataset

Type	Training image	Testing image	Validation image
Apple	2218	489	474
Cherry	1333	288	285
Corn	2695	575	582
Grape	2843	608	611
Potato	1506	363	364
Strawbery	1095	234	236
Tomato	6542	1400	1408
Total	18232	3957	3960

Table 4 Precision, recall, specificity, and F1-score for the plantvillage dataset

Models	Precision%	Recall%	Specificity%	F1-score%
VGG16	98.73	98.63	99.96	98.67
Xception	99.48	99.50	99.99	99.49
EfficientNet B0	99.15	99.25	99.98	99.19
NASNet mobile	96.94	95.69	96.15	96.07
MobileNet	99.50	99.41	99.99	99.45
ResNet50	97.29	96.86	99.90	98.98

4.2 Performance Matrix

To quantitatively measure the proposed model's performance, a number of assessment metrics were used. These metrics include accuracy (ACC), precision (P), recall (R), F1-score, specificity, and area under the curve (AUC) are present in Table 4. Figures 6 and 7 present the performance metric for our used models.

4.3 Classification on PlantVillage Dataset

4.3.1 Accuracy-Baseded Performance

The effectiveness of the models was assessed through a series of experiments using images of the plantvillage dataset, comparing VGG16, Xception, EfficienetNet, NASNet Mobile, MobileNet, and ResNet50. The accuracy graphs for these models are shown in Figs. 8 and 9, respectively. The figures present the model accuracy of validation, training also model loss for validation and training.

As shown in Table 5, the average validation accuracies achieved by the VGG16, Xception, EfficienetNet B0, NASNet Mobile, MobileNet, and ResNet50 models are 99.03%, 99.39%, 99.33%, 95.93%, 99.13%, and 96.39%, respectively. The pre-trained model performs better in terms of accuracy, precision, recall, and F1-score. The pre-trained model successfully classified 25 plant categories from a huge plantvillage dataset, demonstrating considerable performance increase. Table 6 shows the number of trainable and non-trainable parameters, as well as the computing time for our models.

4.3.2 ROC Curve Analysis

Figure 10 displays the ROC curve, a critical parameter for evaluating and enhancing DL model performance. Figure 11 shows the precision vs. recall curves for all six models.

4.4 Comparison of Our Work with Different Existing Works

This subsection compares and describes existing work on the Plantvillage dataset to our own. Table 7 compares our model to alternative models.

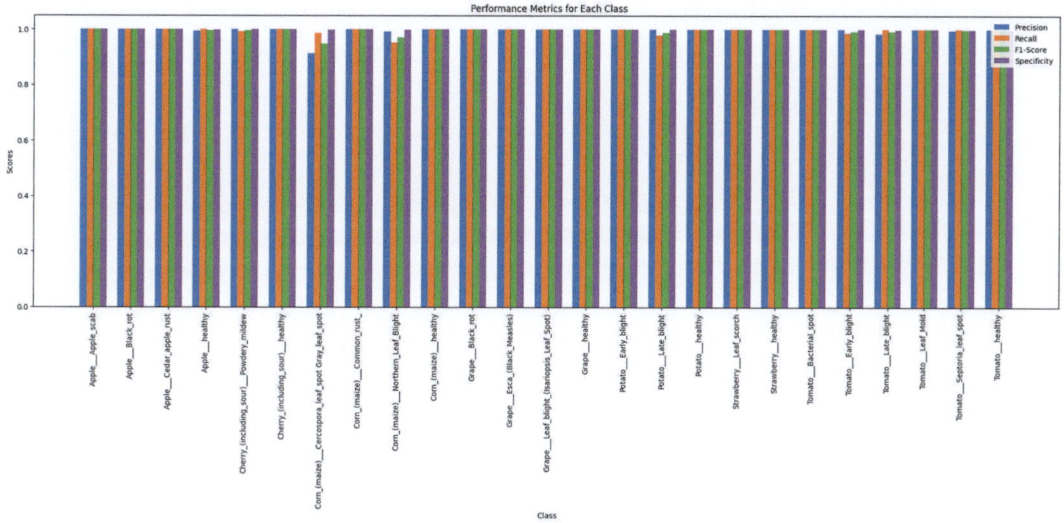

(1) Performance Metrics for Xception

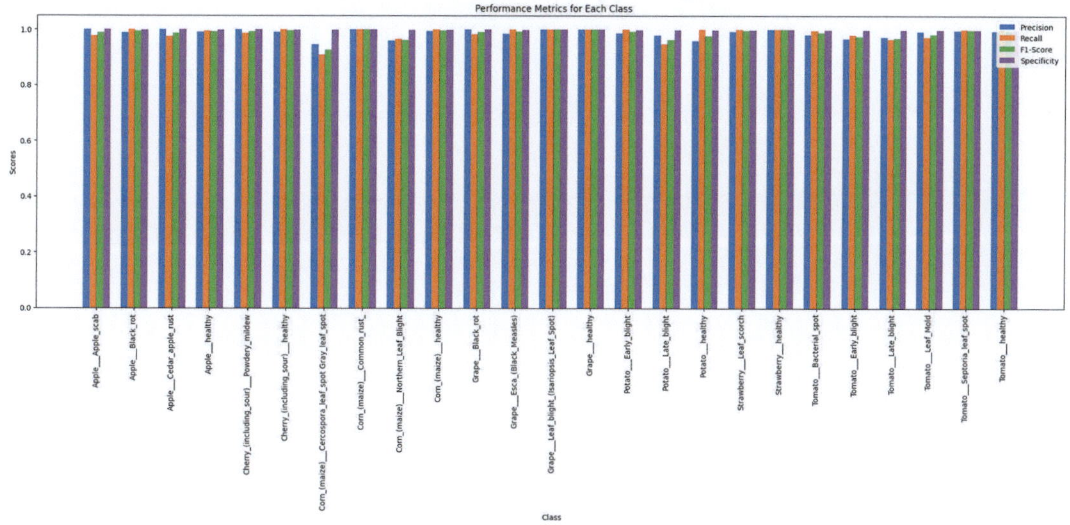

(2) Performance Metrics for VGG16

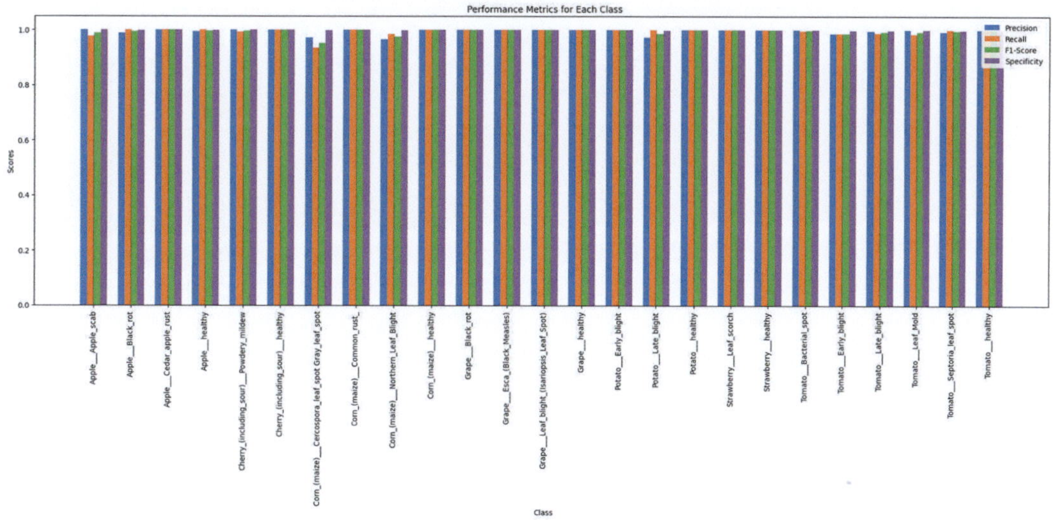

(3) Performance Metrics for MobileNet

Fig. 6 The performance matrix for (1) Xception, (2) VGG16, (3) MobileNet

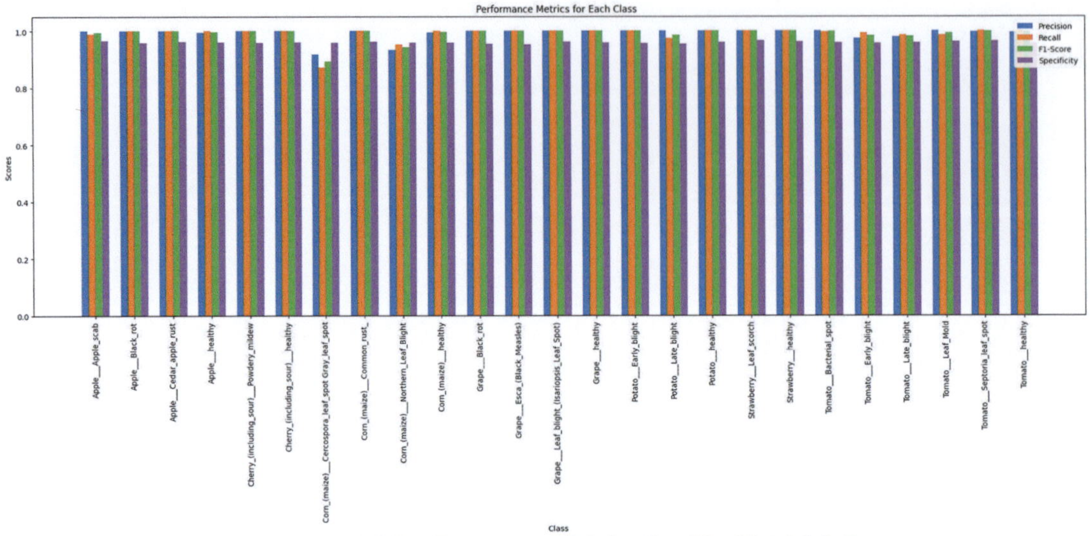

(4) Performance Metrics for NasNet Mobile

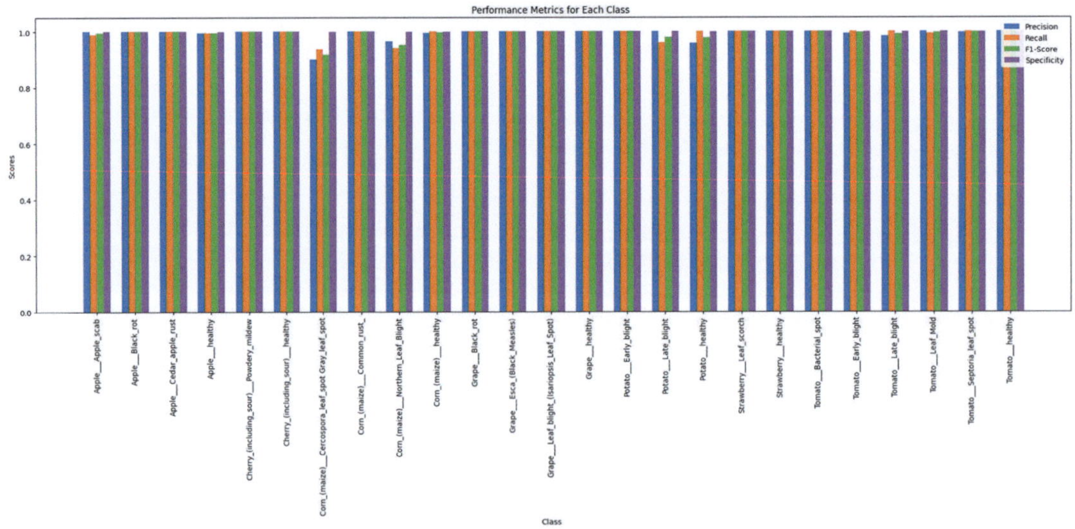

(5) Performance Metrics for EfficientNet B0

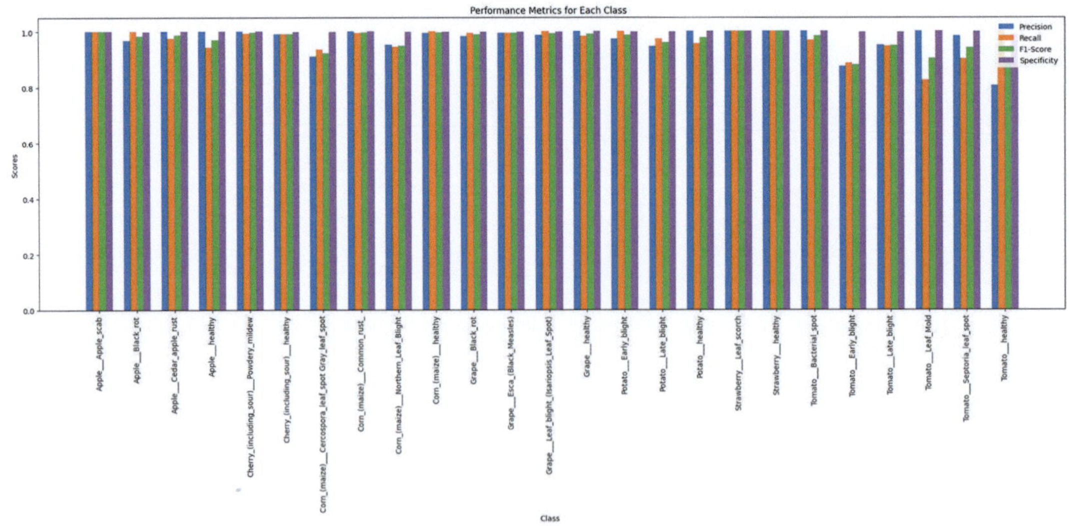

(6) Performance Metrics for ResNet50

Fig. 7 The performance matrix for (4) NASNet mobile, (5) EfficientNetB0, (6) ResNet50

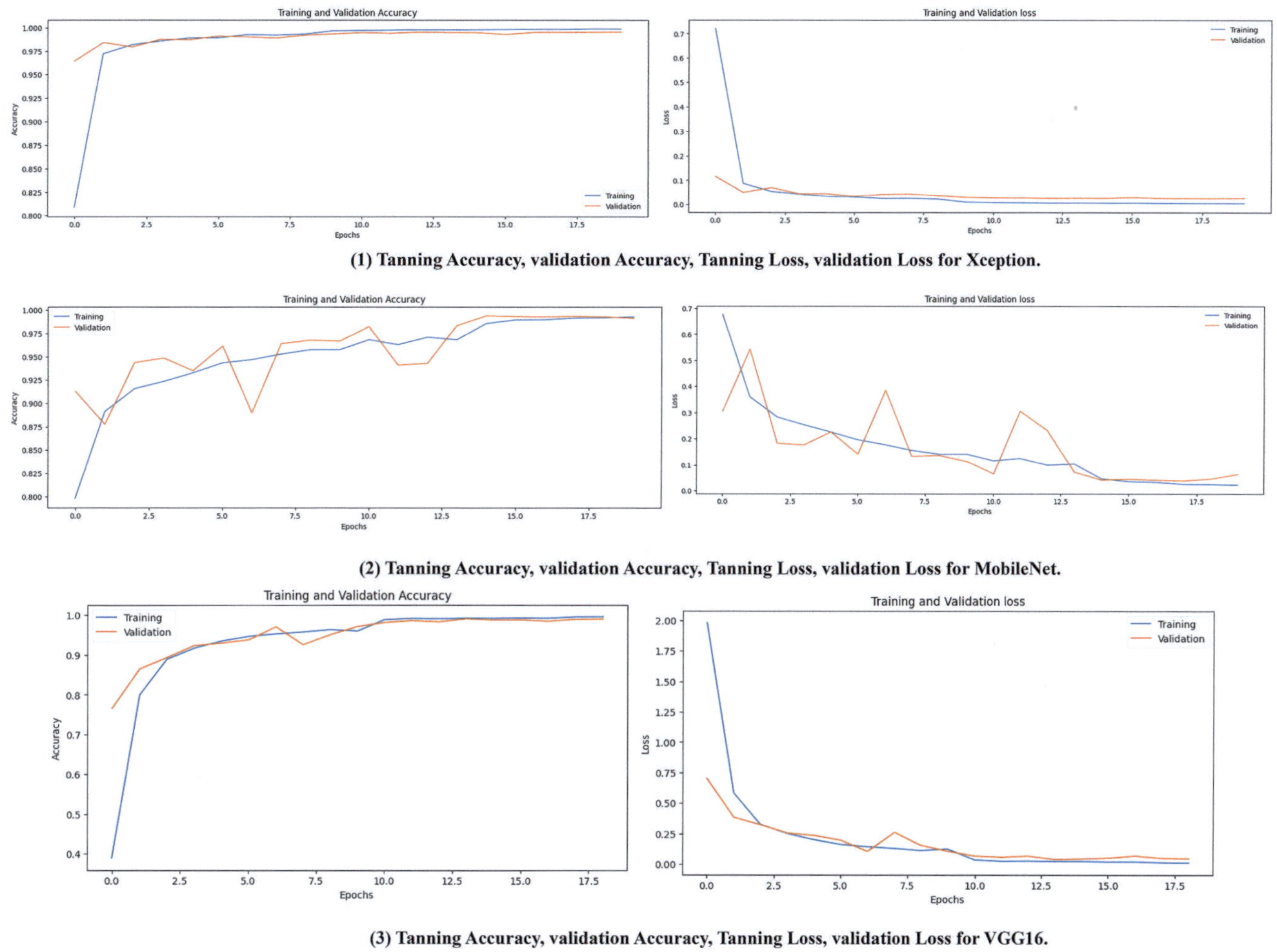

Fig. 8 Graph for tanning, validation accuracy, tanning, validation loss for (1) Xception, (2) MobileNet, and (3) VGG16

5 Conclusion

This study presents deep learning (DL) to develop an automated plant disease detection system using a straightforward classification approach that capitalizes on the capabilities of CNNs. The model incorporates fully connected layers for the final predictions. This research utilized a publicly available dataset of plantvillages. The system achieved an impressive 99.39% accuracy on the public dataset. These results demonstrate the CNN's highly suitable for automatic plant disease detection and diagnosis. The system can be integrated into mini-drones for real-time disease detection in cultivated areas. Although the model was trained on the PlantVillage dataset with 25 classes, it can still identify diseases due to common symptoms across various plants. Future improvements can include adding more real-world images to enhance accuracy and expand the range of plant and disease classifications. Additionally, a three-layer approach can be implemented, where the first layer detects the presence of a plant, the second layer identifies the plant type, and the third layer determines the presence and type of disease.

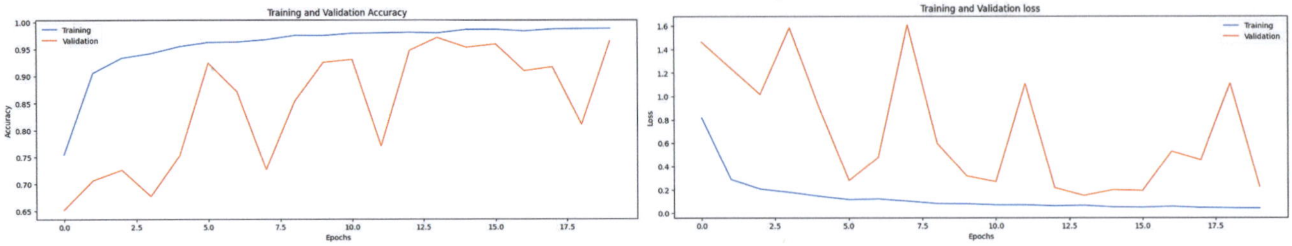

(4) Tanning Accuracy, validation Accuracy, Tanning Loss, validation Loss for ResNet50.

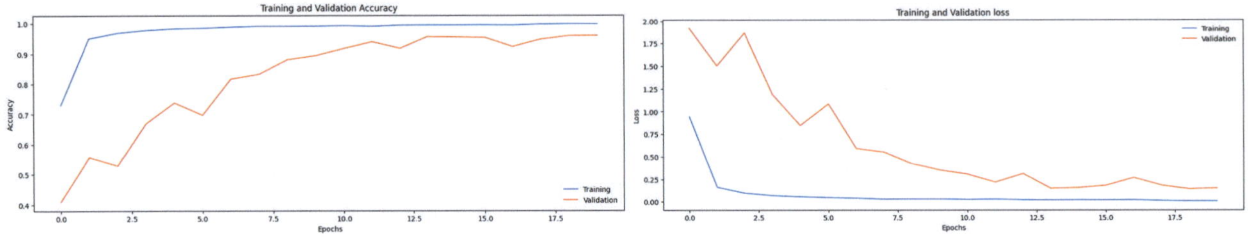

(5) Tanning Accuracy, validation Accuracy, Tanning Loss, validation Loss for NASNet Mobile.

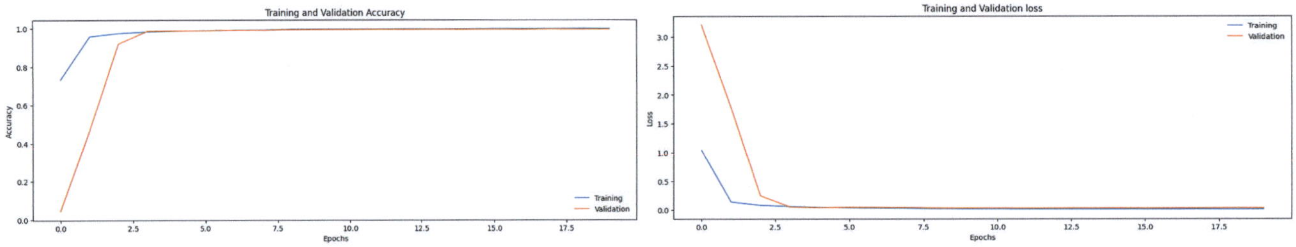

(6) Tanning Accuracy, validation Accuracy, Tanning Loss, validation Loss for EfficientNet.

Fig. 9 Graph for tanning, validation accuracy, tanning, validation loss for (1) ResNet50, (2) NASNet Mobile, and (3) EfficientNet

Table 5 Training and validation accuracy for plantvillage dataset

Models	Training accuracy(%)	Validation accuracy(%)
VGG16	99.53	99.03
Xception	99.92	99.39
EfficientNet B0	99.56	99.33
NASNet mobile	99.81	95.93
MobileNet	99.32	99.13
ResNet50	98.78	96.39

Table 6 Number of parameters, the computational time, train and non-trainable params for our models

Models	Total params	Trainable params	Non-trainable params	Time required (s)
VGG16	134,362,969	134,362,969	0	6,789
Xception	21,923,393	21,868,865	54,528	3,602
EfficientNet B0	4,718,268	4,676,245	42,023	4,533
NASNet mobile	4,823,725	4,786,987	36,738	6,103
MobileNet	3,766,489	3,744,601	21,888	5,961
ResNet50	24,649,625	24,596,505	53,120	4,285

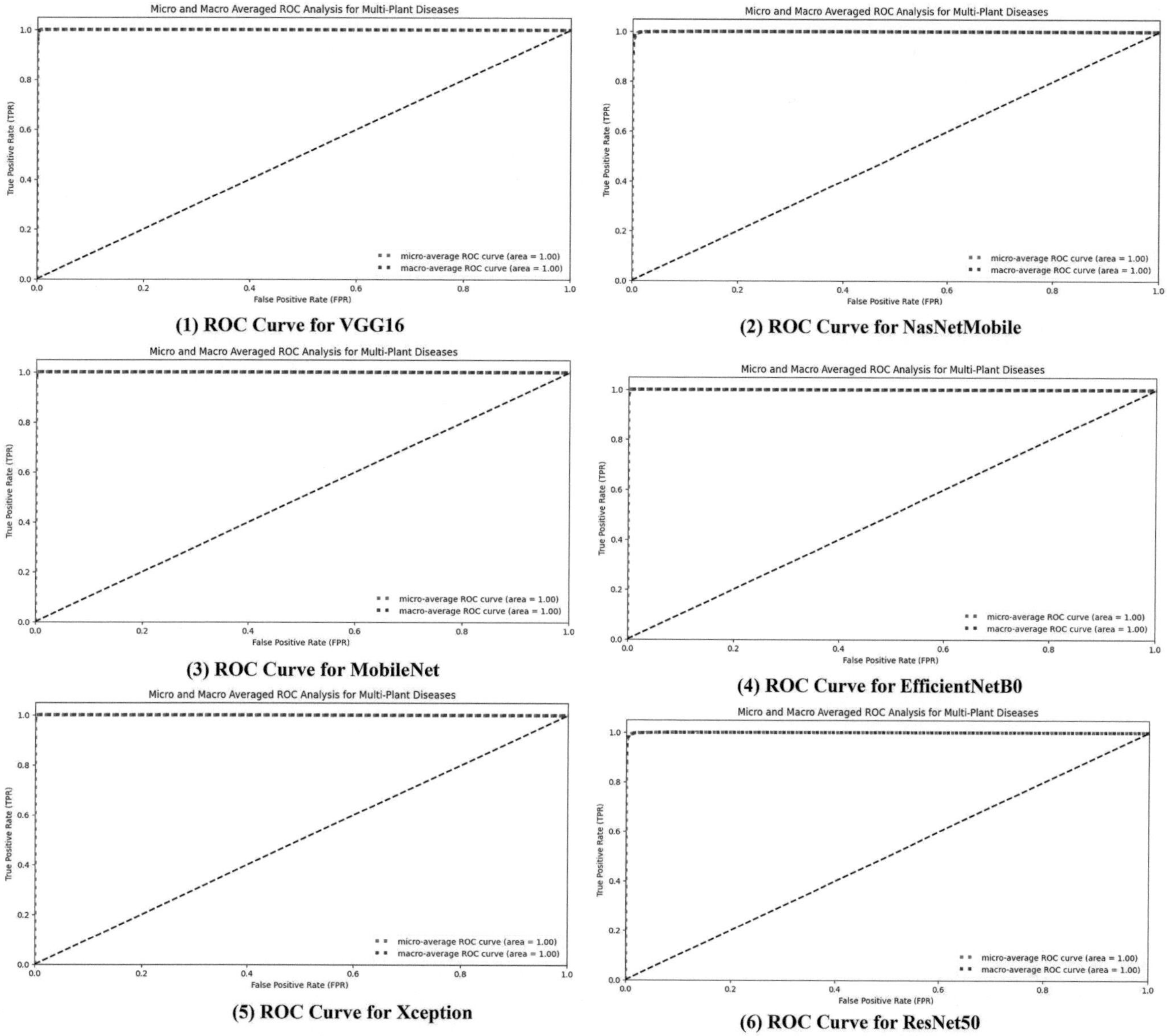

Fig. 10 ROC curve analysis for (1) VGG16, (2) NASNet mobile, (3) MobileNet, (4) EfficientNetB0, (5) Xception, (6) ResNet50

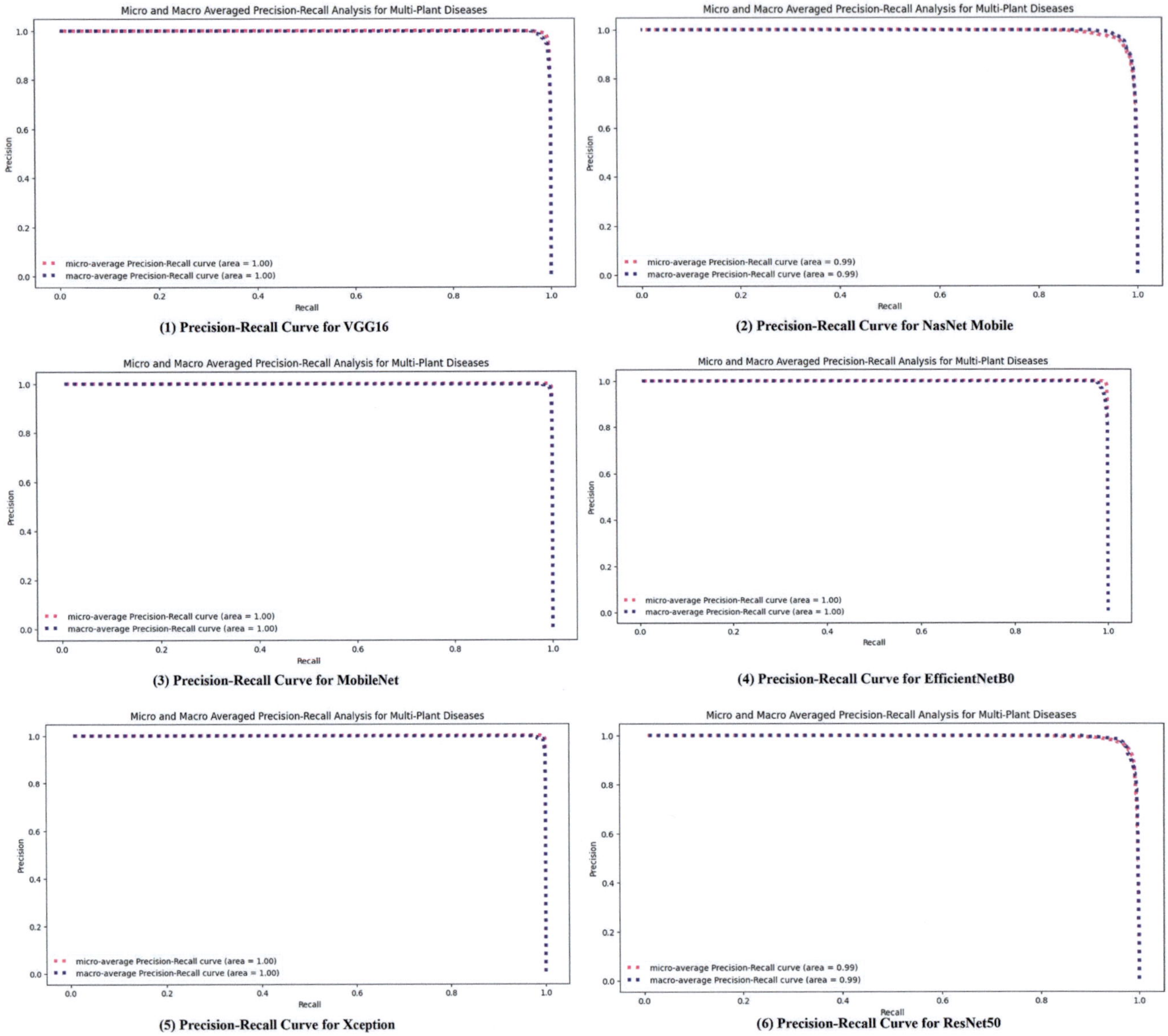

Fig. 11 Precision versus recall for (1)VGG16, (2) NASNet mobile, (3) MobileNet, (4) EfficientNetB0, (5) Xception, (6) ResNet50

Table 7 Comparison with our work with existing work on the Plantvillage dataset

References	Model	Accuracy (%)	F1-score (%)	Precision (%)
Dey et al. [39]	AlexNet	96.63	91	92
Bhagat et al. [40]	VGG-16	98.42	100	99.75
Shafik et al. [41]	PDDNet-LVE	97.79	94.4	93
Bouacida et al. [42]	Inception	94.04	94.50	95.05
Yang et al. [43]	aECA-ResNet34	98.5	98.5	–
Lachure et al. [44]	CNN	98.46	–	–
Ours	Xception	99.39	99.49	99.48

References

1. Ahmed, I., Yadav, P.K.: Plant disease detection using machine learning approaches. Expert Syst. **40**(5), 13136 (2023)
2. Kumar, A., Pant, S.: Analytical hierarchy process for sustainable agriculture: an overview. MethodsX **10**, 101954 (2023)
3. Chin, R., Catal, C., Kassahun, A.: Plant disease detection using drones in precision agriculture. Precis. Agric. **24**(5), 1663–1682 (2023)
4. Ahmad, A., Saraswat, D., El Gamal, A.: A survey on using deep learning techniques for plant disease diagnosis and recommendations for development of appropriate tools. Smart Agric. Technol. **3**, 100083 (2023)
5. Demilie, W.B.: Plant disease detection and classification techniques: a comparative study of the performances. J. Big Data **11**(1), 5 (2024)
6. Sharma, V., Tripathi, A.K., Mittal, H.: Dlmc-net: Deeper lightweight multi-class classification model for plant leaf disease detection. Ecol. Inform. **75**, 102025 (2023)
7. Singh, P., Singh, P., Farooq, U., Khurana, S.S., Verma, J.K., Kumar, M.: Cottonleafnet: cotton plant leaf disease detection using deep neural networks. Multimedia Tools Appl. **82**(24), 37151–37176 (2023)
8. Sahu, S.K., Pandey, M.: An optimal hybrid multiclass SVM for plant leaf disease detection using spatial fuzzy c-means model. Expert Syst. Appl. **214**, 118989 (2023)
9. Sarkar, C., Gupta, D., Gupta, U., Hazarika, B.B.: Leaf disease detection using machine learning and deep learning: review and challenges. Appl. Soft Comput. **145**, 110534 (2023)
10. Shovon, M.S.H., Mozumder, S.J., Pal, O.K., Mridha, M.F., Asai, N., Shin, J.: Plantdet: a robust multi-model ensemble method based on deep learning for plant disease detection. IEEE Access **11**, 34846–34859 (2023)
11. Kaur, P., Pannu, H.S., Malhi, A.K.: Plant disease recognition using fractional-order Zernike moments and SVM classifier. Neural Comput. Appl. **31**, 8749–8768 (2019)
12. Islam, M.R., Oliullah, K., Kabir, M.M., Alom, M., Mridha, M.: Machine learning enabled iot system for soil nutrients monitoring and crop recommendation. J. Agric. Food Res. **14**, 100880 (2023)
13. Abbas, A., Jain, S., Gour, M., Vankudothu, S.: Tomato plant disease detection using transfer learning with C-GAN synthetic images. Comput. Electron. Agric. **187**, 106279 (2021)
14. Agarwal, M., Gupta, S.K., Biswas, K.K.: Development of efficient cnn model for tomato crop disease identification. Sustain. Comput.: Inform. Syst. **28**, 100407 (2020)
15. Waibel, A., Hanazawa, T., Hinton, G., Shikano, K., Lang, K.J.: Phoneme recognition using time-delay neural networks. In: Backpropagation, pp. 35–61. Psychology Press (2013)
16. Patil, N., Ali, R., Wankhedkar, V., Nayak, D.: Crop disease detection using deep convolutional neural networks. Int. J. Eng. Res. Technol. (IJERT) **8**(03) (2019)
17. Ahmed, S.F., Alam, M.S.B., Hassan, M., Rozbu, M.R., Ishtiak, T., Rafa, N., Mofijur, M., Shawkat Ali, A., Gandomi, A.H.: Deep learning modelling techniques: current progress, applications, advantages, and challenges. Artif. Intell. Rev. **56**(11), 13521–13617 (2023)
18. Borhani, Y., Khoramdel, J., Najafi, E.: A deep learning based approach for automated plant disease classification using vision transformer. Sci. Rep. **12**(1), 11554 (2022)
19. Shahi, T.B., Xu, C.-Y., Neupane, A., Guo, W.: Recent advances in crop disease detection using UAV and deep learning techniques. Remote Sens. **15**(9), 2450 (2023)
20. Arun, R.A., Umamaheswari, S.: Effective multi-crop disease detection using pruned complete concatenated deep learning model. Expert Syst. Appl. **213**, 118905 (2023)
21. Haridasan, A., Thomas, J., Raj, E.D.: Deep learning system for paddy plant disease detection and classification. Environ. Monitor. Assess. **195**(1), 120 (2023)
22. Jung, M., Song, J.S., Shin, A.-Y., Choi, B., Go, S., Kwon, S.-Y., Park, J., Park, S.G., Kim, Y.-M.: Construction of deep learning-based disease detection model in plants. Sci. Rep. **13**(1), 7331 (2023)
23. Meena, S.D., Susank, M., Guttula, T., Chandana, S.H., Sheela, J.: Crop yield improvement with weeds, pest and disease detection. Procedia Comput. Sci. **218**, 2369–2382 (2023)
24. Zanzaney, A.U., Hegde, R., Jain, L., Choudhuri, S.S., Sharma, C.K.: Crop disease detection using deep neural networks. In: 2023 International Conference on Network, Multimedia and Information Technology (NMITCON), pp. 1–5 (2023). IEEE
25. Chug, A., Bhatia, A., Singh, A.P., Singh, D.: A novel framework for image-based plant disease detection using hybrid deep learning approach. Soft Comput. **27**(18), 13613–13638 (2023)
26. Mallick, M.T., Biswas, S., Das, A.K., Saha, H.N., Chakrabarti, A., Deb, N.: Deep learning based automated disease detection and pest classification in Indian mung bean. Multimedia Tools Appl. **82**(8), 12017–12041 (2023)
27. Jha, P., Dembla, D., Dubey, W.: Crop disease detection and classification using deep learning-based classifier algorithm. In: International Conference on Emerging Trends in Expert Applications & Security, pp. 227–237. Springer (2023)
28. Adem, K., Ozguven, M.M., Altas, Z.: A sugar beet leaf disease classification method based on image processing and deep learning. Multimedia Tools Appl. **82**(8), 12577–12594 (2023)
29. https://www.kaggle.com/datasets/abdallahalidev/plantvillage-dataset/data
30. Karthik, R., Hariharan, M., Anand, S., Mathikshara, P., Johnson, A., Menaka, R.: Attention embedded residual CNN for disease detection in tomato leaves. Appl. Soft Comput. **86**, 105933 (2020)
31. LeCun, Y., Boser, B., Denker, J.S., Henderson, D., Howard, R.E., Hubbard, W., Jackel, L.D.: Backpropagation applied to handwritten zip code recognition. Neural Comput. **1**(4), 541–551 (1989)
32. Simonyan, K., Zisserman, A.: Very deep convolutional networks for large-scale image recognition (2014). arXiv:1409.1556
33. Chollet, F.: Xception: Deep learning with depthwise separable convolutions. In: Proceedings of the IEEE Conference on Computer Vision and Pattern Recognition, pp. 1251–1258 (2017)
34. Tan, M.: Efficientnet: Rethinking model scaling for convolutional neural networks (2019). arXiv:1905.11946
35. Zoph, B., Vasudevan, V., Shlens, J., Le, Q.V.: Learning transferable architectures for scalable image recognition. In: Proceedings of the IEEE Conference on Computer Vision and Pattern Recognition, pp. 8697–8710 (2018)
36. Howard, A.G.: Mobilenets: Efficient convolutional neural networks for mobile vision applications (2017). arXiv:1704.04861
37. He, K., Zhang, X., Ren, S., Sun, J.: Deep residual learning for image recognition. In: Proceedings of the IEEE Conference on Computer Vision and Pattern Recognition, pp. 770–778 (2016)
38. https://colab.research.google.com/
39. Dey, P., Mahmud, T., Nahar, S.R., Hossain, M.S., Andersson, K.: Plant disease detection in precision agriculture: Deep learning approaches. In: 2024 2nd International Conference on Intelligent Data Communication Technologies and Internet of Things (IDCIoT), pp. 661–667. IEEE (2024)
40. Bhagat, M., Kumar, D., Kumar, S.: Optimized transfer learning approach for leaf disease classification in smart agriculture. Multimedia Tools Appl. **83**(20), 58103–58123 (2024)

41. Shafik, W., Tufail, A., De Silva Liyanage, C., Apong, R.A.A.H.M.: Using transfer learning-based plant disease classification and detection for sustainable agriculture. BMC Plant Biol. **24**(1), 136 (2024)
42. Bouacida, I., Farou, B., Djakhdjakha, L., Seridi, H., Kurulay, M.: Innovative deep learning approach for cross-crop plant disease detection: a generalized method for identifying unhealthy leaves. Inf. Process. Agric. (2024)
43. Yang, W., Yuan, Y., Zhang, D., Zheng, L., Nie, F.: An effective image classification method for plant diseases with improved channel attention mechanism aECAnet based on deep learning. Symmetry **16**(4), 451 (2024)
44. Lachure, J., Doriya, R.: Designing of lightweight deep learning framework for plant disease detection. SN Comput. Sci. **5**(6), 761 (2024)

Improving Bean Disease Detection with Explainable Vision Transformers and Machine Learning Classifiers

Md. Polash Islam, Maharin Afroj, Azrin Nahar Anika, Hanif Bhuiyan, and Md. Rajibul Islam

Abstract

Bean plants are highly vulnerable to various diseases that threaten crop yields and global food security. Accurate and timely disease detection is essential for effective management and the prevention of significant crop losses. Traditional methods, such as manual visual inspection, are inefficient, labor-intensive, and prone to inaccuracies due to human error and subjectivity. This study presents a AI-powered solution that utilizes the complementary strengths of a pre-trained Vision Transformer (ViT) model and a Support Vector Machine (SVM) classifier. The ViT model is employed to extract intricate and high-dimensional features from bean leaf images, providing a rich representation of disease characteristics. These features are then classified using a finely tuned SVM, achieving an outstanding accuracy of 95.00% in disease identification. To enhance interpretability and user trust, Shapley Additive exPlanations (SHAP) is integrated, offering insights into the model's decision-making process. A comparative evaluation with other classifiers, including NB 78%, RF 94.12%, LR 90.5%, and KNNs 86%, underscores the superior performance of the ViT-SVM combination. This approach not only outperforms traditional methods but also introduces a scalable, interpretable, and efficient solution, paving the way for advancements in agricultural disease management, improved crop protection, and a sustainable global food supply.

Keywords

Bean • Deep learning • CNN • Image classification • Leaf disease • ViT • SVM • SHAP • Explainable AI

1 Introduction

Agriculture is regarded as one of the most significant economic resources for both farmers and nations. Beans are among the most vital agricultural products, ranking as the second and third most significant sources of dietary fiber and calories for humans. Unfortunately, a variety of diseases, including bean rust and angular leaf spots, affect these crops and can seriously harm them as well as reduce their yield. Plant diseases that affect beans, such as angular leaf spot and bean rust, lower crop yields [1]. To address this issue in its early stages, a proper diagnosis of this crop disease is necessary. Diseases such as bean rust and angular leaf spot greatly affect bean production. Various pesticides are used to kill pathogens. For example, fungicides are used to control bean rust and angular leaf spot, and biological control and cultural practices such as intercropping, proper plant spacing, and the use of soil amendments to improve soil health and plant nutrition can be used to control these diseases [2]. Nonetheless, the identification and categorization of plant leaf diseases are vital in agriculture, as the overuse and widespread application of pesticides can pose significant risks to human health and the environment. Therefore, an automated system must be developed to control this disease as soon as possible. Identifying crop diseases through automated techniques reduces the amount of work that must be done by supervisors, particularly in large production fields [3].One effective approach is the automated classification of bean leaf diseases

Md. P. Islam · M. Afroj (✉) · A. N. Anika
Department of CSE, Bangladesh University of Business and Technology, Dhaka, Bangladesh
e-mail: maharinshova@gmail.com

M. Afroj
Advanced Machine Intelligence Research Lab (AMIR Lab), Dhaka, Bangladesh

H. Bhuiyan
Monash Data Futures Institute, Monash University, Melbourne, Australia
e-mail: hanif.bhuiyan@monash.edu

Md. R. Islam
Hong Kong Polytechnic University CN, Hong Kong, China
e-mail: mdrajibul.islam@polyu.edu.hk

M. F. Mridha and N. Dey (eds.), *Machine Vision in Plant Leaf Disease Detection for Sustainable Agriculture*, Studies in Computational Intelligence 1202, https://doi.org/10.1007/978-981-96-4520-6_10

through deep learning models. This is an important area of research that will benefit farmers by helping to control large fields of crops quickly and at an early stage which will assist farmers in identifying plant diseases, helping them treat those diseases early on, enhancing the number and quality of crops, and ultimately raising their profit margin. For effective crop management to be effective, an autonomous method for early disease detection must be developed. The need for continual supervision can be greatly decreased by using automatic methods to detect crop illnesses, particularly in fields with extensive production. The use of deep learning models to automatically classify bean leaf diseases is a promising approach. This approach not only makes it easier to identify illnesses quickly and effectively, but it also makes it easier to manage [4]. Therefore, it is critical to conduct research on the automatic classification of illnesses of bean leaves. It facilitates the prompt detection and treatment of plant diseases by farmers, improving crop quality and yields and increasing profitability. We can offer sustainable solutions that result in healthier crops and more affluent farming communities by utilizing cutting-edge technology. These technologies can also contribute to more sustainable farming practices. We can lessen the detrimental impact of chemical treatments on the environment and human health by using less of them. Farming may become more adept at managing disease outbreaks as we continue to create and refine these automatic classification systems, guaranteeing food security and economic stability for communities everywhere. It is very promising to use cutting edge technologies such as deep learning models in agriculture. They facilitate sustainable farming methods, lessen the need for hazardous pesticides, and enable early and accurate disease diagnosis. This strategy benefits farmers by increasing food yields and earnings, but it also enhances environmental quality and worldwide human well-being.

Our work offers several vital contributions:

- By integrating a ViT for feature extraction and a SVM for classification among other machine learning classifiers, we combine the strengths of both models.
- It has employed SHAP to interpret the decisions of our hybrid ViT-SVM model. SHAP offers localized explanations for each prediction, highlighting the specific features influencing the model's decisions.

The remainder of this paper structure: The paper follows the following structure: In Sect. 2, a thorough exploration of recent research findings on the topic is provided, offering a comprehensive overview. Section 3 outlines the detailed methodology proposed for the model development. The results of the proposed models and experimental findings of our research are presented in Sect. 4. Further, several critical points have been discussed in Sects. 5 and 6 summarizes the significant conclusions drawn from the study.

2 Related Work

Crop loss due to diseases is a global issue that significantly affects food security, international trade, and the economy [5]. Early disease detection often involves the examination of the leaves of various crops. Numerous studies have utilized ML models to classify crop diseases, encompassing a wide range of crops such as coffee [6, 7], maize [8, 9], beans [8, 10–13], rice [7, 8, 14], guinea corn [8], sunflower [8], and wheat [7, 14] based on disease-infected leaf images. Study [14] incorporated the "Plant Village" dataset, which encompasses 14 distinct crop species. Additionally, study [10] addresses the common limitation of dataset size by employing a dataset of 1295 images containing both healthy and diseased bean leaves. Initially, supervised machine learning methods such as NB, Decision Tree (DT), KNN, SVM, and RF algorithms were predominant [9]. In study [14], image processing technology is employed for disease detection and classification, which includes modifying the color composition of plant images and extracting features such as color and texture details. Subsequent studies have utilized artificial neural networks (ANN) [6], linear SVMs [9, 12], CNNs [6, 8, 10–12, 14, 15], and genetic algorithms for similar tasks related to leaf disease detection and classification. The process of selecting and extracting pertinent features from images significantly influences the performance of models. Insufficient representation of features can result in suboptimal detection and classification of diseases. Traditional methods face challenges, including high computational demands and vulnerability to environmental variations, particularly when dealing with novel samples. Consequently, the majority of research endeavors transitioned to deep learning [8, 10, 11, 14] approaches, where feature extraction proves to be more robust. Plant disease detection using leaf images employs various advanced techniques, including the application of ViT [6, 7, 14] alongside traditional methods. For example, in detecting diseases on bean leaves, color-based methods such as color histograms [16] and co-occurrence matrices capture changes in leaf coloration indicative of diseases like rust or blight. Texture analysis [16, 17] helps in identifying textural anomalies caused by pathogens. Shape-based approaches such as elliptic Fourier descriptors quantify deformities in leaf structure. Transfer learning from pre-trained ViT models, trained on diverse datasets, improves accuracy in distinguishing between healthy and diseased bean leaves.

Table 1 Summary of related research and identified limitations

References	Method	Dataset	Accuracy (%)	Limitations
[9]	ML models: NB, DT, KNN, RF	3823 images	79.23	Feature Engineering
[19]	CNN, SVM	9629 images	98.8–98.9	Limited generalization, interpretability, data imbalance
[14]	Hybrid ViT-CNN	54,306 images (Plant Village dataset)	95.99	Generalization to new diseases, optimization and hyperparameters, interpretability
[8]	CNN	268 images	94.03	Accuracy variation, data variety, and single leaf classification which indicates at the time of the study, the model could only categorize single leaves against a consistent background. This limitation restricts the model's applicability to real-world scenarios where diseases may affect different parts of the plant or appear on both sides of the leaves
[10]	ResNet50 architecture	1295 images	98.33	Domain shift, overfitting, limited adaptability, data quality and quantity, interpretability

The hybrid approach in the study [13] combines the strengths of traditional feature extraction with the capabilities of ViT, demonstrating its potential in advancing plant disease detection systems for agricultural management. Finally, for classification, Support Vector Machines (SVMs) are highly regarded for their capability to manage high-dimensional data and address nonlinear classification tasks effectively using kernel methods. Studies [9, 12, 18, 19] highlight SVMs effectiveness in identifying plant diseases based on features extracted from images, such as color, texture, and shape descriptors. Research has made significant strides in plant disease detection, particularly through the use of SVMs and CNNs. SVMs are effective in managing high-dimensional and nonlinear data, while CNNs are adept at extracting complex patterns from images, such as leaf texture, shape, and color, achieving accuracy rates between 93 and 97% [12, 14, 18, 19]. Despite challenges such as limited datasets, environmental variability, and model interpretability, methods like transfer learning and data augmentation have enhanced model performance. Future research aims to improve dataset quality, combine advanced ML techniques with remote sensing for real-time disease monitoring, and create robust decision support systems for precision agriculture, offering sustainable solutions to improve crop yields and ensure global food security. Table 1 presents key challenges and gaps identified in related research.

3 Methods and Materials

The approach to fruit classification is built on a series of carefully designed experiments. In this section, we detail the main aspects of our methodology Fig. 1, which includes the data description, preprocessing steps, and the architecture of the model.

3.1 Dataset Description

The dataset [20] consists of images of legume leaves, each representing one of three different lesion conditions. It is splited into two parts: a training set and a calibration set, to ensure the creation and validation of a reliable model. It comprises three categories of lesions: healthy leaves, angular bean rust, and leaf spot (Fig. 2).

3.2 Data Pre-processing

Our preprocessing workflow employed a number of cutting-edge methods to improve the quality of our dataset. To fit the numbers inside a given range while preserving the original distribution, we employed the Min–Max Scaler. This was very useful considering the various magnitudes and abnormalities found in our dataset. The Min-Max formula is expressed by the equation below, where m represents the transformed value, x is the original value in the cell, xmin is the minimum value in the column, and xmax is the maximum value in the column.

$$m = \frac{(x - x\min)}{(x\max - x\min)} \tag{1}$$

To tackle the issue of missing values, particularly in time series data, we used many techniques, such as mean imputation and standardization (Fig. 3).

3.3 Proposed Architecture

This model utilizes the feature extraction capabilities of ViT and the classification strengths of SVM to ensure effective

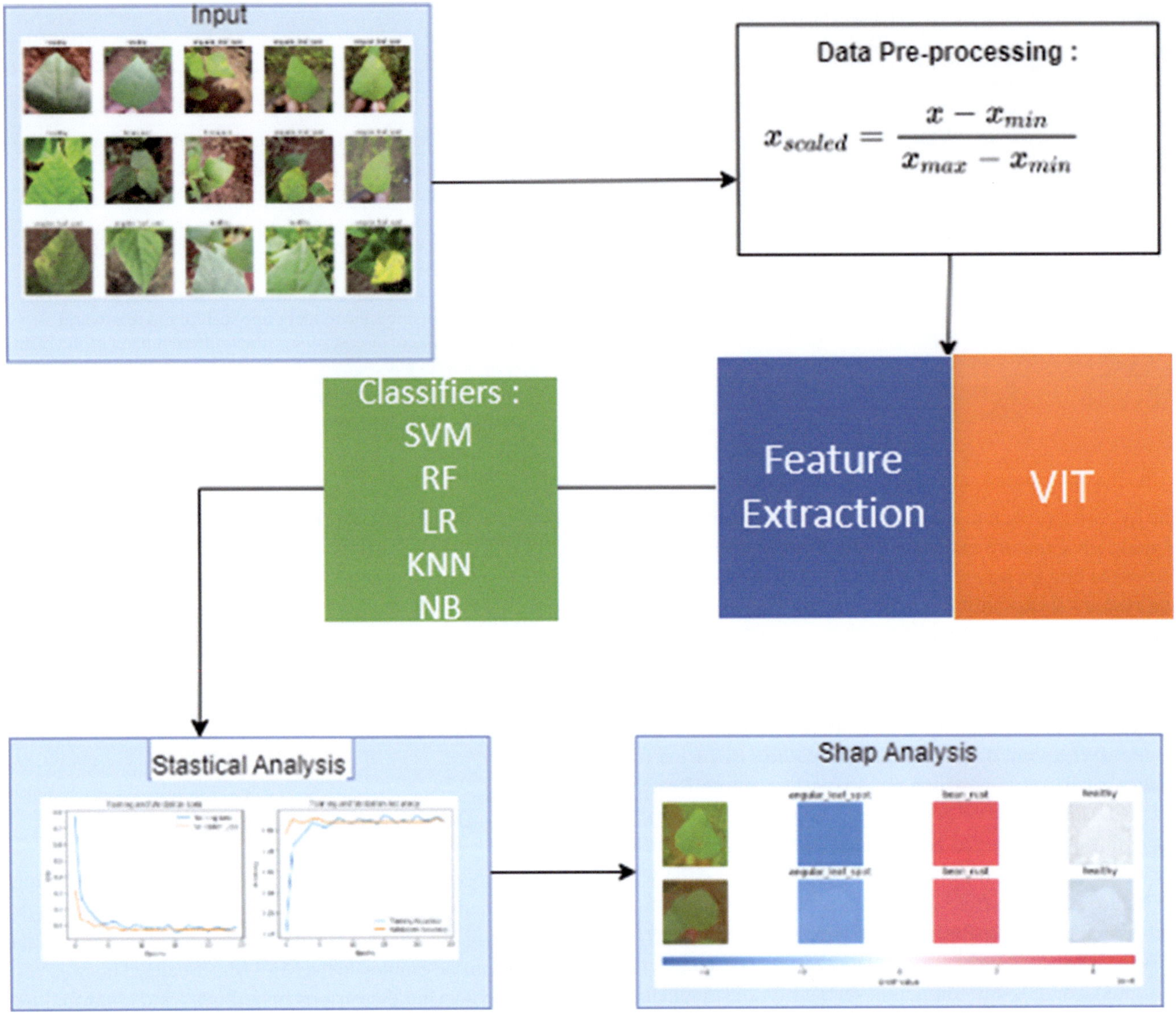

Fig. 1 Workflow of bean leaf disease classification

performance. This process aims to improve image classification accuracy by using the feature representation of ViT alongside the classification capability of SVM (Fig. 4).

3.4 VIT

The Vision Transformer (ViT) is an image categorization tool that utilizes the concepts of the Transformer model, which was initially created for natural language processing. The design of this model is distinguished by a number of crucial phases, which include positional embedding, picture patch embedding, processing via the Transformer encoder layers, and classification head processing.

Patch Embedding

Initially, the input image of size $H \times W \times C$ is divided into a sequence of fixed-size non-overlapping patches, each of size $P \times P$. The total number of patches N is calculated as

$$N = \frac{H \times W}{P^2}$$

Each patch is flattened into a vector and projected into a lower-dimensional embedding space of dimension D. The patch embeddings are

$$\mathbf{z}_0 = [\mathbf{z}_0^{[\mathrm{CLS}]}; \mathbf{z}_0^1; \mathbf{z}_0^2; \ldots; \mathbf{z}_0^N] + \mathbf{E}_{\mathrm{pos}}$$

Fig. 2 Sample images of dataset

where $\mathbf{E}_{\text{pos}}$ represents the positional embeddings.

Multi-head Self-attention (MHSA)

The embedded patches, with added positional information, are processed through several Transformer encoder layers, each containing MHSA and a FFN. The MHSA mechanism is defined as

$$\text{Attention}(\mathbf{Q}, \mathbf{K}, \mathbf{V}) = \text{softmax}\left(\frac{\mathbf{QK}^T}{\sqrt{D_k}}\right)\mathbf{V}$$

where $\mathbf{Q}$, $\mathbf{K}$, $\mathbf{V}$ are the query, key, and value matrices, and D_k is the dimension of the key vectors.

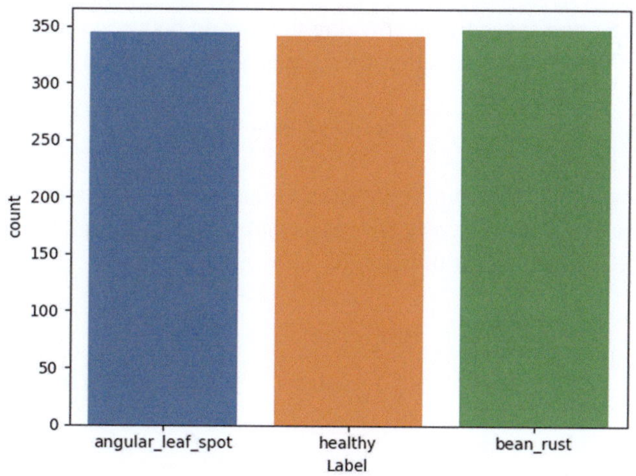

Fig. 3 Training set distribution of data

Transformer Encoder Layer

Each Transformer encoder layer applies MHSA followed by a Feed-Forward Neural Network. The operations in each encoder layer are

$$\mathbf{z}'_l = \text{LayerNorm}(\mathbf{z}_{l-1} + \text{MHSA}(\mathbf{z}_{l-1}))$$

$$\mathbf{z}_l = \text{LayerNorm}(\mathbf{z}'_l + \text{FFN}(\mathbf{z}'_l))$$

where l denotes the layer index, and FFN is the feed-forward network.

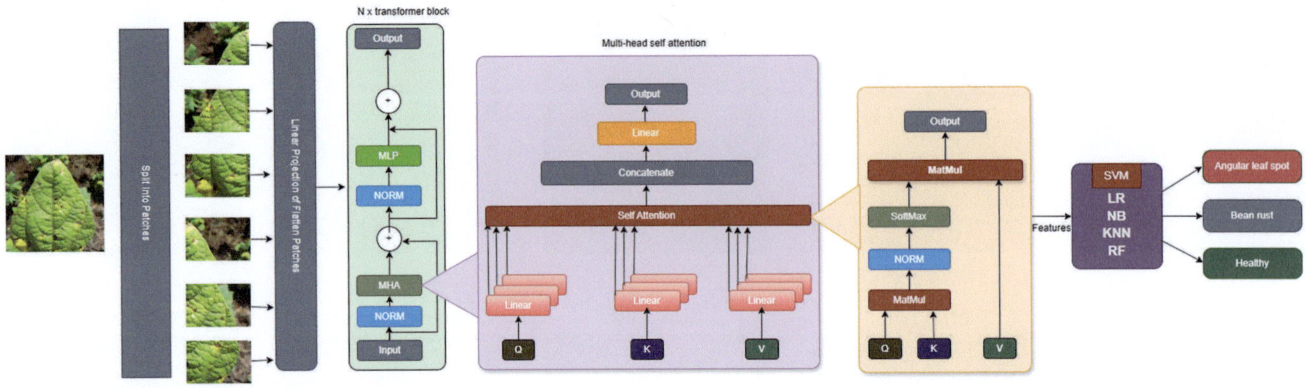

Fig. 4 Architecture of proposed model

Classification Head

A [CLS] token is added to the patch embeddings sequence for classification. The final encoder layer's output for this token is passed through a fully connected layer to generate the classification result. The classification head is as follows:

$$\mathbf{y} = \text{Softmax}(\text{Linear}(\mathbf{z}_L^{[\text{CLS}]}))$$

3.5 SVM

SVMs are supervised learning models used for classification, focusing on finding the hyperplane that best divides the classes in feature space by maximizing the margin between them.

The decision function for a linear SVM is defined as

$$f(\mathbf{x}) = \mathbf{w} \cdot \mathbf{x} + b$$

where $\mathbf{w}$ is the weight vector, $\mathbf{x}$ is the input vector, and b is the bias term. The objective is to find $\mathbf{w}$ and b that minimize the following cost function:

$$\min_{\mathbf{w},b} \frac{1}{2}\|\mathbf{w}\|^2 + C\sum_{i=1}^{N} \max(0, 1 - y_i(\mathbf{w} \cdot \mathbf{x}_i + b))$$

where C is the regularization parameter, N is the number of training samples, and y_i is the class label for the i-th sample.

4 Experimental Analysis

This section presents a comprehensive analysis of the results obtained from the rigorous testing and evaluation performed on deep learning models specifically developed for the classification of Bean leaf diseases.

The Table 2 provides a comparative analysis of existing models and the proposed Hybrid ViT-SVM model for a specific task. It highlights the performance of models such as CNN by Ibrahim et al. [8] 94.03%, ResNet50 by Serttas et al. [10] 91.67%, CNN K-means by Sahu et al. [11] 93–97%, Hybrid ViT-CNN by Borhani et al. [14] 95.99%, and CNN-SVM by Chaganti et al. 93%. Among these, the Hybrid ViT-CNN by Borhani et al. achieves the highest accuracy of 95.99%. The proposed Hybrid ViT-SVM model demonstrates competitive performance with an accuracy of 95%, showcasing its effectiveness compared to others.

The accuracy of several models is contrasted in Table 3 with our suggested Hybrid VIT-SVM model. Our Hybrid VIT-SVM and ResNet50 both reach the best accuracy of 95%. The accuracy of MobileNetV2 is lower, at 92.70%. Following closely with accuracies of 94.67% and 94.17%, respectively, are EfficientNet and DenseNet121. This comparison demonstrates that our Hybrid VIT-SVM model outperforms MobileNetV2, EfficientNet, and DenseNet121 while

Table 2 Comparison of existing works with our proposed model

Existing work	Model	Accuracy (%)
Ibrahim et al. [8]	CNN	94.03
Serttas et al. [10]	ResNet50	91.67
Sahu et al. [11]	CNN K-means	93–97
Borhani et al. [14]	Hybrid Vit-CNN	95.99
Chaganti et al. [18]	CNN-SVM	93
Our proposed	Hybrid ViT-SVM	95

Table 3 Comparison of traditional models with our proposed model

Model	Accuracy (%)
ResNet50	95
MobileNetV2	92.70
Effecientnet	94.67
DenseNet121	94.17
Our proposed (Hybrid ViT-SVM)	95

Table 4 Comparison of different classifiers pairing with ViT as feature extractor

Classifier	Accuracy (%)	Remarks
NB	78	Lowest accuracy
RF	94.12	High accuracy
LR	90.5	Good accuracy
KNN	86	Moderate accuracy
SVM	95	Best performance

performing on par with the best available model, ResNet50 (Table 4).

4.1 Performance Metrics

Table 5 provides a classification report for three categories: Angular Leaf Spot, Bean Rust, and Healthy are given in the table. With 20 samples as support, the model gets an F1-score of 0.89, recall of 0.80, and precision of 0.99 for Angular Leaf Spot. With 20 samples, it achieves 0.83 precision, 0.95 recall, and 0.88 F1-score for Bean Rust. With 20 samples, healthy leaves had precision of 0.95, recall of 0.99, and F1-score of 0.98. With all things considered, the model obtains an accuracy of 0.95.

Accuracy

Accuracy refers to the ratio of accurately predicted data points to the total number of data points. The accuracy is calculated according to the method shown in Eq. 2.

$$\text{Accuracy} = \frac{\text{TP} + \text{TN}}{\text{TP} + \text{TN} + \text{FP} + \text{FN}} \tag{2}$$

FP refers to instances where normal behavior is mistakenly identified as an attack, while FN occurs when a false attack is incorrectly classified as normal. TP represents cases correctly identified as normal, and True Negative TN refers to cases accurately classified as attacks.

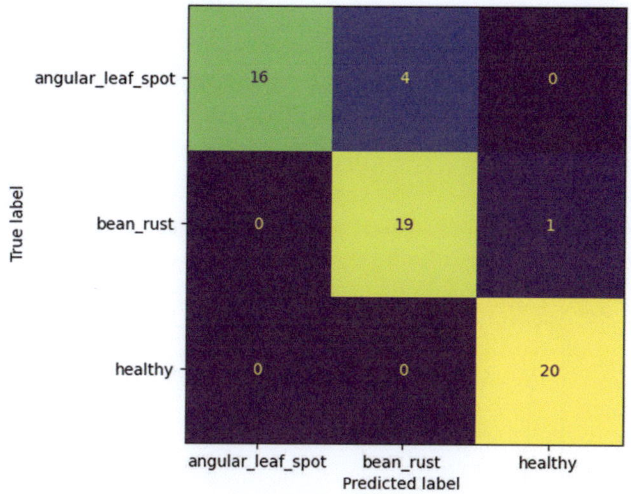

Fig. 5 Visual depiction illustrating the evolving performance metrics

Precision

The formula that is shown in Eq. 3, is a probability measure that evaluates the probability of a positive prediction being accurate.

$$\text{Precision} = \frac{\text{TP}}{\text{TP} + \text{FP}} \tag{3}$$

Recall

The term TPR refers to the proportion of correctly predicted positive occurrences inside the positive class Eq. 4.

$$\text{Recall} = \frac{\text{TP}}{\text{TP} + \text{FN}} \tag{4}$$

F1-score

The score is calculated by considering the precision (P) and recall (R) of the exam (Figs. 5 and 6).

$$f1\ \text{score} = 2 * \frac{\text{Precision} * \text{Recall}}{\text{Precision} + \text{Recall}} \tag{5}$$

4.2 Explainability with SHAP

SHAP has been employed in Fig. 7 which approach has been incorporated into our ViT-SVM model for bean leaf disease classification to enhance interpretability. SHAP provides a quantitative assessment of the contribution of individual pixels or regions within the input image, offering essential insights into the underlying mechanisms of the model's predictive decisions. By offering localized explanations for

Table 5 Classification Report

	Precision	Recall	F1-score	Support
Angular leaf spot	0.99	0.80	0.89	20
Bean Rust	0.83	0.95	0.88	20
Healthy	0.95	0.99	0.98	20
Accuracy	0.95			

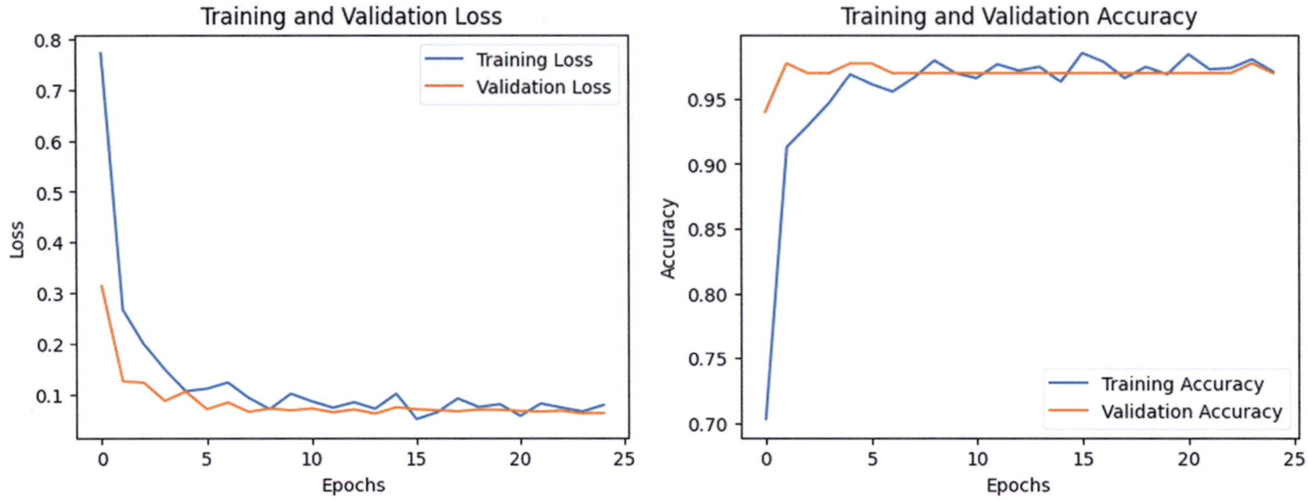

Fig. 6 Evaluating model performance: accuracy and loss curves for bean disease detection

each prediction, SHAP aids in visualizing and interpreting the decision-making process of the ViT-SVM model. This transparency provides farmers and other agricultural stakeholders with valuable information, helping them to understand and trust the model's accuracy.

5 Discussion

This study reveals the transformative power of AI in tackling crucial agricultural issues, particularly in detecting diseases on bean leaves. Farmers often struggle with identifying diseases early, and when left unchecked, these conditions can devastate crops. The SVM classifier was selected because of its superior performance compared to other classifiers like NB, RF, LR, and KNNs. The ability of SVM to handle complex, high-dimensional data and create clear decision boundaries makes it particularly effective for disease classification tasks. The use of an ML classifier ensures efficient and interpretable classification by utilizing the deep features extracted from the ViT. With a promising accuracy of 95%, the suggested approach, which combines a pre-trained ViT model with a SVM classifier, marks a significant advance over conventional human inspection techniques. With such great precision, the model can accurately distinguish healthy and diseased leaves, which makes it an invaluable tool for agronomists and farmers alike. Another significant challenge in agriculture is the reliance on time-consuming and labor-intensive visual inspections, which are often subjective and prone to human error. The use of SHAP to guarantee the model's interpretability is one of its main advantages. Although end users might otherwise be wary of a "black-box" AI model [21], SHAP fosters confidence by revealing which features go into classification decisions. The interpretability of AI technologies in agriculture is of utmost importance as it facilitates the comprehension and validation of stakeholders of the rationale underlying the predictions of the model [22]. This work has important practical implications. Since beans are widely grown as a staple crop, especially in poor countries, an automated and trustworthy disease detection system can result in early intervention, minimizing crop losses and improving food security. This approach can decrease human error, cut expenses, and save time by substituting an automated solution for the subjective and labor-intensive visual inspection procedure. With the widespread cultivation of beans as a staple food crop [23], particularly in developing regions, the need for an automated, reliable, and scalable disease detection system is critical to ensuring food security and reducing economic losses. For a number of reasons, we chose to use an SVM classifier as a ML model in this work rather than a deep learning model. Even though deep-thought models are renowned [24] for their strong fea-

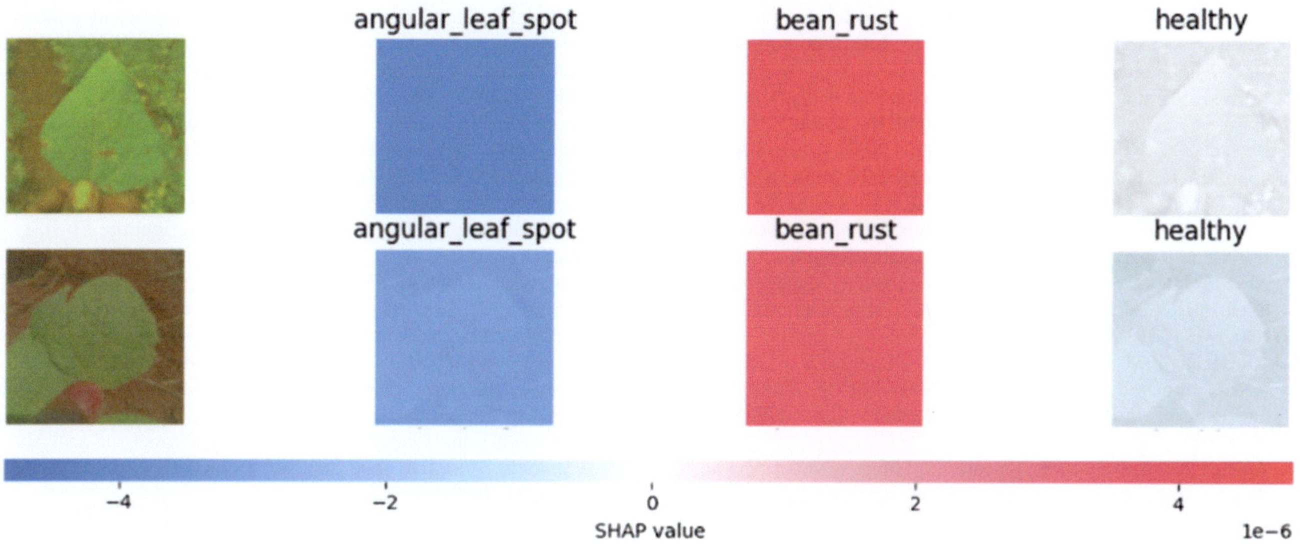

Fig. 7 Visualization of SHAP-enhanced interpretability in the ViT-SVM model for bean leaf disease classification

ture extraction abilities and high task performance, they frequently call for substantial computational resources and a lot of labeled data. ViT's feature extraction capabilities combined with the SVM's ease of use and interpretability provided a well-balanced method that produced excellent results without requiring a lot of data or processing power. This increases the model's usability and practicality for agricultural applications in the real world, where resources could be few.

Though the results are encouraging, there are a few areas that still need to be looked into more. The current model's generalizability to other crops or to different varieties of beans still needs to be confirmed. It was evaluated on a particular dataset of bean leaf photos. Subsequent investigations ought to delve into the degree to which this methodology may be adjusted to diverse farming environments, with the possibility of integrating supplementary data sources like environmental elements or multi-spectral imaging to enhance its resilience.

6 Conclusions

Our study offers a novel AI-driven method for automating bean leaf disease detection and classification, tackling important issues with food security and agricultural production. Our hybrid model achieves a noteworthy accuracy of 95% by utilizing the ViT for reliable feature extraction and incorporating the SVM for accurate classification. This is superior to eye examination techniques, providing farmers and agronomists with a quicker and more dependable option. By adding SHAP, the model's decision-making process becomes more transparent and easier to grasp. Our research paves the way for future developments in AI applications in agriculture, including scalability and integration with technology for precision farming continued enhancement of model interpretability to support sustainable crop management practices and ensure global food security.

References

1. Singh, V., Chug, A., Singh, A.P.: Classification of beans leaf diseases using fine tuned CNN model. Procedia Comput. Sci. **218**, 348–356 (2023)
2. Singh, S.P., Schwartz, H.F.: Breeding common bean for resistance to diseases: a review. Crop Sci. **50**(6), 2199–2223 (2010)
3. Abed, S.H., Al-Waisy, A.S., Mohammed, H.J., Al-Fahdawi, S.: A modern deep learning framework in robot vision for automated bean leaves diseases detection. Int. J. Intell. Robot. Appl. **5**(2), 235–251 (2021)
4. Elfatimi, E., Eryigit, R., Elfatimi, L.: Beans leaf diseases classification using MobileNet models. IEEE Access **10**, 9471–9482 (2022)
5. Savary, S., Ficke, A., Aubertot, J.-N., Hollier, C.: Crop losses due to diseases and their implications for global food production losses and food security. Food Secur. **4**(4), 519–537 (2012)
6. Pham, T.C., Le, C.H., Packianather, M., Hoang, V.-D., et al.: Artificial intelligence-based solutions for coffee leaf disease classification. In: IOP Conference Series: earth and Environmental Science, vol. 1278, p. 012004. IOP Publishing (2023)
7. Li, G., Wang, Y., Zhao, Q., Yuan, P., Chang, B.: PMVT: a lightweight vision transformer for plant disease identification on mobile devices. Front. Plant Sci. **14**, 1256773 (2023)
8. Ibrahim, H.B., Benjamin, O.O.: Plant disease detection on some selected plants using convolutional neural network (CNN)
9. Panigrahi, K.P., Das, H., Sahoo, A.K., Moharana, S.C.: Maize leaf disease detection and classification using machine learning algorithms. In: Progress in Computing, Analytics and Networking: proceedings of ICCAN 2019, pp. 659–669. Springer (2020)
10. Serttas, S., Deniz, E.: Disease detection in bean leaves using deep learning (2023)
11. Sahu, P., Chug, A., Singh, A.P.: Crop leaf disease detection for beans using ensembled-convolutional neural networks. Int. J. Food Eng. **19**(11), 521–537 (2023). https://doi.org/10.1515/ijfe-2023-0055
12. Banerjee, D., Kukreja, V., Yadav, R., Joshi, K., Singh, A.: Effective disease detection in bean leaves using deep CNN and SVM

ensemble. In: 2023 3rd Asian Conference on Innovation in Technology (ASIANCON), pp. 1–6 (2023). https://doi.org/10.1109/ASIANCON58793.2023.10270779

13. Önler, E.: Feature fusion based artificial neural network model for disease detection of bean leaves. Electron. Res. Arch. (2023)
14. Borhani, Y., Khoramdel, J., Najafi, E.: A deep learning based approach for automated plant disease classification using vision transformer. Sci. Rep. **12**(1), 11554 (2022)
15. Poovidha, S., Thaiyalnayaki, S., Sampath, A.K.: Leaf disease detection using image processing. In: 2023 International Conference on New Frontiers in Communication, Automation, Management and Security (ICCAMS), vol. 1, pp. 1–7 (2023). https://doi.org/10.1109/ICCAMS60113.2023.10526087
16. Ali, H., Lali, M., Nawaz, M.Z., Sharif, M., Saleem, B.: Symptom based automated detection of citrus diseases using color histogram and textural descriptors. Comput. Electron. Agric. **138**, 92–104 (2017)
17. Ahmad, N., Asif, H.M.S., Saleem, G., Younus, M.U., Anwar, S., Anjum, M.R.: Leaf image-based plant disease identification using color and texture features. Wirel. Personal Commun. **121**(2), 1139–1168 (2021)
18. Chaganti, S.Y., Nanda, I., Pandi, K.R., Prudhvith, T.G., Kumar, N.: Image classification using SVM and CNN. In: 2020 International Conference on Computer Science, Engineering and Applications (ICCSEA), pp. 1–5. IEEE (2020)
19. Kibriya, H., Abdullah, I., Nasrullah, A.: Plant disease identification and classification using convolutional neural network and SVM. In: 2021 International Conference on Frontiers of Information Technology (FIT), pp. 264–268. IEEE (2021)
20. Marquis03: Bean Leaf Lesions Classification Dataset (2021). https://www.kaggle.com/datasets/marquis03/bean-leaf-lesions-classification. Accessed 01 Sept. 2024
21. Tugrul, B., Elfatimi, E., Eryigit, R.: Convolutional neural networks in detection of plant leaf diseases: a review. Agriculture **12**(8), 1192 (2022)
22. Costa, C., Antonucci, F., Pallottino, F., Aguzzi, J., Sun, D.-W., Menesatti, P.: Shape analysis of agricultural products: a review of recent research advances and potential application to computer vision. Food Bioprocess Technol. **4**(5), 673–692 (2011)
23. Álvarez Sánchez, D.-E., Arévalo Rodríguez, A., Benavides Martínez, I.F., Salazar González, C., Betancourth García, C.: Use of trained convolutional neural networks for analysis of symptoms caused by botrytis Fabae Sard. Revista de Ciencias Agrícolas **40**(1) (2023)
24. Mahajan, M., Sharma, R., Singh, A., Kukreja, V., Vats, S.: Towards smart agriculture: Automated detection and diagnosis of bean leaf diseases using resnext50 model. In: 2023 14th International Conference on Computing Communication and Networking Technologies (ICCCNT), pp. 1–6. IEEE (2023)

Deploying CNN-ResNet50-BiLSTM for Paddy Leaf Disease Detection

Md Basitur Rahman Bappi, S. M. Masfequier Rahman Swapno, Sumiya Akhter, and M. M. Fazle Rabbi

Abstract

Research in agriculture, specifically on paddy plants, is needed to accurately categorize the diseases that affect rice crops in their early stages. This is achievable if automated technologies are capable to support farmers in identifying rice diseases based on photographs of paddy leaves. Applying deep learning (DL) techniques to identify diseases in farmed plants can reduce farmers dependency on the protection of paddy crop production. In this study, we compiled a dataset comprising five distinct classes: bacterial leaf blight, bacterial leaf streak, bacterial panicle blight, downy mildew, and normal. We leverage a variety of DL models to check the system validity in the plant disease detection field. Our hybrid model, CNN-ResNet50-BiLSTM, exhibited exceptional performance, boasting an extraordinary training accuracy of 98.67% and a validation accuracy of 99.93%. On top, we evaluate the precision, recall, F1 score, and anticipated results to determine the effectiveness of the proposed system. This promising system has the potential to support agricultural professionals in the early and precise diagnosis of paddy leaf disease, ultimately improving crop outcomes and alleviating the burden on agricultural systems.

Md Basitur Rahman Bappi · S. M. Masfequier Rahman Swapno (✉) · S. Akhter · M. M. F. Rabbi
Department of Computer Science and Engineering, Bangladesh University of Business and Technology, Mirpur, Dhaka, Bangladesh
e-mail: masfequier.cse.bubt@gmail.com

M. M. F. Rabbi
e-mail: rabbi@bubt.edu.bd

Keywords

Agriculture • Deep learning • Paddy leaf • CNN-ResNet50-BiLSTM • CNN • Leaf disease • Valuable crop • Effective classification

1 Introduction

Agriculture [1, 2] engages a significant role in the livelihoods of people in Bangladesh. The agricultural sector prioritizes the production of high-quality food while minimizing costs. Precise identification and a fast solution to these issues are crucial for effective crop management. Identifying [3, 4] a problem manually is a highly intricate undertaking that involves multiple criteria. Progress in image processing technology offers a more convenient solution to these issues. Rice is the predominant crop in more than 75% of the world's population. Rice growing [5] regions are commonly affected by various diseases affecting rice plants, such as false smut, leaf blast, brown spot, sheath blight, leaf streak, and bacterial blight. The proposed method that has been suggested can identify and categorize specific illnesses that affect paddy plants and provide farmers with immediate information regarding the disease stage.

An essential factor [6] for mitigating the decline in agricultural product quantity and productivity is the accurate identification of plant diseases. Disease recognition and monitoring of plant wealth have negative impacts on sustainable farming. The term "recognition of plant diseases" refers to the identification and observation of visual patterns on plants that indicate the presence of illnesses. Traditionally, the detection of paddy diseases [7, 8] has been performed manually by experts who rely solely on visual sight. This process is both time-consuming and costly, particularly for large farms. Quantifying the type of disease can be challenging and occasionally results in errors. Due to the lack of proper management to address leaf diseases in rice plants, paddy

M. F. Mridha and N. Dey (eds.), *Machine Vision in Plant Leaf Disease Detection for Sustainable Agriculture*, Studies in Computational Intelligence 1202, https://doi.org/10.1007/978-981-96-4520-6_11

production has recently declined. Efficient and prompt identification methods must be implemented to effectively diagnose diseases influencing paddy leaves.

1.1 The Purpose and Significance of this Research

The reason behind this study is the significance of maintaining food security and livelihoods. Rice, a staple crop for billions of people worldwide, was the main focus of this study. A wide range of diseases can affect paddy plants, significantly affecting crop quantity and quality. Timely intervention and abatement of these diseases depends on accurate identification and classification. Furthermore, manually identifying plant illnesses is laborious and frequently necessitates specialized expertise, making difficulty to farmers in efficiently oversee their crops. This project attempts to improve illness identification in agriculture by utilizing automated technologies, specifically DL models, to evaluate photos of paddy leaves. The work aims to create a robust dataset and utilize advanced DL techniques to offer farmers a dependable tool for quickly and precisely identifying paddy illnesses. This study aims to improve crop yields, minimize losses, and reduce the pressure on agricultural systems, thereby promoting global food security and sustainability.

Our research primarily focused on creating a sophisticated method for identifying diseases affecting paddy leaves. This study aimed to identify five distinct diseases that affect rice plants. Figure 1 shows a graphical representation of the five distinct classes that represent various diseases affecting paddy leaves. We collected a dataset of 6405 images, carefully classified into five categories, to improve the accuracy and effectiveness of paddy leaf disease detection. Our approach combines a hybrid CNN-ResNet50-BiLSTM model that utilizes the strength of each component, resulting in a robust and efficient model that accurately identifies diseases. By building our hybrid model exceptional training and validation accuracy were achieved. A remarkable 98.67% training accuracy and an even more remarkable 99.37% validation accuracy were attained by the model. These remarkable accuracy metrics demonstrate the usefulness of our developed system, highlighting its ability to accurate detect paddy leaf disease. The precision of our paddy leaf disease detection criterion exceeded that of industry benchmarks, making our discoveries even more critical. This highlights our hybrid model's

Fig. 1 Visual representation of the paddy plant affected by a disease characterized by the presence of five leaves. Present an exhibition the pathology of paddy leaf disease

intricacy and holds promise for improving precision farming techniques and lessening the effects of illnesses on paddy crops. The primary aspects of this study are as follows:

- Concentrating particular benchmark techniques to guarantee the effective identification of paddy leaf disease to boost agriculture by means of enhanced automatic detection of paddy leaf disease.
- Utilizing sophisticated machine learning (ML) and deep learning (DL) methods to accurately identify paddy leaf disease.
- Developing a hybrid CNN-ResNet50-BiLSTM architecture dedicated to detecting paddy leaf disease.

The remainder of this paper is structured as follows. The pertinent literature is summarized in Sect. 2. The methodology is explained in Sect. 3, and the analytical results are presented and contrasted with previous approaches in Sect. 4. A succinct summary of the study is given in Sect. 5.

2 Relevant Work

Healthy paddies enhance the export of premium rice. Rice is a primary food source in many rural areas and the second most abundant cereal. Rice serves as a primary and essential source of nutrition in Bangladesh. Papademetriou (2000) state that rice disease is responsible for 10–15% of crop losses in Asia. Agricultural crops are cultivated in all five continents: Asia, Africa, America, Europe, and Oceania. FAOSTAT data reveal that humans account for 91.05% of global rice consumption. The distribution of rice production throughout the different regions of the world was as follows: Africa accounted for 2.95% of the population, America accounted for 5.19%, Asia accounted for 91%, Europe accounted for 0.67%, and Oceania accounted for 0.15%. This information is presented in Fig. 2.

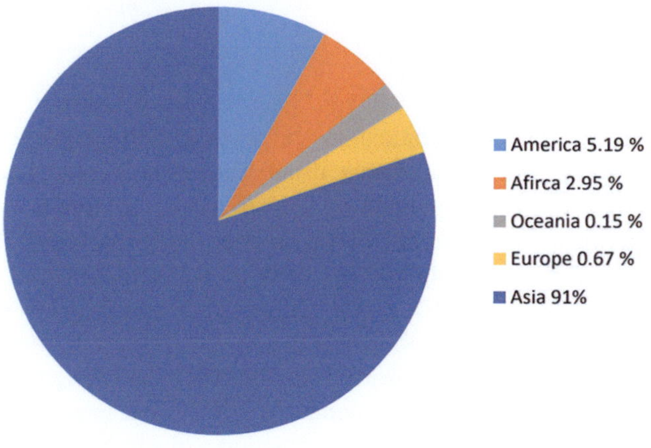

Fig. 2 Rice production worldwide is obtained from the FAOSTAT source

Machine learning (ML) [9–15] and deep learning (DL) [16–22] have recently been increasingly employed to detect diseases in the agricultural field. These advanced technologies enhance the precision and efficiency of plant disease identification, which are essential for ensuring healthy crop yields. Additionally, ML and DL techniques are widely used for classification and segmentation, and automation processes that traditionally require manual work. By applying these technologies, the agricultural industry can achieve better disease management, optimize crop production, and lessen reliance on chemical treatments, thus supporting more environmentally friendly agricultural methods. Patidar et al. [23] used a Residual Neural Network for disease classification, proving that it is faster and more effective than traditional convolutional neural networks (CNN) and classifiers such as support vector machines (SVMs). It mitigates saturation in vast datasets or complex networks, resulting in superior outcomes with a precision of 95.83%. Shahidur et al. [24] performed preprocessing and feature extraction using clear photos of healthy and sick rice leaves captured against a white background. We constructed an image categorization model using random Forest is one ML approach. Random Forest stood out as the highest achiever, attaining an outstanding accuracy rate of 97.50%. Our edge gadget has a model that effectively detects diseases in the paddy leaves. The CNN architecture proposed by Sagarika et al. [25] has ten layers and is trained for a maximum of 20 epochs. The learning rate during the training period was set as 0.0001. A great accuracy rate of 94.12% is indicated by the model's results. Published on the Internet of Things (IoT) platform are these findings. This technology provides rapid and accurate safety recommendations and assists in disease control by offering suggestions for pesticides, thereby improving crop productivity.

Acharya et al. [26] utilized GoogLeNet, ResNet, ShuffleNet, ResNeXt, and Wide ResNet as distinct models to identify diseases in paddy leaves. These models were trained individually and combined using different weights. The performance study, which considered the loss and accuracy indicators, revealed a remarkable accuracy of 95.54% for the final method. An ensemble technique improves the identification of illnesses in paddy leaves by utilizing the strengths of many models. Burhan et al. [27] gathered field data and subjected it to preprocessing, which involved removing the background and shadow, before model evaluation. ResNet50 achieved a top accuracy of 75.0% on the synthetic dataset, however, using our dataset, ResNet101V2 achieved an accuracy of 86.99%, outperforming it. The results emphasize the effectiveness of ResNet models in detecting harmful conditions in agricultural photography. Pothen et al. [28] utilized Otsu's methodology to partition images of diseased rice leaves into discrete classifications: bacterial leaf blight, leaf smut, and brown spot. The procedure entails the extraction of features using Local

Binary Patterns (LBP) and Histogram of Oriented Gradients (HOG), followed by classification using a SVM. The polynomial Kernel SVM and HOG combination yielded an accuracy of 94.6%. The utilization of the mask recurrent convolutional neural network (Mask RCNN) by Mashroor et al. [29] allows for accurate classification, resulting in an astounding accuracy rate of nearly 99%. The segmentation results exhibit minimal loss values, precisely 0.09 for the class, 0.29 for the bounding box, and 0.30 for the mask. Furthermore, the segmentation model exhibited strong performance with an approximate mean average precision (mAP) of 89%.

The primary issue is the challenging detection of paddy leaf disease. In addition, the current methods exhibit significant underperformance. The utilization of our DL approach, specifically the CNN-ResNet50-BiLSTM algorithm, to effectively tackle this challenge and generate the interest of researchers in this particular area of research has received much recognition in the past.

3 Research Methodology

Our research efforts were significantly advanced by successfully creating a hybrid CNN-ResNet50-BiLSTM model to effectively detect diseases in paddy leaves. By successfully addressing multiple obstacles, we established a flawless, resilient detection system. Figure 3 shows the structure of our hybrid model, illustrating the incorporation of the CNN, ResNet50, and BiLSTM components. This complicated amalgamation not only symbolizes a technological milestone but also demonstrates the efficacy of our method in tackling the intricacies linked to the detection of paddy leaf disease.

3.1 Analysis and Discussion of the Dataset

A dataset from Kaggle was used, which contained images of paddy leaf diseases and the size of dataset is 320 MB. There are 6405 paddy leaf disease images in this dataset. The images presented here can be classified into five distinct categories: standard, bacterial panicle blight, bacterial leaf streak, bacterial leaf blight, and bacterial leaf blight. We obtained historical findings from this dataset by incorporating them into our model. The quantity of photos for every class is listed in Table 1.

3.2 Data Preprocessing

We utilized various strategies throughout the data preprocessing stage to optimize model performance and address problems such as overfitting. First, in order to provide a uniform analysis format, we make sure that our picture inputs are resized to 224 * 224 pixels. Subsequently, three subsets are created from the dataset: The dataset was divided into three sets: a testing set that included the remaining 10% of the data, a training set that included 80% of the data, and a validation set that included 10% of the data. This division ensures that we have enough data for training, optimizing the hyperparameters, and evaluating the model's functionality. We used data augmentation techniques to address class representation discrepancies and strengthen our model's robustness. To add variations to the training data, these methods use transformations including rotation, flipping, and scaling. To optimize the usefulness of our dataset, we also apply a fourfold cross-validation technique. The suggested approach iteratively splits the dataset into four equal parts: three for training and one for validation. Four iterations of this procedure are performed to guarantee that each subset is used as training and validation data. Each subset was used as the training and validation data, and this procedure was carried out four times. Additionally, we apply noise removal techniques to exclude data points that contain noise, which improves the model's generalization and strengthens the dataset's balance. To maximize the computational resources and improve the model's efficacy, we also use image-resizing algorithms. We ensure that our dataset is balanced, noise-free, and suitably scaled by using these preprocessing procedures, which lowers overfitting and enhances our model's overall performance.

3.3 Proposed Hybrid CNN-ResNet50-BiLSTM Model Implmenetation

Convolutional layers are mostly used to establish localized relationships between features from preceding layers and assign their data to visible feature maps. The mathematical process that joins the filter F $\left(F \epsilon \mathbb{R}^{2a_1+2a_2}\right)$ and input I is denoted as a convolution.

$$(I * F)_{n,m} = \sum_{k=-a_1}^{a} 1 \sum_{1=-a_1}^{a} 2F_k, I_{n-k}, m - l \tag{1}$$

A non-linear activation function is a Rectified Linear Unit (ReLU), which is applied to the feature maps generated by the convolutional layers using the activation function $(g(z) = \max(0, z))$. Uniting similar traits communicated from the previous layer is the main goal of the max-pooling layers. By locating the highest value on the feature map that falls under the bounds of the filter, the max-pooling layers accomplish downsampling. CNNs and multilayer perceptron (MLPs) often have comparable architectures, spanning from a detection layer to a fully connected (FC) layer. In an MLP, the FC layers perform the same function as the hidden layers. The architecture of a CNN can contain one or more FC layers. The fully linked layer connects all the neurons in

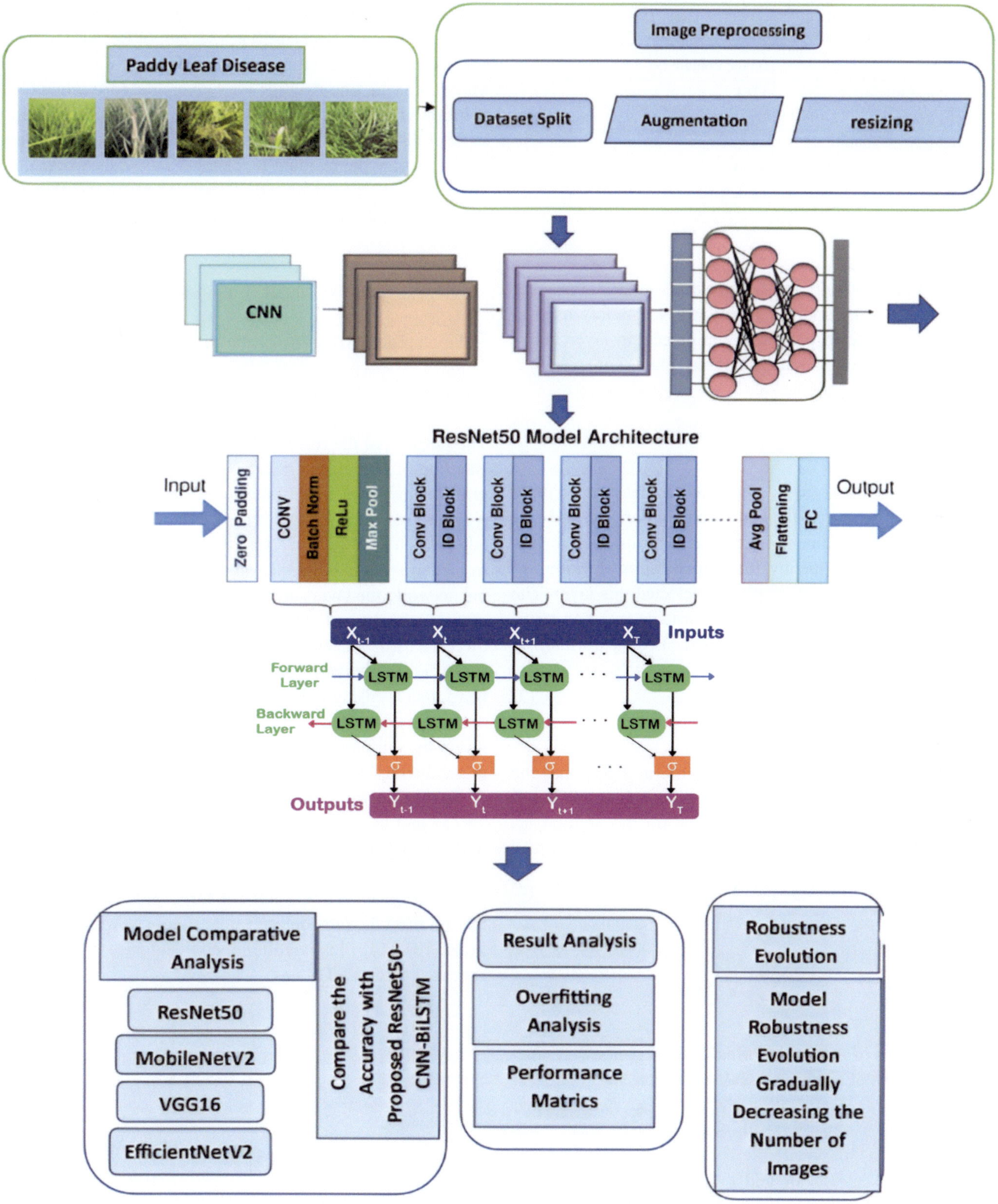

Fig. 3 Detecting paddy leaf disease by using the CNN-ResNet50-BiLSTM model architecture. Showcase the distinct architectures of the CNN, ResNet50, and BiLSTM models that were created especially for the prediction of paddy leaf disease

Table 1 The distribution of the datasets by class. The images per class are displayed as five classes

Class	Image contain
Bacterial leaf blight	1181
Bacterial leaf streak	1056
Bacterial panicle blight	1230
Downy mildew	9950
Normal	1988

the preceding layer and all of the neurons in the subsequent layer. In CNNs, the softmax function is commonly used to convert unnormalized data from the preceding layer into a probability distribution across predicted class scores.

$$\sigma(x_i) = \frac{e^{x_i}}{\sum_{j=1}^{k} e^{xj}}, \; j = 1, \ldots, K \tag{2}$$

Here, σ denotes the softmax output, and x_i denotes the input vector values. To reduce the CNN training duration and network initialization vulnerabilities, batch normalization layers were used. Consequently, the normalization algorithm of the proposed CNN design explicitly selected this layer. The input variable x_i, mini-batch mean variable m_b, and mini-batch variance variable v_b were used to calculate normalized activation.

$$\hat{x}_i = \frac{x_i - m_b}{\sqrt{v_b^2 + \epsilon}} \tag{3}$$

Given that ϵ is a constant, it plays a role in determines the numerical state when the value of v_b is extremely small. Equations (5) and (6) show the calculations for the m_b and v_b, respectively.

$$m_b = \frac{1}{q}\sum_{p=1}^{q} z_p \tag{4}$$

$$v_b = \frac{1}{q}\sum_{i=1}^{n} (z_p - m_b)^2 \tag{5}$$

The activations of the batch normalization layer's were fine-tuned by means of a shift and scale operation.

$$y_p = r\hat{t_i} + s \tag{6}$$

where, is the scale factor and r is the balance. Throughout the training process, these modifiable variables were updated to the most suitable values. The softmax function applied element-by-element to the vector z is mathematically described as follows:

$$\text{softmax}(z)_i = \exp(y_i) \tag{7}$$

where exp is the exponential function and i ranges over all sentiment classes. ResNet-50 is a variant of ResNet, which is a CNN architecture with 50 layers. The model included one MaxPool layer, one Average Pool layer, and forty-eight Convolution layers. The reason behind its performance is still a topic of discussion, but the most direct method to understand it is to clarify the residual blocks and how they work. Let's analyze a neural network block designed to process the input p and learn the true distribution $s(r)$. Let us denote the difference as:

$$T(p) = \text{output} - \text{input} = S(r) - r \tag{8}$$

Upon reorganizing the given information, we obtain:

$$H(p) = S(r) + x \tag{9}$$

The purpose of the residual block is to gain a precise understanding of output, $H(p)$. Upon careful analysis of the diagram, shows that the existence of an identity connection, which arises from x, enables the layers to acquire the knowledge residual, $H(p)$. In a traditional network, the layers are trained to acquire the precise output $s(r)$; However, in a residual network, the layers are trained to acquire the discrepancy between the desired output and the present output $s(r)$. Furthermore, it has been noted that acquiring knowledge about the difference between output, and input is more straightforward than focusing uniquely on the input. The identity residual model enables the reuse of activation functions from prior levels by skipping them, thereby avoiding additional architectural complications.

To regulate the updating and retaining of the cell state, the LSTM unit uses input, forget, and output gates. Certain elements of the newly preserved data are chosen by the input gate. While the output gate chooses which parts of the new cell state are given to the cell, the forget gate decides if information from the previous cell state should be retained. Equations (10)–(14) illustrate the precise sequence of operations executed by an LSTM unit.

$$d_z = \xi\left(K_i * \left[m_{z-1}, \acute{l}_z\right] + b_q\right) \tag{10}$$

$$e_z = \xi\left(K_f * \left[m_{z-1}, \acute{l}_z\right] + b_r\right) \tag{11}$$

$$f_z = \xi\left(K_o * \left[m_{z-1}, \acute{l}_z\right] + b_s\right) \tag{12}$$

$$G_z = f_t * C_{z-1} + i_z * \tanh\left(K_c * \left[m_{z-1}, \acute{l}_z\right] + b_v\right) \tag{13}$$

$$m_z = o_z * \tanh(C_z) \tag{14}$$

In this case, the output of the input gate, forget gate, output gate, and cell at time z are denoted by the variables d_z, e_z, f_z, and G_z, respectively. The input vector and hidden layer vector at time z are equivalently represented by $\acute{l}_t$ and m_z, respectively. The symbol ξ stands for the sigmoid activation function. The characters k and b stand for the weight matrix and bias vector, respectively. The subscripts k and b denote the categories k_i and b_q, which stand for the weight matrix and bias vector associated with the input gate construction, respectively. By using the LSTM unit, Bi-LSTM efficiently retains contextual information.

Because they are at the beginning and end of a sentence, they can retain information from both ends. The Web service description document's previous and subsequent contextual data, the content that comes before and after captured and stored by Bi-LSTM. The Web service description document $l = [l_1, l_2, ..l_d]$ feature vector transforms each feature word l_i into its matching word embedding representation, a data format that the neural network model can easily process. Two parallel LSTM layers examined the input l_z in opposite directions during the zth time step. The forward layer deals with the input coming from the front direction, whereas the reverse layer processes the input coming from behind. The cumulative total number of hidden state vectors was the outcome.

$$m_z = \acute{k}_1 \overrightarrow{m_z} + \acute{k}_2 \overleftarrow{m_z} + b_{d2} \tag{15}$$

The output consists of two LSTM layers, running in parallel, which produce the vectors $\overrightarrow{m_z}$ and $\overleftarrow{m_z}$. These layers operate in both forward and reverse orientations. The weight parameters for these parallel LSTM layers are labelled as $\acute{k}_1$ and $\acute{k}_2$, whereas the bias is denoted as b_{d2}. Therefore, by applying formula (15), we can obtain the anticipated depth detection vector as $d_2 = [m_1, m_2, ...m_n]$.

4 Result Analysis

We developed a paddy leaf disease detection system using the CNN-ResNet50-BiLSTM architecture and achieved exceptional results.

4.1 Evaluation Standards and Outcome Presentation

To determine whether our system was accurate in identifying paddy leaf disease, we employed the following key metrics: true positive (TP), true negative (TN), false positive (FP), and false negative (FN). The model obtained 98.67% for training and 99.93% for validation. This accuracies are showed Fig. 4. When conducting a DL experiment, it is crucial to assess the determine accuracies to evaluate the model's performance and capacity for generalization. Training accuracy measures the

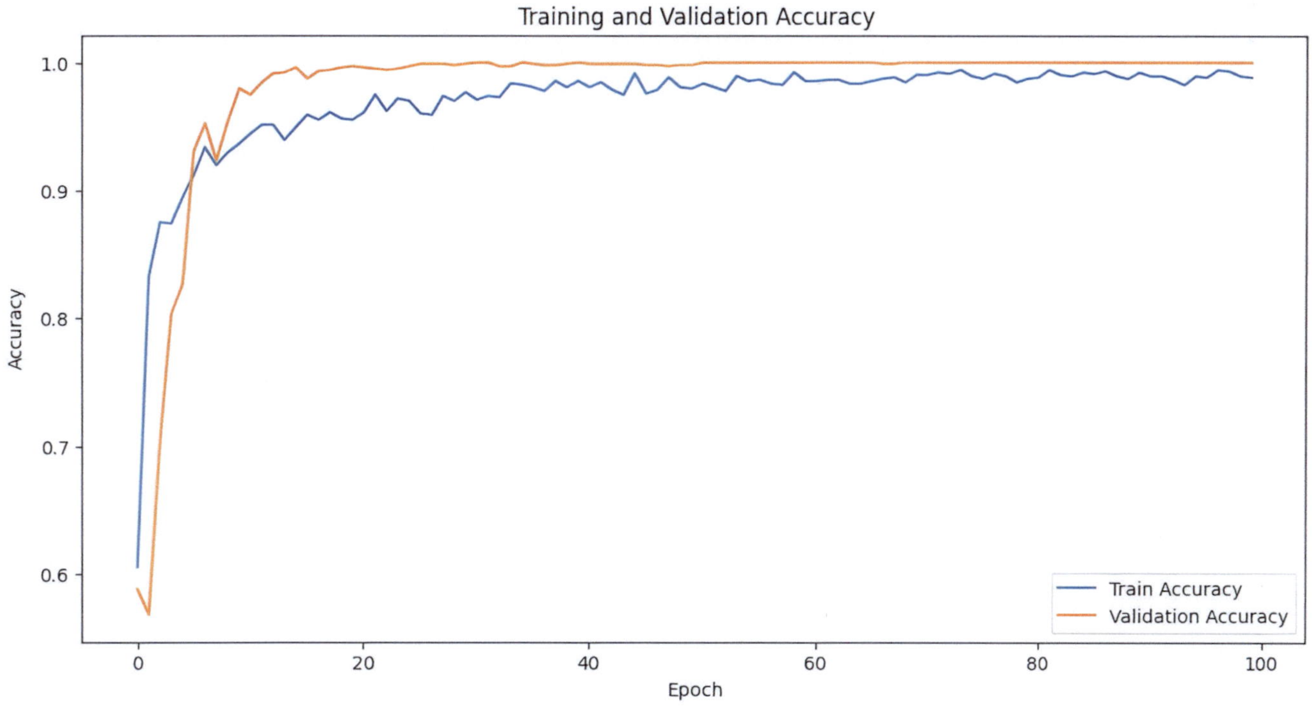

Fig. 4 The proof of accuracy in training and validation as a way to gauge the model's performance. Additionally expressed is the detection accuracy in this system

degree to which the model accurately captures the patterns in the training data that it has encountered. Nevertheless, a model that exhibits high performance on training data may need to be more balanced, which refers to its tendency to incorporate noise and specific information that does not apply to fresh, unknown data. A metric called validation accuracy was used to evaluate a model's performance on a dataset that was not utilized for training. This is a rough estimate of the capacity of the model to generalize new data. By comparing training and validation accuracy, one can identify the presence of overfitting or underfitting. If the model's accuracy during training is high, but the accuracy during validation is low, it suggests that overfitting is occurring in the model. On the other hand, if the accuracies of both models are low, the model may be underfitting and not effectively learning from the training data. Ensuring a balance between these accuracy levels is essential to construct a resilient model that demonstrates high performance when applied to real-world data.

A key role in assessing the efficacy of DL models and promoting additional improvements is outlined by the training and validation loss. The training loss metric quantifies the degree to which the model accurately captures the patterns in the training data, and supports the optimization parameters to minimize discrepancies. Additionally, validation loss assesses the way the model performs on data that it has yet to be trained on, guaranteeing that it can apply its knowledge effectively and does not excessively adapt to the training data. By observing both loss functions, it is possible to assess the effectiveness of the learning process of the model and determine whether tweaks such as regularization or early halting are necessary to improve generalization and prevent overfitting. This two fold assessment offers a thorough comprehension of model behavior and aids in creating strong, and precise models. After computing the model's loss, 0.13 was obtained for training and 0.07 for validation. The training and validation losses are shown in Fig. 5.

Fundamental evaluation metrics including precision, recall, and F1 score are frequently employed in DL research, especially for jobs where precise categorization is essential, such as paddy leaf disease detection. Precision gauges the capacity of the model to prevent false positives by calculating the percentage of true positive predictions—that is, correctly detected disease cases—among all positive predictions. The high precision of the model means that it can accurately detect the presence of a disease without incorrectly classifying healthy leaves as infected. Recall, sometimes referred to as sensitivity, quantifies the model's capacity to identify every instance of a disease by dividing the number of true positive predictions by the total number of actual positive instances. In agriculture, where early diagnosis can stop extensive crop damage, high recall makes sure that sick leaves are not missed. The F1 score is a balanced statistic that consider both false positives and false negatives. It is calculated as the harmonic mean of the precision and recall. When there is an imbalance in the dataset for example, when some diseases are more prevalent than others—this score becomes even more significant. In paddy leaf disease detection, these metrics guarantee that the model not only meets the unique requirements of agricultural applications, where false positives (unnecessary treatment) can result in resource waste and false negatives (missed diseases) can cause large crop losses. All of these criteria work together to provide a thorough assessment framework that helps researchers improve and optimize their models for accurate and dependable illness detection in practical settings. The metrics for the five classes in proposed system are listed in table below. This study provides a thorough insight into how well the system performs across a number of evaluation criteria and is a crucial indicator of the system's efficacy. The

Fig. 5 Our model's training and validation losses quantify the amount of inaccuracy the model produces

Table 2 Precision, recall, and f1 score—all computed for five distinct classes—showcase the effectiveness of our proposed approach

Class	Precision	Recall	F1 score
Bacterial leaf blight	0.93	0.91	0.92
Bacterial leaf streak	0.92	0.94	0.93
Bacterial panicle blight	0.86	0.83	0.85
Downy mildew	0.92	0.81	0.86
Normal	0.80	0.96	0.87

performance matrix for each of the five classes is presentedin Table 2.

4.2 Showcase a Confusion Matrix to Calculate the Actual and Expected Value

Behind DL experiments, a confusion matrix (CM) is a crucial tool because it thoroughly assesses the model's performance across several classes. It provides useful data regarding the precision of forecasts for every class, including false positives, false negatives, true positives, and true negatives. This aids in the understanding of the strengths and weaknesses of the model. It also makes it possible to evaluate the model's performance in detail for different classes. In order to ensure the accuracy and reliability of DL models in classification tasks, these insights are essential for their improvement and optimization. We provide a confusion matrix that shows the performance of the algorithm for five different classes. The classification results are shown in the figure below, which demonstrates how well the model can distinguish and diagnose paddy disease across five different categories. In Fig. 6, the confusion matrix for the five classes is displayed.

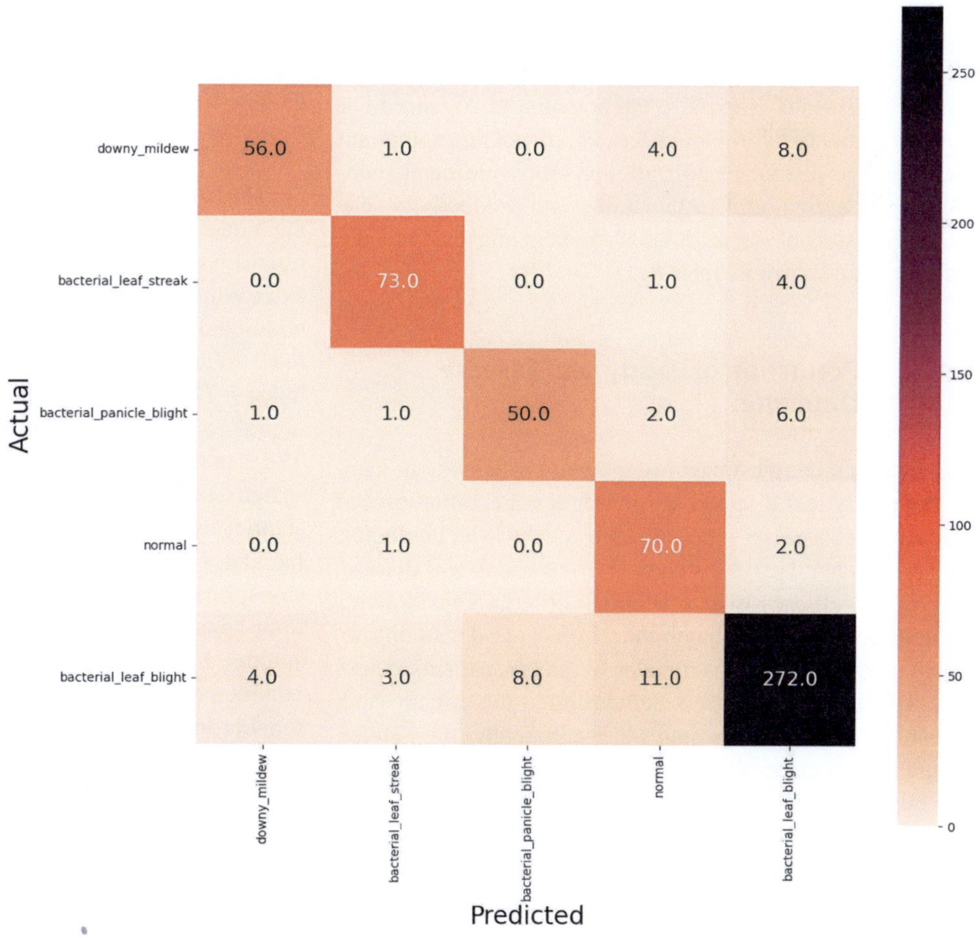

Fig. 6 The CM measures paddy leaf diseases by displaying the expected and actual merit for five distinct class photos

Table 3 Implemented models training and validation to measure the effectiveness of different models in detecting paddy leaf disease

Model	Training accuracy (%)	Validation accuracy (%)
VGG16	94.22	92.17
ResNet50	96.25	93.26
DenseNet121	97.11	95.16
MobileNetV2	94.55	92.26
CNN-ResNet50-BiLSTM (Proposed)	98.67	99.93

4.3 Performance Demonstration of the Different Models We Assessed

A range of DL models has been used extensively in our investigation to discover the optimal methods for detecting paddy leaf disease. By outline the table, the results demonstrate the validation and training accuracy for each employed model. Our hybrid CNN-ResNet50-BiLSTM model demonstrated exceptional performance by outperforming all the other models in terms of accuracy. The integrated model demonstrated superior precision in detecting paddy leaf disease, highlighting its effectiveness in agricultural disease detection. The combination of CNNs, ResNet50 architecture, and BiLSTM networks plays a significant role in its exceptional performance. The results emphasize the potential of advanced DL methods, namely hybrid architectures, in tackling important issues such as disease identification in crops, ultimately contributing to agricultural sustainability and productivity. The accuracy results of the model that evedient suggested model efficiency are shown in Table 3.

4.4 Prediction of Paddy Leaf Disease Detection

Our approach demonstrated exceptional proficiency in identifying paddy leaf diseases, exhibiting a remarkable level of efficacy that surpasses other currently available technologies. We able to identify five different types of paddy leaf disease, proving the efficacy of our detection criteria. Our system's amazing performance is highlighted by its great detection rate, which places it as a top contender for paddy leaf ease detection. The accuracy numbers demonstrated the effectiveness of our technology, surpassing other competing alternatives. Figure 7 shows a detailed picture of our accurate forecasting abilities, demonstrating the expected categories of the diseases affecting paddy leaves. The anticipated labels accurately represented the range of the illnesses. This comprehensive study emphasizes the capacity of the system to differentiate between different diseases and its dependability to carry accurate and actionable data for efficient agricultural management.

4.5 Improvement and Findings of our System

This work presents the development of an automated system designed to identify diseases in paddy leaves. The system aims to solve the crucial requirements of keep healthy paddy leaves for optimal rice production. In order to preserve agricultural sustainability and provide food security, paddy is a vital crop. We examined five prevalent diseases affecting paddy leaves. These diseases have the potential to propagate quickly, resulting in substantial reductions in agricultural yield. To tackle this issue, we created the Hybrid CNN-ResNet50-BiLSTM, a complex deep learning model. This model utilizes the capabilities of CNN, ResNet50, and Bidirectional BiLSTM networks to detect and categorize paddy leaf illnesses precisely. By utilizing these sophisticated neural network designs, our system can efficiently capture the disease symptoms' spatial and temporal characteristics. The training results showed a remarkable accuracy of 98.67%, while the validation accuracy achieved an astonishing 99.93%. These results demonstrate the system's excellent effectiveness in detecting illnesses in paddy leaves. Aside from assessing accuracy, We also looked at precision, recall, and F1 score to assess our model's performance. These measures are essential for assessing the model's resilience and dependability. A confusion matrix, which provides a thorough analysis of the actual classifications in comparison to the predicted classifications, was also used to assess the model's performance. This study facilitates the identification of particular domains in which the model demonstrates extraordinary performance and regions that may necessitate additional enhancement. The performance metrics substantiate the accuracy of the detection findings from our technology. The results of this investigation emphasize our technique's efficacy compared to currently available solutions. The accuracy of our automated paddy leaf disease detection system exceeds that of prior technologies and offers substantial advancements in the agriculture sector. Our approach facilitates the early and accurate identification of leaf diseases, enabling farmers to address crop damage promptly, improving rice output and contributing to food security. In summary, our hybrid CNN-ResNet50-BiLSTM model is a notable breakthrough in agricultural technology, providing a reliable and efficient technique for identifying paddy leaf diseases. This breakthrough can transform disease management procedures in rice farming, resulting in improved crop health and increased yields.

Fig. 7 Five classes of disease detection have been predicted, demonstrating the projected class level of paddy leaf disease

4.6 Comaprison of Existing and Proposed Model

We carefully examined some conventional models from existing research and our innovative CNN-ResNet59-BiLSTM architecture in our comparative evaluation. Although the established models yielded satisfactory results, our comprehensive analysis unequivocally showcases the remarkable superiority of our proposed approach. The accompanying table illustrates this discernible disparity, underscoring the exceptional effectiveness of our method compared to existing models. Our novel architecture incorporated CNNs, ResNet50, and BiLSTM networks, leveraging their collective strengths to achieve unprecedented performance. Through this thorough examination, we affirm our model's significant advancement in disease detection, particularly in agricultural contexts such as identifying paddy leaf disease. These results high point of developing the novel DL architectures to transform disease detection techniques and improve the productivity and sustainability of agriculture. A comparison of the various models and the suggested model is shown in Table 4.

4.7 Advantages of Paddy Leaf Disease Detection

Detecting diseases in paddy leaves provides several benefits that greatly influence agricultural production and sustainability. Timely identification of disease in paddy field is essential for efficient crop administration and can

Table 4 Analyze the current models utilized in the study of paddy leaf disease and show how the proposed model outperforms them

References	Model	Accuracy (%)
[30]	KNN	97
[31]	CNN	70
[32]	CNN	90
[33]	K-Nearest	65
[34]	YOLOv3	79.19
Proposed	CNN-ResNet50-BiLSTM	99.93

avert substantial reductions in production. One of the main advantage is the ability to produce and perform accurate interventions. Through early disease detection, farmers can administer suitable interventions such as pesticides or biological agents, thereby reducing disease transmission and protecting the health of crops. Moreover, precise disease identification leads to decreased reliance on pharmaceutical substances. Conventional approaches frequently employ the indiscriminate application of pesticides, which can result in environmental contamination and the emergence of insect strains resistant to these chemicals. Accurate disease diagnosis enables the treatment of only affected regions, resulting in more focused and eco-friendly pest management methods. This level of precision minimizes the total expenditure on crop care by reducing the quantity of pesticides required and the effort required for their application. Another notable benefit is the improvement in crop quality and output. Robust plants are more prone to quit superior-grade grains, consequently augmenting the market worth of the agricultural supply. Disease detection systems, particularly modern technologies such as remote sensing and DL, offer practical and valuable information to farmers. These observations can enhance crop management decisions by increasing the efficiency with which labor, fertilizer, and water are used. Furthermore, the implementation of early and precise detection technologies can significantly contribute to adopt sustainable farming practices. These technologies guarantee long-term soil fertility and environmental balance by preserving the health in paddy fields. This sustainable method not only enhances food security but also fosters the well-being of farming communities. The benefits of detecting paddy leaf diseases include economic, environmental, and social aspects, making them a crucial element of contemporary agriculture.

5 Conclusion

Plants often show signs of disease on their leaves, which makes them a prominent site of disease. Various diseases exhibit unique effects on leaves, setting them separately. Rice plants constitute the majority of the world's food; approximately 50% of people consider rice plants to be extremely important. Paddy plant infections significantly affect the yield and quality of rice. Paddy diseases are estimated to cause a yearly reduction of 20–40% in crop production. The manual identification of these diseases necessitates the understanding of farmers. This requires labor-intensive efforts to physically inspect large agricultural areas with individual rice crops for early detection. Our technology guarantees the automated identification of paddy leaf disease. These findings are accurate, and demonstrate the validity of the proposed method. The proposed technology monitors and predicts future trajectories of the agricultural sector.

References

1. Ahmed, K., Shahidi, T.R., Alam, S.M.I., Momen, S.: Rice leaf disease detection using machine learning techniques. In: 2019 International Conference on Sustainable Technologies for Industry 4.0 (STI), pp. 1–5. IEEE (2019)
2. Devi, T.G., Srinivasan, A., Sudha, S., Narasimhan, D.: Web enabled paddy disease detection using compressed sensing. Math. Biosci. Eng. **16**(6), 7719–7733 (2019)
3. Bashir, K., Rehman, M., Bari, M.: Detection and classification of rice diseases: an automated approach using textural features. Mehran Univ. Res. J. Eng. Technol. **38**(1), 239–250 (2019)
4. Phadikar, S., Sil, J.: Rice disease identification using pattern recognition techniques. In: 2008 11th International Conference on Computer and Information Technology, pp. 420–423. IEEE (2008)
5. Mishra, B., Kumar, N., Mukhtar, M.S.: Systems biology and machine learning in plant–pathogen interactions. Mol. Plant-Microbe Interact. **32**(1), 45–55 (2019)
6. Grinblat, G.L., Uzal, L.C., Larese, M.G., Granitto, P.M.: Deep learning for plant identification using vein morphological patterns. Comput. Electron. Agric. **127**, 418–424 (2016)
7. Hesami, M., Alizadeh, M., Jones, A.M.P., Torkamaneh, D.: Machine learning: its challenges and opportunities in plant system biology. Appl. Microbiol. Biotechnol. **106**(9–10), 3507–3530 (2022)
8. Singh, A., Ganapathysubramanian, B., Singh, A.K., Sarkar, S.: Machine learning for high-throughput stress phenotyping in plants. Trends Plant Sci. **21**(2), 110–124 (2016)
9. Swapno, S.M.R., Nobel, S.N., Ramachandra, A.C., Islam, M.B., Haque, R., Rahman, M.M.: Traffic light control using reinforcement learning. In: 2024 International Conference on Integrated Circuits and Communication Systems (ICICACS), pp. 1–7. IEEE (2024)
10. Swapno, S.M.R., Chhabra, G., Kaushik, K., Nobel, S.N., Islam, M.B., Shahiduzzaman, M.: An adaptive traffic signal management system incorporating reinforcement learning. In: 2023 Annual International Conference on Emerging Research Areas: international Conference on Intelligent Systems (AICERA/ICIS), pp. 1–6. IEEE (2023)
11. Nobel, S.N., Swapno, S.M.R., Shajeeb, H.H., Islam, M.B., Haque, R.: Hybrid CNN LSTM Approach for sentiment analysis of bengali language comment on facebook. In: 2024 International Conference on Integrated Circuits and Communication Systems (ICICACS), pp. 1–8. IEEE (2024)
12. Nobel, S.N., Swapno, S.M.R., Islam, M.R., Safran, M., Alfarhood, S., Mridha, M.F.: A machine learning approach for vocal fold segmentation and disorder classification based on ensemble method. Sci. Rep. **14**(1), 14435 (2024)

13. Swapno, S.M.R., Nobel, S.N., Islam, M.B., Haque, R., Meena, V.P., Benedetto, F.: A novel machine learning approach for fast and efficient detection of mango leaf diseases. In: 2024 IEEE 3rd International Conference on Computing and Machine Intelligence (ICMI), pp. 1–7. IEEE (2024)
14. Nobel, S.N., Swapno, S.M.R., Islam, M.B., Meena, V.P., Benedetto, F.: Performance improvements of machine learning-based crime prediction, a case study in Bangladesh. In: 2024 IEEE 3rd International Conference on Computing and Machine Intelligence (ICMI), pp. 1–7. IEEE (2024)
15. Swapno, S.M., Nobel, S.M., Meena, P., Meena, V.P., Azar, A.T., Haider, Z., Tounsi, M.: A reinforcement learning approach for reducing traffic congestion using deep Q learning. Sci. Rep. **14**(1), 1–20 (2024)
16. Nobel, S.N., Swapno, S.M.R., Islam, M.B., Azad, A.K.M., Alyami, S.A., Alamin, M., Liò, P., Moni, M.A.: A novel mixed convolution transformer model for the fast and accurate diagnosis of glioma subtypes. Adv. Intell. Syst. 2400566 (2024)
17. Bappi, B.R., Swapno, S.M.R., Chhabra, G., Kaushik, K., Nobel, S.N., Islam, M.B.: Deep learning based tuberculosis and pneumonia disease detection using CNN. In: 2023 Seventh International Conference on Image Information Processing (ICIIP), pp. 670–676. IEEE (2023)
18. Bappi, M.B.R., Swapno, S.M.R., Rabbi, M.F.: Skin cancer disease detection using MCD-GRU: a deep learning approach. In: 2024 6th International Conference on Electrical Engineering and Information & Communication Technology (ICEEICT), pp. 445–450. IEEE (2024)
19. Rahman, S., Haque, R., Swapno, S.M.R., Islam, M.B., Nobel, S.N., et al.: Deep learning-based left ventricular ejection fraction estimation from echocardiographic videos. In: 2023 International Conference on Evolutionary Algorithms and Soft Computing Techniques (EASCT), pp. 1–6. IEEE (2023)
20. Bappi, M.B.R., Swapno, S.M.R., Akhter, S., Rabbi, M.F.: Deploying hybrid VGG19-BiGRU model for kidney disease segmentation. In: Intelligent Systems Conference, pp. 47–61. Springer (2024)
21. Bappi, M.B.R., Swapno, S.M.R., Rabbi, M.F.: Deploying densenet for cotton leaf disease detection on deep learning. In: International Conference on Trends in Electronics and Health Informatics, pp. 485–498. Springer (2023)
22. Imran, M.A., Swapno, S.M.R., Chhabra, G., KaushiK, K., Mahi, M.A.A., Islam, M.B., Haque, R.: IoT-enabled smart manhole management system for real-time status, water level, and gas detection. In: 2024 International Conference on Intelligent Systems for Cybersecurity (ISCS), pp. 01–07. IEEE (2024)
23. Patidar, S., Pandey, A., Shirish, B.A., Sriram, A.: Rice plant disease detection and classification using deep residual learning. In: Machine Learning, Image Processing, Network Security and Data Sciences, volume 1240 of Communications in Computer and Information Science. Springer, Singapore (2020)
24. Rumy, S.S.H., Hossain, M.I.A., Jahan, F., Tanvin, T.: An IoT based system with edge intelligence for rice leaf disease detection using machine learning. In: 2021 IEEE International IOT, Electronics and Mechatronics Conference (IEMTRONICS), pp. 1–6 (2021)
25. Sagarika, G.K., Prasad, S.K., Kumar, S.M.: Paddy plant disease classification and prediction using convolutional neural network. In: 2020 International Conference on Recent Trends on Electronics, Information, Communication & Technology (RTEICT), pp. 208–214 (2020)
26. Acharya, A., Muvvala, A., Gawali, S., Dhopavkar, R., Kadam, R., Harsola, A.: Plant disease detection for paddy crop using ensemble of CNNS. In: 2020 IEEE International Conference for Innovation in Technology (INOCON), pp. 1–6 (2020)
27. Burhan, S.A., Minhas, S., Tariq, A., Hassan, M.N.: Comparative study of deep learning algorithms for disease and pest detection in rice crops. In: 2020 12th International Conference on Electronics, Computers and Artificial Intelligence (ECAI), pp. 1–5 (2020)
28. Pothen, M.E., Pai, M.L.: Detection of rice leaf diseases using image processing. In: 2020 Fourth International Conference on Computing Methodologies and Communication (ICCMC), pp. 424–430 (2020)
29. Mashroor, F., Ishrak, I.F., Alvee, S.M., Jahan, A., Suvon, M.N.I., Siddique, S.: Rice paddy disease detection and disease affected area segmentation using convolutional neural networks. In: TENCON 2021–2021 IEEE Region 10 Conference (TENCON), pp. 891–896 (2021)
30. Ahmed, K., Shahidi, T.R., Alam, S.M.I., Momen, S.: Rice leaf disease detection using machine learning techniques. In: 2019 International Conference on Sustainable Technologies for Industry 4.0 (STI), pp. 1–5 (2019)
31. Swathika, R., Srinidhi, S., Radha, N., Sowmya, K.: Disease identification in paddy leaves using cnn based deep learning. In: 2021 Third International Conference on Intelligent Communication Technologies and Virtual Mobile Networks (ICICV), pp. 1004–1008 (2021)
32. Cherukuri, N., Kumar, G.R., Gandhi, O., Thotakura, V.S.K., NagaMani, D., Basha, C.Z.: Automated classification of rice leaf disease using deep learning approach. In: 2021 5th International Conference on Electronics, Communication and Aerospace Technology (ICECA), pp. 1206–1210 (2021)
33. Bharanidharan, N., Chakravarthy, S.S., Rajaguru, H., Kumar, V.V., Mahesh, T.R., Guluwadi, S.: Multiclass paddy disease detection using filter-based feature transformation technique. IEEE Access **11**, 109477–109487 (2023)
34. Kiratiratanapruk, K., Temniranrat, P., Kitvimonrat, A., Sinthupinyo, W., Patarapuwadol, S.: Using deep learning techniques to detect rice diseases from images of rice fields. In: Fujita, H., Fournier-Viger, P., Ali, M., Sasaki, J. (eds.) Trends in Artificial Intelligence Theory and Applications, vol. 12144. Springer, Cham (2020)

Advancing Agriculture Through Irrigation Automation: Smart Management for Sustainable Farming

Md. Khaled, Kazi Atik Foysal, Saida Binta Alam, Md. Eshmam Rayed, Jamin Rahman Jim, and Md. Zakir Hossain Zamil

Abstract

Efficient water management in agriculture is increasingly critical due to the growing scarcity of water resources and the impact of climate change. Traditional irrigation methods, characterized by uneven water distribution and high labor costs, are insufficient to meet these challenges. Smart irrigation systems, leveraging automation technologies, present a viable solution to optimize water usage, enhance crop yields, and reduce environmental impact. This paper explores the role of advanced technologies such as the Internet of Things (IoT), Artificial Intelligence (AI), and machine learning in revolutionizing irrigation practices. Through sensor-based monitoring and data-driven decision-making, these technologies enable precision irrigation, ensuring optimal water delivery to crops. The study highlights the economic, environmental, and social benefits of irrigation automation, including increased efficiency, reduced water usage, and improved profitability. Moreover, the paper explores the obstacles and potential dangers related to adopting these technologies, underscoring the necessity for sustained research and advancement. The findings underscore the transformative potential of smart irrigation systems in achieving sustainable agricultural practices.

Keywords

Smart agriculture • Irrigation automation • Internet of things • Sustainable agriculture • Machine learning

1 Introduction

Agriculture, a critical sector for global food security, faces challenges such as inefficient water usage and climate change, which are exacerbated by traditional irrigation methods. Smart agriculture, incorporating technologies like IoT, AI, and machine learning, offers solutions to these issues. Kabir et al. [1] demonstrated the potential of AI in plant disease diagnosis, while Prodeep et al. [2] applied deep CNNs for identifying diseases from leaf images. Islam et al. [3] developed an IoT-enabled soil nutrient monitoring and crop recommendation system, showcasing data-driven resource optimization approaches. Additionally, Nobel et al. [4] emphasize the importance of precise monitoring and management in improving crop health and yields. While irrigation automation offers advantages like improved efficiency and lower labor costs, it still faces challenges, including significant upfront expenses and technical difficulties. Thus, continued research and innovation are crucial to addressing these issues and encouraging the broader implementation of smart irrigation systems for sustainable farming.

Irrigation plays a vital role in supporting agricultural crops, preserving landscapes, and restoring degraded soils, especially in arid regions and during times of low rainfall. It entails the controlled application of water to the soil using various systems, including pipes, pumps, and sprinklers.

Md. Khaled · K. A. Foysal
Department of CSE, Bangladesh University of Business and Technology, Mirpur, Dhaka, Bangladesh

S. B. Alam (✉) · Md. E. Rayed · J. R. Jim
Department of Computer Science, American International University-Bangladesh, Dhaka, Bangladesh
e-mail: saidabintaalam@gmail.com

Md. Z. H. Zamil
School of Computer Sciences, Western Illinois University, Macomb, Illinois, USA

M. F. Mridha and N. Dey (eds.), *Machine Vision in Plant Leaf Disease Detection for Sustainable Agriculture*, Studies in Computational Intelligence 1202, https://doi.org/10.1007/978-981-96-4520-6_12

However, conventional watering techniques encounter several challenges, including uneven water distribution, pooling in uneven terrain, and inefficiency due to reliance on manual and human labor. Moreover, these methods often require significant investment in initial costs, resulting in environmental challenges like excessive water consumption and pollution.

Adopting smart management practices in agriculture and leveraging automation technologies presents an opportunity to overcome these challenges and enhance agricultural sustainability. By incorporating automation into irrigation processes, it becomes possible to optimize water usage, improve efficiency, and reduce negative environmental impacts. Smart irrigation offers significant potential for transforming agriculture with its sensor-based monitoring, data-driven decision-making, and precision irrigation techniques.

Irrigation is essential for agricultural production, especially in areas of limited water availability. According to the Food and Agriculture Organization of the United Nations (FAO), irrigation accounts for 70 of all water used for human activities and is responsible for 40 of the world's food supply [5]. Irrigation increases crop yields by providing plants with the water and nutrients they need to grow. The use of irrigation has been linked to several positive outcomes in agriculture, including higher crop yields, better soil fertility, and improved crop quality. The use of irrigation systems can lower the chances of crop loss caused by drought or flooding and enable farmers to cultivate a wider range of crops. In addition to its positive impacts on agriculture, irrigation can also have many negative environmental effects. These include waterlogging and salinization of soils, increased water consumption, and the pollution of rivers, lakes, and groundwater sources. For these reasons, it is essential for farmers to practice water conservation, and use efficient irrigation techniques [6].

Farmers have been distributing water equitably among themselves for years using a proportionate division mechanism. These structures can be unreliable from a hydraulic standpoint, and it was calculated that their performance criteria depend not only on the notch width but also on other hydraulic design parameters [7]. Drip irrigation works well for applying irrigation water directly to the soil around plant roots. It lessens frequent losses such as runoff, deep percolation, and soil erosion. The drip irrigation technique is thought to provide the solution to a number of problems in dry-land agriculture [8, 9]. The initial investment for drip irrigation systems is relatively high. Additionally, maintaining and replacing equipment requires ongoing capital expenditures. The emitters used in drip irrigation are also susceptible to clogging and failure. The amount of water that evaporates from the spray nozzle to the crop canopy is called ACSEL. The irrigation systems with a traveling gun and a center pivot produced ACSEL, the most consistent across all nozzles. Whether or if water droplets moved into the wind or with the wind affected the ACSEL [10]. Sprinklers should not be used on soils that readily produce a crust. Light, fine sprays should be utilized if sprinkler watering is the sole option. Avoid using bigger sprinklers that produce larger water droplets.

This paper aims to provide an organized overview of the role of irrigation automation in smart agriculture management. It explores the benefits and challenges of irrigation automation, emphasizing the potential for addressing the limitations of traditional methods and promoting sustainable practices. The economic, environmental, and social advantages of irrigation automation include increased crop yield, water savings, reduced labor costs, improved profitability, and positive environmental impacts such as reduced water waste, minimized nutrient leaching, and mitigated soil erosion.

Advanced technologies like the Internet of Things (IoT), Artificial Intelligence (AI), and machine learning algorithms are also highlighted for their role in irrigation automation. These technologies enable the integration of sensors, weather data, and predictive analytics to optimize irrigation scheduling, monitor soil moisture levels, and minimize water consumption. However, along with the benefits, there are challenges associated with irrigation automation, such as high initial costs, technological complexity, and concerns regarding data privacy.

The key contributions of this research are summarized as follows:

- Comprehensive analysis of irrigation automation, including the challenges of traditional methods, the potential benefits of automation, and the integration of advanced technologies.
- Evaluation of irrigation automation's economic, environmental, and social impacts, highlighting increased crop yield, reduced water consumption, improved profitability, and positive environmental outcomes.
- Identification of challenges and recommendations for successful implementation, addressing issues such as high costs, technological complexity, and data privacy concerns while emphasizing robust systems, capacity building, and knowledge dissemination.

2 Releted Work

The author accurately predicted that varying watering methods would lead to higher growth depending on the fertility and moisture level of the soil. This system's motor and solenoid valves are controlled by sensors that measure the water level and soil moisture. Water delivery to the field is facilitated by this. An MQTT server monitors and controls the entire system through the [11] internet.

Irrigated dairy farming is the main agricultural activity in the Shepparton Irrigation Region of Northern Victoria. High-quality irrigation is necessary for increased pasture quality and quantity. IoT-based smart farming systems may prove to be very useful to agriculturalists because fewer water systems are worthless for production. Based on the regional ecology of that area, edge values for climatic factors like stickiness, warmth, and wetness may be computed. Using data from the climate store and field-observed constant information, this framework develops a water system [12, 13] plan.

A group of academics has built a model of an autonomous irrigation system that uses solar energy as its only electrical source. The water level is continuously monitored by sensors, which send the information to the farmer's cell phone. The farmer uses a cellphone to control the engine from outside the paddy [14] field. Zig-bee and internet technologies were used to create a concept that allows farms to become self-sufficient in terms of water usage. Farms have sensors installed that continually perceive and gather data. The data is kept at the central monitor, transferred to the interface for data collection, and then [15] transmitted to the wireless sensor node. demonstrates a cheap, intelligent irrigation system for agriculture. In today's farming, automated, remote monitoring, and management systems that are connected to cloud, fog, and IoT networks are [16] used. This paper presents an innovative approach for automated water management and a pest detection platform, designed to regulate water levels and irrigate crops in agricultural fields efficiently. The proposed algorithm leverages features extracted from plant leaves, as outlined in [17]. In Bangladesh and India, the majority of irrigation systems still rely on manual operation, although automated and semi-automated technologies are gradually replacing traditional farming methods. There is a pressing need for systems capable of interpreting agricultural data intelligently and analyzing it to predict outcomes effectively. Researchers, as noted in [18], have developed a system that utilizes wireless sensor network technology to optimize water use for crops. The automation process is triggered by data collected from three soil moisture sensors, which activate or deactivate the solenoid valve for irrigation. The system has been shown to function effectively, with real-time data transmission from the field [19]. The implementation of automatic control systems in irrigation has significantly enhanced irrigation efficiency and the sustainable use of water resources. According to the World Bank, these systems can boost the productivity and water efficiency of rice paddies [20].

3 System Architecture

The smart management system for agriculture using irrigation automation is composed of various components that work together to optimize irrigation practices and promote sustainable agriculture. This section provides a comprehensive overview of the proposed system, including the circuit diagram Fig. 1, pin diagram Fig. 2, block diagram Fig. 3, and the execution logic Algorithm 1.

Algorithm 1 Process execution logic

$Water\ pump \leftarrow 0$
if $(no\ rain\ next\ 2\ hours)\ And(moisture < required - level)\ And\ (water - level < required - level)$ **then**
  $Water\ pump \leftarrow 1$
else
  if $(rain\ next\ 2\ hours)\ And(moisture > required - level)\ And\ (water - level > required - level)$ **then**
    $Water\ pump \leftarrow 0$
  end if
end if

3.1 Technical Details

The technical details of irrigation automation in agriculture involve the use of various sensors, data analysis techniques, and automation systems. The following are some of the technical details involved in the implementation of irrigation automation:

- Sensors: There are different types of sensors used in irrigation automation, including soil moisture sensors, weather sensors, and plant sensors. Soil moisture sensors can be installed at different depths in the soil to measure moisture levels, while weather sensors can measure temperature, humidity, and solar radiation. Plant sensors, such as leaf sensors, can measure the water content of plants to help determine their water needs.
- Data analysis: The sensor data is processed and evaluated to identify the precise amount of water required for the crops. This data analysis may involve using algorithms and machine learning techniques to predict crop water needs based on various factors, including soil moisture, weather conditions, and crop type.
- Automation systems: Automated irrigation systems can be used to deliver water to the crops based on the data analysis. These systems may include drip irrigation, center pivot irrigation, or micro-irrigation. Automated irrigation systems may be controlled by a central computer or through mobile devices, allowing farmers to adjust irrigation settings remotely.
- Communication systems: Communication systems, such as wireless networks, can be used to transmit data from the sensors to the central computer or mobile device used to control the irrigation system.
- Power supply: The operation of irrigation automation systems relies heavily on a consistent power supply. Solar-

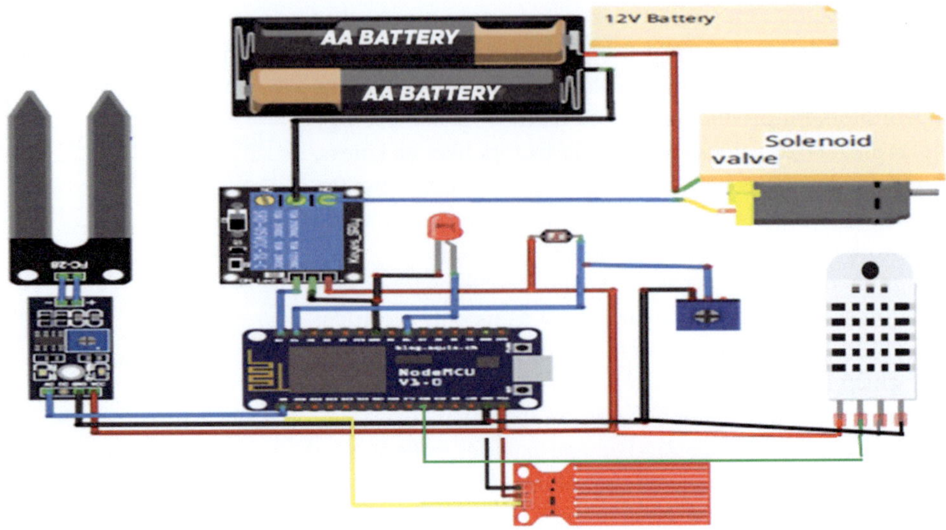

Fig. 1 Circuit diagram of the smart agriculture management system

powered systems can be used in areas with limited access to electricity.

- Maintenance: Proper maintenance is critical for the effective operation of irrigation automation systems. This includes regular calibration of sensors, inspection of irrigation equipment, and software updates to ensure that the system is working correctly.

These technical details are essential for the successful implementation of irrigation automation in agriculture. By integrating these components, farmers can optimize their irrigation practices and improve the sustainability and profitability of their farms.

3.1.1 Connectivity

Take a breadboard, NodeMCU, DHT22 temperature and Humidity sensor, Soil Moisture sensor, Water lavel sensor, LDR sensor, 3v water pump, 9volt battery and jumper wires. Connect the Node MCU to the breadboard at b16–i16 × b30–i30 Fig. 2.

Power Supply Connections

- Connect the 3.3 V pin to the positive rail (15) on the breadboard.
- Connect the GND pin to the negative rail (13) on the breadboard.

DHT11 Sensor Connections

- Connect the positive pin of the DHT11 sensor to the positive rail (7) of the breadboard.
- Connect the output pin of the DHT11 to the D4 pin of the NodeMCU.
- Connect the negative pin of the DHT11 sensor to the negative rail (7) of the breadboard.

Relay Module Connections

- Connect the VCC pin of the relay module to the Vin pin of the NodeMCU.
- Connect the GND pin of the relay module to the GND pin of the NodeMCU.
- Connect the IN pin of the relay module to the D0 pin of the NodeMCU.
- Connect the screw terminal of the relay to the positive terminal of the battery, and then to the solenoid valve. Ensure the negative terminal of the solenoid valve is connected back to the negative terminal of the battery.

Soil Moisture Sensor Connections

- Connect the VCC pin of the soil moisture sensor to the positive rail (13) of the breadboard.
- Connect the GND pin of the soil moisture sensor to the negative rail (11) of the breadboard.
- Connect the D0 pin of the soil moisture sensor to the D3 pin of the NodeMCU.
- Connect the A0 pin of the soil moisture sensor to the D3 pin of the NodeMCU as well.

Immerse the soil moisture and water level sensor into the agricultural field or plant pot, and connect the water supply to the solenoid valve. An IoT-enabled Smart Agriculture System has been designed for real-time monitoring of soil moisture, temperature, and humidity using NodeMCU integrated with various sensors and cloud computing. This system is highly efficient and reliable, delivering precise live data on

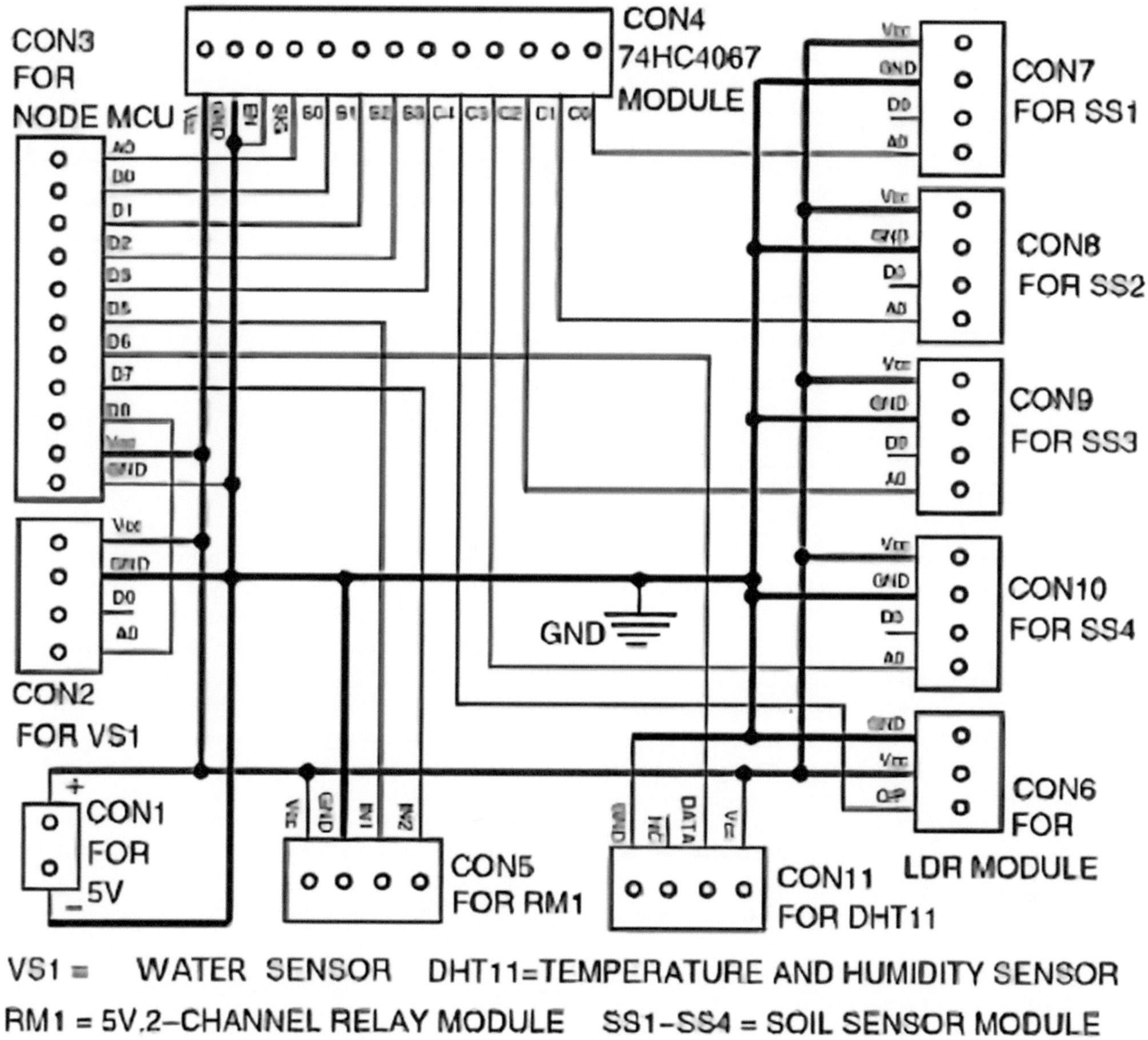

Fig. 2 Pin diagram showing the connectivity

temperature, soil moisture, water level, and light intensity (via LDR). The proposed IoT-driven smart farming system aims to support farmers in enhancing crop productivity and managing food production more effectively by providing accurate real-time environmental data, including temperature, humidity, soil moisture, water level, and light intensity, with over 99% accuracy.

3.1.2 Data Collection

The data collection process in this research was comprehensive and carefully planned to guarantee accurate tracking and regulation of irrigation. Sensors deployed at various strategic points in the field collected real-time data on essential factors including soil moisture, water levels, temperature, and humidity. The NodeMCU microcontroller played a central role in processing this data, interfacing with the relay module to manage the activation of the water pump. Data was continuously transmitted to the ThingSpeak server, an IoT platform utilized for data collection, analysis, and visualization. This setup allowed for remote monitoring and real-time adjustments based on sensor feedback. The system's automation logic ensured that the water pump was activated only when necessary, based on predefined thresholds for soil moisture and water levels, and considering weather forecasts to avoid unnecessary irrigation during impending rainfall. This data-driven approach facilitated precision irrigation, optimizing water usage, enhancing crop yields, and promoting sustainable agricultural practices. Regular calibration and mainte-

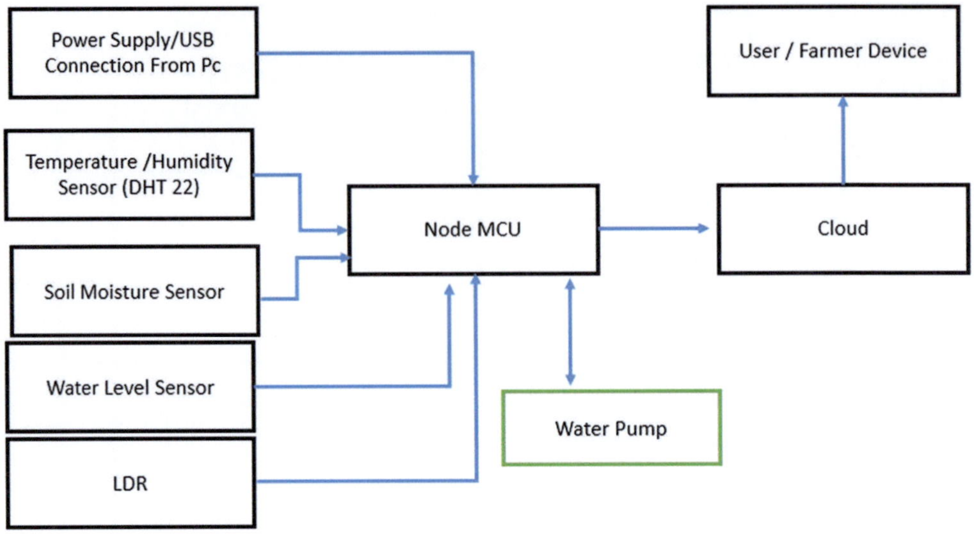

Fig. 3 Block diagram of the smart system

nance of the sensors ensured the accuracy and reliability of the collected data, which was crucial for the system's overall efficacy.

3.1.3 Used Components

- **NodeMCU**: The NodeMCU ESP8266 development board features the ESP-12E module, which includes a Tensilica Xtensa 32-bit LX106 RISC microprocessor. This processor supports RTOS and operates within an adjustable clock frequency range of 80MHz to 160MHz. Its high processing capabilities make it suitable for IoT applications.
- **DHT22 sensor**: The DHT22 is an affordable digital sensor used to measure temperature and humidity. Humidity refers to the amount of water vapor present in the air, and the level of humidity can impact various chemical and biological processes used in agriculture.
- **Relay Module** A DC relay is an electromagnetic component that electrically isolates two circuits while connecting them magnetically. It is commonly used to connect a low-voltage electronic circuit to a high-voltage electrical circuit, such as using a 5V DC battery to control a 230V AC mains circuit.
- **DC Motor**: DC motors, which are a type of rotary electrical machine, convert electrical energy into mechanical energy. These motors were among the first widely used because they could be powered by direct-current lighting power systems. The speed of a DC motor can be varied across a broad range by adjusting the supply voltage or changing the current strength in the field windings.
- **Soil Moisture**: Soil moisture sensors play a vital role in determining water levels, making it easier for farmers to manage irrigation and reduce costs. These sensors allow farmers to assess soil moisture without needing to be physically present in the field. The output voltage from the sensor correlates with the soil's moisture or dryness level.
- **Water level**: Measuring water levels in deep tanks or reservoirs helps control the water depth and prevents overflow in industrial settings. In residential applications, such as aquariums, this sensor can also be used to monitor water levels. The sensor detects the water level in the field and sends data to the NodeMCU, which helps determine when the pump should be activated or deactivated.
- **ThingSpeak**: To connect the soil moisture sensor, water level sensor, and DHT22, use the D4 pin on the NodeMCU. The 5V pin of the NodeMCU can be used to power the motor and relay. Smart farming systems use these sensors to automate irrigation and monitor agricultural fields efficiently.

4 Result and Discussion

The implementation of smart management in agriculture using irrigation automation has shown promising results in improving crop growth and reducing water consumption. In this section, we will discuss some of the results and findings related to the use of irrigation automation in agriculture.

- Improved Crop Yields: The use of irrigation automation has been shown to significantly improve crop yields in several studies. By delivering the optimal amount of water to crops, smart irrigation systems can promote healthy plant growth, leading to increased yields. Research indicates that smart irrigation systems have the potential to boost crop yields by as much as 20%, depending on the type of crop and the methods used for irrigation management.

- Water Savings: Another significant benefit of irrigation automation is the significant water savings it can offer. By utilizing sensors to track soil moisture and weather conditions, farmers can deliver the optimal amount of water to their crops, reducing water waste and promoting resource conservation. Research has demonstrated that smart irrigation systems can cut water usage by as much as 50% compared to conventional irrigation techniques.
- Cost Savings: Irrigation automation can also result in cost savings for farmers by reducing labor and energy costs. By automating the irrigation process, farmers can minimize the need for manual labor and reduce energy consumption, leading to lower operational costs and increased profitability.
- Environmental Benefits: Smart irrigation systems can also have significant environmental benefits. By minimizing water waste, farmers contribute to the conservation of natural resources and lessen the environmental footprint of agriculture. Furthermore, by enhancing crop growth and reducing the need for chemical fertilizers and pesticides, smart irrigation systems support sustainable farming practices and help safeguard the environment.
- Scalability: Smart irrigation systems are versatile and can be adapted to farms of various sizes. Whether for small family-owned farms or large commercial enterprises, all farmers can experience the advantages of implementing smart irrigation solutions.

In conclusion, the use of irrigation automation has shown promising results in improving crop yields, reducing water and operational costs, promoting sustainable agriculture, and protecting the environment. By optimizing irrigation practices and adopting smart management in agriculture, farmers can enhance their productivity, profitability, and sustainability while conserving natural resources and reducing the environmental impact of agriculture.

Correlation Matrix A correlation matrix is a table that displays the correlation coefficients between variables, as shown in Fig. 4. Each cell in the matrix represents the correlation between two variables. This matrix is used to summarize data, serve as input for more complex analyses, and act as a diagnostic tool for advanced methods. In this case, we present Fig. 4 the correlation matrix for Temperature, Humidity, Soil Moisture, and Water level.

Graph Details: In Figs. 5 and 6 Time series px line graph is being used here. The X-axis represents time as well as the motor, while the Y-axis represents soil moisture. The motor on and off represents the values O and 1. The value of soil moisture and water level depends on the motor. At this point, the motor is initially off, and we see variations in the water level and soil moisture. Once we've been using the motor for a while, we can see how the curve has evolved. In the scenario, we'll check to see if rain is forecast for the following two hours.

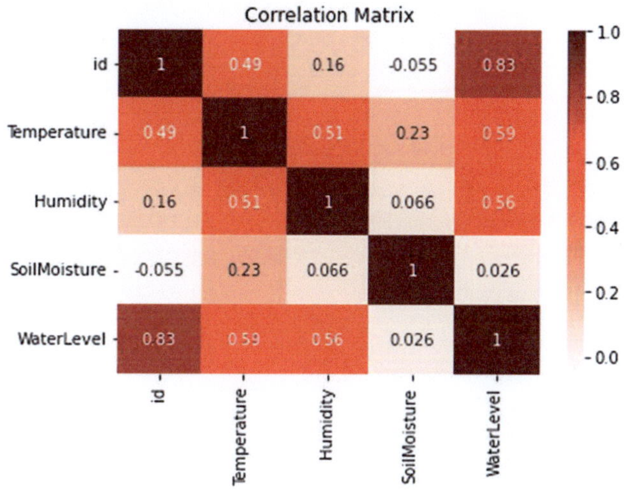

Fig. 4 Correlation matrix

In Figs. 7 and 8 Forecasting rain over the next two hours, at which point the pump would turn off. The scenario will be shown on the graph.

This study aims to achieve three key objectives: promoting sustainable growth in agricultural productivity and income, enhancing resilience to climate change, and minimizing or eliminating greenhouse gas emissions where possible. Along with increasing agricultural output, it has the potential to enhance product quality, reduce labor expenses, raise farmers' income, and support the advancement of agricultural modernization and smart farming. The paper offers an in-depth review of the latest research on the Internet of Things (IoT) in agriculture.

4.1 Test Bade Details

The water pump must be completely submerged in water, with the outlet pipe positioned in the field for irrigation. Likewise, soil moisture sensors and water level sensors are placed into the soil

Once the device is powered on, ThingSpeak will begin displaying real-time data for soil moisture, water level, LDR, humidity, and temperature. When the soil moisture and water levels drop, the water pump will activate and irrigate the field until the desired moisture level is reached. You can view the entire process in the picture.

Additionally, you can access the data remotely from anywhere in the world through the ThingSpeak server. Simply visit the private view on ThingSpeak to monitor the soil moisture, water level, LDR, humidity, temperature, and relay status.

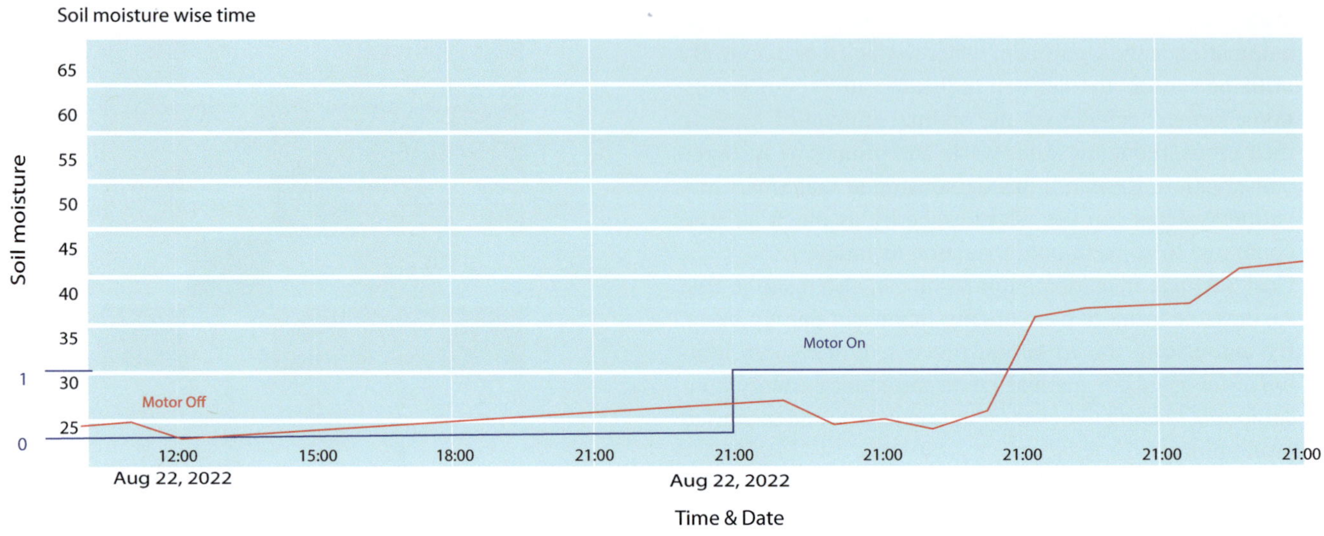

Fig. 5 Soil moisture

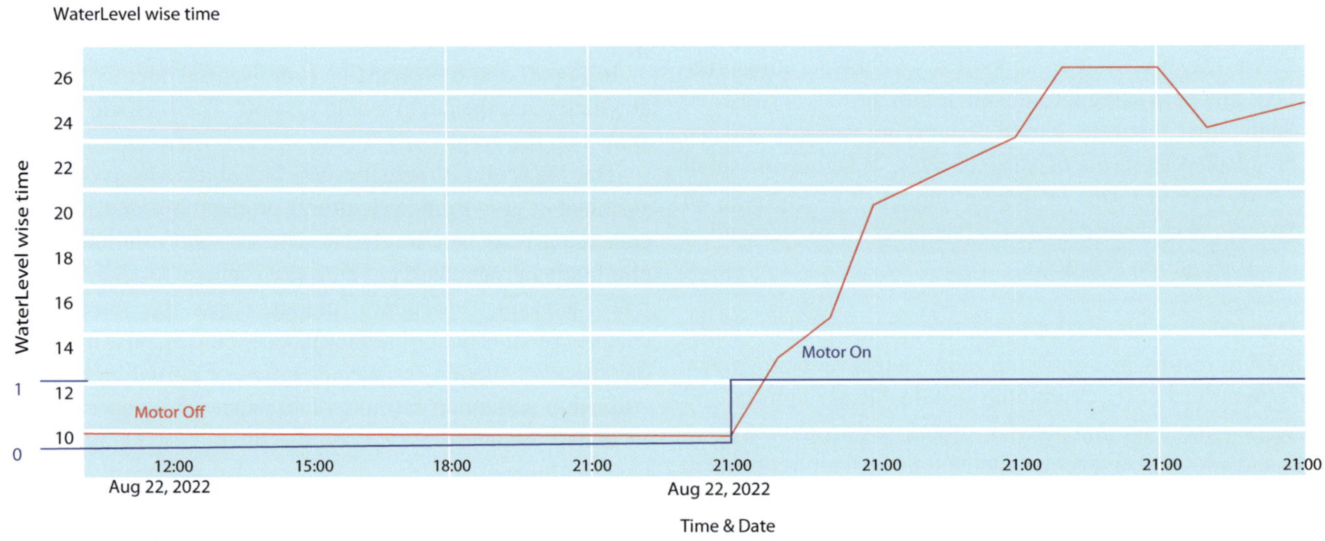

Fig. 6 Water level

5 Evaluations and Discussion

Irrigation automation has become increasingly important in modern agriculture, and the implementation of smart irrigation systems has the potential to transform traditional farming practices. By automating the irrigation process, farmers can optimize crop growth and yields, reduce water consumption and costs, and minimize labor requirements. In this section, we will discuss some of the key evaluations and discussions related to smart management in agriculture using irrigation automation.

- Water Savings: A key advantage of smart irrigation systems is their ability to significantly reduce water usage. Farmers can optimize water usage and conserve resources by using sensors to monitor soil hydration and weather conditions, ensuring that crops receive the precise amount of water needed while minimizing waste. Research indicates that smart irrigation systems can lower water consumption by as much as 50.
- Increased Yields: Beyond saving water, smart irrigation systems can enhance crop yields by optimizing irrigation practices. By supplying the right amount of water at the right time, farmers ensure that plants receive the optimal conditions for growth, which can result in higher yields.
- Reduced Costs: Smart watering systems can also lower labor and energy costs for farmers. With the automation of the irrigation process, the need for manual labor is reduced, and energy consumption is minimized, leading to decreased operational expenses and higher profitability.

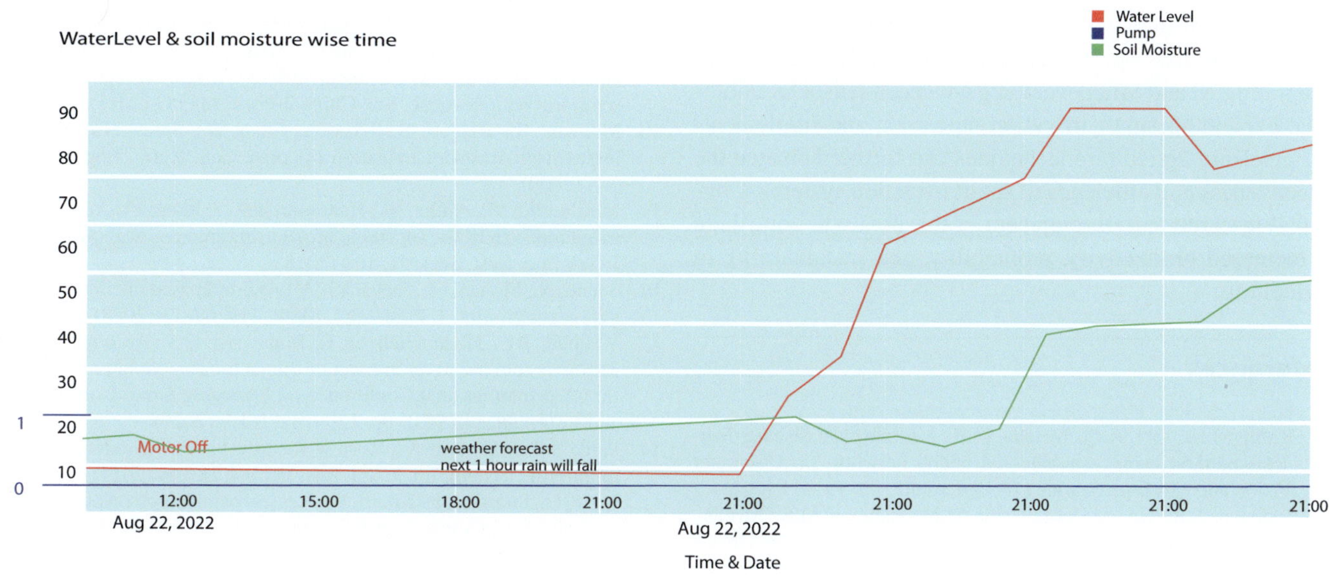

Fig. 7 Soil moisture and water level when rain fall

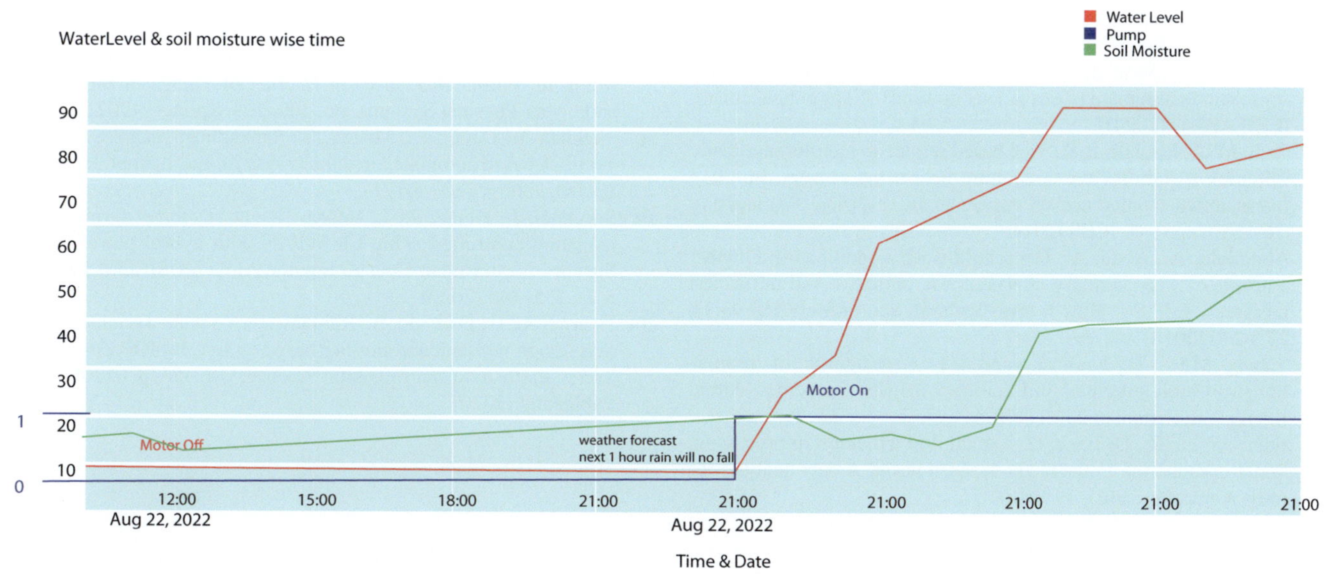

Fig. 8 Soil moisture and water level when rain no fall

- Scalability: Smart irrigation systems are highly adaptable and can be deployed on farms of all sizes. Whether on small family-owned farms or large commercial agricultural operations, farmers of all scales can benefit from the implementation of smart irrigation systems.

In conclusion, smart management in agriculture using irrigation automation has significant potential to revolutionize traditional farming practices. By optimizing irrigation practices, farmers can increase yields, reduce costs, and promote sustainable agriculture while conserving natural resources and protecting the environment.

6 Conclusion

In conclusion, the use of smart management in agriculture through irrigation automation holds great promise for revolutionizing traditional farming practices. By automating irrigation processes, farmers can optimize crop growth and yields, reduce water consumption and costs, and minimize labor requirements. Smart irrigation systems also support sustainable farming practices, conserving natural resources and reducing the use of chemicals. Future advancements in integrating AI and ML, developing advanced sensor technologies, expanding smart irrigation networks, and implementing precision agriculture techniques can further enhance the effectiveness and efficiency of smart irrigation systems. Continued research and development in this area will contribute to increased productivity, profitability, and sustainability in agriculture.

References

1. Kabir, M.M., Ohi, A.Q., Mridha, M.F.: A multi-plant disease diagnosis method using convolutional neural network. In: Computer Vision and Machine Learning in Agriculture, pp. 99–111 (2021)
2. Prodeep, A.R., Hoque, A.M., Kabir, M.M., Rahman, M.S., Mridha, M.: Plant disease identification from leaf images using deep CNN's efficientnet. In: 2022 International Conference on Decision Aid Sciences and Applications (DASA), pp. 523–527. IEEE (2022)
3. Islam, M.R., Oliullah, K., Kabir, M.M., Alom, M., Mridha, M.: Machine learning enabled iot system for soil nutrients monitoring and crop recommendation. J. Agric. Food Res. **14**, 100880 (2023)
4. Nobel, S.N., Imran, M.A., Bina, N.Z., Kabir, M.M., Safran, M., Alfarhood, S., Mridha, M.: Palm leaf health management: a hybrid approach for automated disease detection and therapy enhancement. IEEE Access (2024)
5. Rani, M.U., Kamalesh, S.: Web based service to monitor automatic irrigation system for the agriculture field using sensors. In: 2014 International Conference on Advances in Electrical Engineering (ICAEE), pp. 1–5. IEEE (2014)
6. Abdullahi, A., Abdu, A.: Household water conditions in Gombe: a profile of water scarcity in Nassarawo, palliative and mitigation measures. Am. Sci. Res. J. Eng. Technol. Sci. (ASRJETS) **43**(1), 227–240 (2018)
7. Yahya, M.H.: Hydraulical approach in designing of proportional division structures in traditional irrigation schemes. Jurnal Keteknikan Pertanian **23**(2) (2009)
8. Francaviglia, R., Di Bene, C.: Deficit drip irrigation in processing tomato production in the mediterranean basin. A data analysis for Italy. Agriculture **9**(4), 79 (2019)
9. Kharrou, M.H., Er-Raki, S., Chehbouni, A., Duchemin, B., Simonneaux, V., Le Page, M., Ouzine, L., Jarlan, L.: Water use efficiency and yield of winter wheat under different irrigation regimes in a semi-arid region. Agric. Sci. China **2**(03), 273–282 (2011)
10. McLean, R., Ranjan, S., Klassen, G., et al.: Spray evaporation losses from sprinkler irrigation systems. Can. Agric. Eng. **42**(1), 1–8 (2000)
11. Marimuthu, M., Ajitha, M., Nandhini, R.P., Scholar, U.: Automatic irrigation system for vegetable crops using internet of things. Res. J. Sci. Eng. Syst. **3**(1), 91–100 (2018)
12. Seenu, N., Mohan, M., Jeevanath, V.: Android based intelligent irrigation system. Int. J. Pure Appl. Math. **119**, 67–71 (2018)
13. Wakhare, P.B., Neduncheliyan, S., Sonawane, G.S.: Automatic irrigation system based on internet of things for crop yield prediction. In: 2020 International Conference on Emerging Smart Computing and Informatics (ESCI), pp. 129–132. IEEE (2020)
14. Patil, S.S., Malviya, A.: Review for arm based agricultural field monitoring system. Int. J. Sci. Res. Publ. **4**(2), 1–4 (2014)
15. Rane, D., Indurkar, P., Khatri, D.: Review paper based on automatic irrigation system based on RF module. Int. J. Adv. Inf. Commun. Technol. **1**(9), 736–738 (2015)
16. Akhund, T.M., Ullah, N., Newaz, N.T., Zaman, Z., Sultana, A., Barros, A., Whaiduzzaman, M.: Iot-based low-cost automated irrigation system for smart farming. In: Intelligent Sustainable Systems, pp. 83–91. Springer, Singapore (2022)
17. Kalpana, A., Chandrasekar, T., Devi, S.R., Jeaunita, T.: Automated irrigation system with pest detection using IoT with OTSU algorithm. In: Further Advances in Internet of Things in Biomedical and Cyber Physical Systems, pp. 209–217. Springer (2021)
18. Shekhar, Y., Dagur, E., Mishra, S., Sankaranarayanan, S.: Intelligent IoT based automated irrigation system. Int. J. Appl. Eng. Res. **12**(18), 7306–7320 (2017)
19. Sitompul, J., Hutauruk, J., Sitorus, R., et al.: Smart farming-drip irrigation controlled using LR-WPAN with hybrid power. Jurnal Sistem Informasi dan Ilmu Komputer Prima (JUSIKOM PRIMA) **5**(2), 135–140 (2022)
20. Noerhayati, E., Dwisulo, B., Rahmawati, A., et al.: Sprinkler irrigation design with microcontroller based on IoT. In: IOP Conference Series: earth and Environmental Science, vol. 456, p. 012063. IOP Publishing (2020)

Deep Learning Algorithms for Detecting Banana Leaf Spot Diseases

Hadiur Rahman Nabil, Md. Golam Rabbani Abir, Mst. Moushumi Khatun, Md. Eshmam Rayed, and Md. Abdul Hamid

Abstract

Banana leaf spot diseases such as Cordana, Pestalotiopsis, and Sigatoka pose significant challenges to the banana sector worldwide, limiting the supply of essential nutrients as well as the stability of the global economy. In this study, we utilize deep learning (DL) models to ensure the precise detection of these diseases in a proper way with better accuracy. Currently, detection methods for these diseases are often labor-intensive, time-consuming, and tending to inaccuracies. A significant gap exists in the development of efficient and precise detection techniques that can be readily deployed in the field. Our research emphasizes the importance of effective management techniques to ensure food security and long-term livelihood. With a subset of the Banana Leaf Spot Disease (BananaLSD) dataset, the experimental findings with YOLOv8 showed promising metrics: a precision rate of 88.9%, and a recall rate of 92%, for the YOLOv8s model. These results have the potential to significantly increase crop yields and allow for more accurate leaf spot disease detection in fields.

Keywords

Leaf spot disease • Disease detection • YOLO • Object detection • YOLOv8s

1 Introduction

Bananas, which have vibrant yellow peels and delicious sweetness, are popular worldwide. However, hidden behind this widely consumed fruit is a silent battle against a powerful enemy known as Banana Leaf Spot Disease (BLSD) [1]. This term includes various fungal infections that quietly drain the life out of banana plants, posing a threat to the quality of the fruit and the livelihoods of millions who rely on its cultivation. Its economic impact is staggering, with annual losses exceeding USD 5 billion, casting a dark cloud over food security and economic stability in vulnerable regions [2, 3].

Recognizing the urgent need to address this critical issue, our study is motivated by the potential to develop more efficient and accurate detection methods for banana leaf spots. Current detection practices are often labor-intensive, time-consuming, and prone to errors, leading to delayed responses and increased crop damage [4]. By leveraging advanced deep learning techniques, we desire to create a powerful and reliable detection system that can be easily implemented in the field. This will help farmers to identify and manage infections earlier enhance crop yields and ensure food security for communities that depend heavily on banana cultivation.

BLSD presents a grim sight on banana leaves, appearing as unattractive spots ranging from small specks to large, merging lesions. These marks disrupt the crucial process of photosynthesis, leaving the plants weak and struggling to produce the plump, vibrant bananas that we adore [5]. Black Sigatoka [6], one of the most common types of Banana Leaf Spot, can drastically reduce yields, pushing already resource-limited farmers into even greater economic hardship. This disease not only reduces their earnings but also presents a danger to the very

H. R. Nabil (✉) · Md. E. Rayed
Computer Science and Engineering, American International University Bangladesh, Dhaka, Bangladesh
e-mail: hadiurrahmannabil@gmail.com

Md. G. R. Abir · Mst. M. Khatun
Computer Science and Engineering, Bangladesh University of Business and Technology, Dhaka, Bangladesh

Md. A. Hamid
Computing & Information Technology, King Abdulaziz University, Jeddah, Saudi Arabia
e-mail: mabdulhamid1@kau.edu.sa

M. F. Mridha and N. Dey (eds.), *Machine Vision in Plant Leaf Disease Detection for Sustainable Agriculture*, Studies in Computational Intelligence 1202, https://doi.org/10.1007/978-981-96-4520-6_13

establishment of their food protection and the well-being of their communities.

Fighting BLSD requires a two-pronged approach: prevention and effective disease management. Early detection is the key to both methods. Similar to catching a cold at its earliest symptom, identifying BLSD in its initial stages allows for targeted application of fungicides, minimizing yield losses and preventing the disease from spreading rapidly through plantations. However, the traditional methods of visual inspection by farmers have limitations. Although trained, the human eye can miss subtle early signs, leading to delayed action and compromised effectiveness. Moreover, visual inspection requires skilled labor, which is often scarce in resource-limited settings, creating a gap in frontline defense against BLSD.

This is where the blessing of technology comes to work. The potential of automated BLSD detection, using the power of image analysis and machine learning, proposes an optimistic future for banana growers. Imagine farmers equipped with user-friendly apps on their smartphones can capture leaf images and, diagnose the presence and type of BLSD in the blink of an eye. Such technology could transform disease management from a reactive scramble to a proactive shield, authorizing agriculturalists to make instructed findings, optimize fungicide usage, and ultimately protect their yields.

This study demonstrates significant improvements in crop leaf detection performance, including Cordana, Pestalotiopsis, and Sigatoka, with our lightweight YOLO models. The usage of machine learning to notice these conditions in banana leaves is becoming more convenient than before. Using machine learning models such as YOLO with high accuracy it has become more efficient. To solve the gap between accuracy and efficiency, this analysis explores the domain of machine learning and agriculture, intending to produce a strong and precise BLSD detection model using image analysis. The overall contributions of this study are as follows:

- Our approach involved implementing the YOLOv8 model to detect and accurately identify different types of Banana Leaf Spot Disease (BLSD). By closely examining subtle visual cues in leaf images, the model guides specific treatment strategies, avoiding unnecessary fungicide application for less severe infections. This precision optimizes resource utilization and minimizes environmental impact, contributing to a more sustainable future in agriculture.
- We implemented the YOLOv8 model to reliably perform under various challenging field conditions, including variations in lighting, leaf maturity, and banana cultivars. This ensures that the model serves as a dependable tool for farmers worldwide, irrespective of plantation size or technical expertise. By comparing YOLOv8 with other YOLO models, we aim to demonstrate its superior resilience and effectiveness, ensuring that both large and small-scale farms can effectively utilize this technology, enhance accessibility, and promote agricultural resilience.

This research further discusses the detection of BLSD using machine learning models in depth. Section 1 introduces the potential risk of BLSD and the solution this research paper aims at Sect. 2 discusses the work that has been conducted related to this research and its significance. Section 3 explains the methods that are being used in detecting BLSD. Section 4 presents the outcome of the research. Lastly, Sect. 7 wraps up the article.

2 Releted Work

The world relies on agricultural growth. Growth is often hampered by plant diseases. Currently, banana farmers are the primary managers of banana orchards. The process of gathering bananas in the orchard also depends heavily on human labor [7]. Early detection is key to the effective management of plant diseases [8–10]. There are several ways to identify these conditions, one of which is through visual inspection. However, this approach is inaccurate and requires a significant amount of resources. Laboratory analysis is an other option for this purpose. But this method takes a lot of time, requires certain knowledge, and requires access to fully furnished laboratories [11]. Unfortunately, many farmers are unable to use these resources because they are frequently in limited availability in developing nations [12]. However, artificial intelligence (AI) presents a possible replacement. AI algorithms can be used for automated disease detection, allowing for the fast and accurate analysis of large datasets for the benefit of agriculture and crop health [13].

Artificial intelligence (AI) should be taken into consideration when it comes to utilizing deep learning models. These models are particularly useful in identifying plant diseases based on the plant's formation and graphic signs. This ability to mimic human behavior is truly remarkable [14]. Deep learning is a novel method for image processing and object detection with greater accuracy in the classification of various crop diseases [15]. Smartphones can help farmers detect plant disease in an instant. With the help of deep learning, this can be easily achieved with high accuracy [16]. Transfer learning is a widely used technique in deep learning, where pretrained models are modified to accomplish updated activities. Deep transfer learning (DTL) introduces a new architecture for analyzing qualitative images and making predictive analyses, offering improved precision and showing great potential in detecting crop diseases. The DTL approach also provides a promising avenue for recognizing diseases in the field, using large datasets of trained images, and offers a convenient solution for overcoming the limitations posed by mobile applications [17]. Suganya Devi et al. introduced a strategy that aims to recognize diseases in groundnut leaves. To accomplish this, they employed the use of HSV segmentation to segment the images and locate the affected areas. They then utilized the Harris corner detector and Histogram of Gradient method to extract relevant features. Disease classification was achieved through the implementation of the K-Nearest Neighbour classifier. The researchers compared the developments obtained using their offered H2K method with those obtained from the current multiclass SVM approach. The comparison demonstrated the robustness and optimality of the proposed method in accurately classifying groundnut leaf diseases [18]. Hassanien et al. successfully developed a more advanced method for detecting diseases on tomato plants, known as the improved moth flame approach. Their system incorporated an algorithm called MFORSFS, which was specifically designed for feature selection. Through their research, they compared the performance of their algorithm with feature sampling methods founded on Particle Swarm Optimization (PSO) and Genetic Algorithms. The results revealed that their proposed algorithm was not only robust but also more efficient in terms of speed and accuracy [19]. Anjna et al. showcased their research on detecting capsicum disease. They utilized the histogram equalization technique to enhance the images. Extracting GLCM features, they employed tree, linear discriminant, SVM, and KNN techniques from Applied Computer Systems for classification. The result of their study revealed that the tree and linear discriminant methods provided an accuracy of 98.4%, while SVM and KNN achieved a perfect accuracy of 100% [20]. A hybrid model was proposed by Bedi et al. to notice peach leaf diseases. The CAE network, assembled by the model, successfully decreased the image size from 256 × 256 to 32 × 32. To achieve this reduction, the CAE network underwent training to minimize reconstruction losses. The loss between the authentic and rebuilt images was computed by employing the Normalized Root Mean Squared Error (NRMSE) function. A CNN classifier was used to classify an image [21].

2.1 Applications Using YOLO

The negative impacts of Fusarium oxysporum on the Philippine banana export market were highlighted by Nionel et al. They suggest employing the YOLOv4 algorithm to automate disease identification, which might lower losses and increase productivity [22]. Models such as YOLO have become more efficient in detecting BLSD. Fu et al. proposed and compared two enhanced YOLOv4 neural network detection models for a banana tree. One is the YOLO-Banana detection model, which examines banana features and network structure to prune the slight necessary network layers. The other is the YOLO-Banana-l4 detection model, which, by adding a YOLO head layer to the pruned network structure, analyzes the impact of a four-scale prediction system on the pruning network. The average precision (AP) values of the YOLO-Banana detection model for banana bunches and stalks were 98.4 and 85.98%, and the mean average precision (mAP) of the detection model was 92.19% [23]. Cai et al. proposed an enhanced YOLOv7 deep learning object detection algorithm, YOLOv7-FM, to detect banana pseudostems under various development prerequisites. The results show that the AP and inference speed of the YOLOv7-FM algorithm are higher than those of the compared models, with an average inference time of 8.0 ms per image containing banana pseudostems and AP of 81.45% [24].

3 Methodology

This part provides a detailed outline of the methodology for the detection of Banana Leaf Spot Disease. The process of this paper will include the overall processes of data collection, data pre-processing, and applying different CNN architectures for disease detection. The training method of this model was conducted using the Google Colabotary platform.

3.1 Dataset

This study used a 200-image dataset from the Banana Leaf Spot Diseases dataset containing photos from the four primary classes of Sigatoka, Cordana, Pestalotiopsis, and Healthy. These accurately annotated photos provide a strong basis for supervised learning algorithms, allowing the accurate detection and identification of diseases affecting banana leaves. Researchers may investigate various deep learning techniques because of the diversity of the dataset, which advances our knowledge and ability to manage Banana Leaf Spot diseases in an agricultural context. Figure 1 shows the example of an image dataset used for this research.

3.2 Data Preproessing

In data pre-processing, we first, completed the annotations of all images. In addition, in pre-processing auto-Orient is used to fix picture orientation issues. Additionally, making every image, even if it implies altering its original shape, fits into a 640 × 640 square box. For augmentation only horizontal flips were applied. After augmentation, the total number of images was approximately 300, where 70% was employed for training, 15% for testing, and another 15% for validation.

(a) Healthy

(b) Cordana

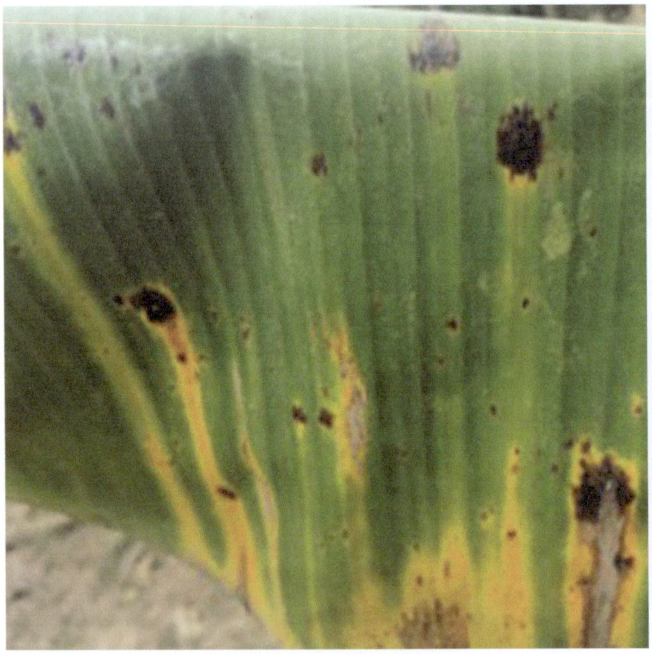

(c) Pestalotiopsis

(d) Sigatoka

Fig. 1 These figures (**a**–**d**) illustrate the images from the dataset

3.3 Applied Model

We conducted an in-depth assessment of the YOLOv5s, YOLOv8s, and YOLOv9 models as part of our study on banana leaf spot disease. Following an in-depth evaluation, the YOLOv8s model was determined to be the best choice because of its excellent accuracy and speed in identifying diseases. Its exceptional performance improves the overall effectiveness of disease management evaluations in banana cultivation and simplifies the diagnosis process. The selection highlights how crucial it is to utilize cutting-edge technology for environmentally friendly agriculture and how essential it is to use advanced machine learning approaches to handle agricultural difficulties.

3.4 Architectural Overview of YOLOv5

YOLOv5 utilizes CSP-Darknet53 as its backbone, which is essentially the convolutional network Darknet53 (originally employed in YOLOv3) enhanced with the cross-stage partial (CSP) network method. YOLOv5 introduced two significant modifications to the model neck: it implemented a variant of Spatial Pyramid Pooling (SPP) and revised the Path Aggregation Network (PANet) by integrating BottleNeckCSP into its structure. The head of YOLOv5 remains the same as in YOLOv3 and YOLOv4, consisting of three convolutional layers liable for indicating the location of the bounding boxes (x, y, height, and width), scores, and object types. Figure 2 shows the architectural overview of YOLOv5.

3.5 Architectural Overview of YOLOv8

The YOLOv8 object detection process is an improved version of the YOLO technique that provides more accurate and quicker object detection in images [25]. By integrating innovative architectural and developmental improvements, it improves upon previous versions and establishes itself as a key algorithm in object detection. Figure 3 shows the architectural overview of YOLOv8.

The main Features of YOLOv8: YOLOv8 surpass previous versions in detecting objects accurately owing to novel approaches and optimizations [26]. Compared with other object detection algorithms, YOLOv8 provides faster inference speeds while maintaining excellent accuracy [27]. Additionally, it offers the freedom to choose the most suitable model for its unique requirements by supporting a variety of backbones, such as EfficientNet, ResNet, and CSPDarknet. YOLOv8 also achieves better model performance by dynamically optimizing the learning rate and adjusting the loss functions during training. The core of YOLOv8 is built on a rebuilt CSPDarknet53 architecture, with 53 convolutional layers and the use of cross-stage fractional connections to improve the layer-to-layer transfer of information [25]. YOLOv8 consists of several convolutional layers at the top, which are followed by fully linked layers. For object detection, this section predicts bounding boxes, objectness scores, and probability of classes. The model may categorize various visual elements according to their importance for the task since YOLOv8 has a unique self-attention mechanism built into its head.

3.6 Architectural Overview of YOLOv9

To handle the issue of knowledge bottlenecks and data loss in the feed-forward approach, the founders of YOLOv9 introduced a unique idea called programmable gradient information (PGI). This prototype renders trustworthy gradients through an auxiliary reversible branch, ensuring that deep components continue performing the target task, whereas the additional branch mitigates semantic loss from multi-path features. The writers obtained optimal training results by applying PGI propagation at different semantic levels. The

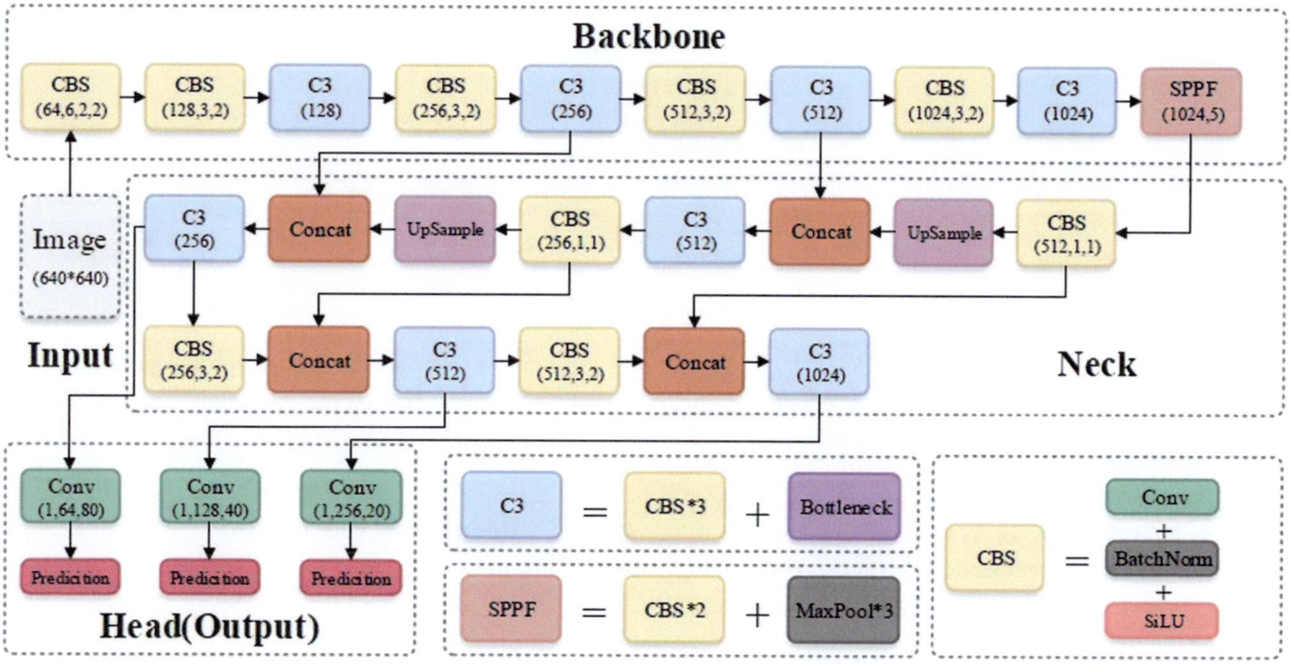

Fig. 2 Architecture of YOLOv5

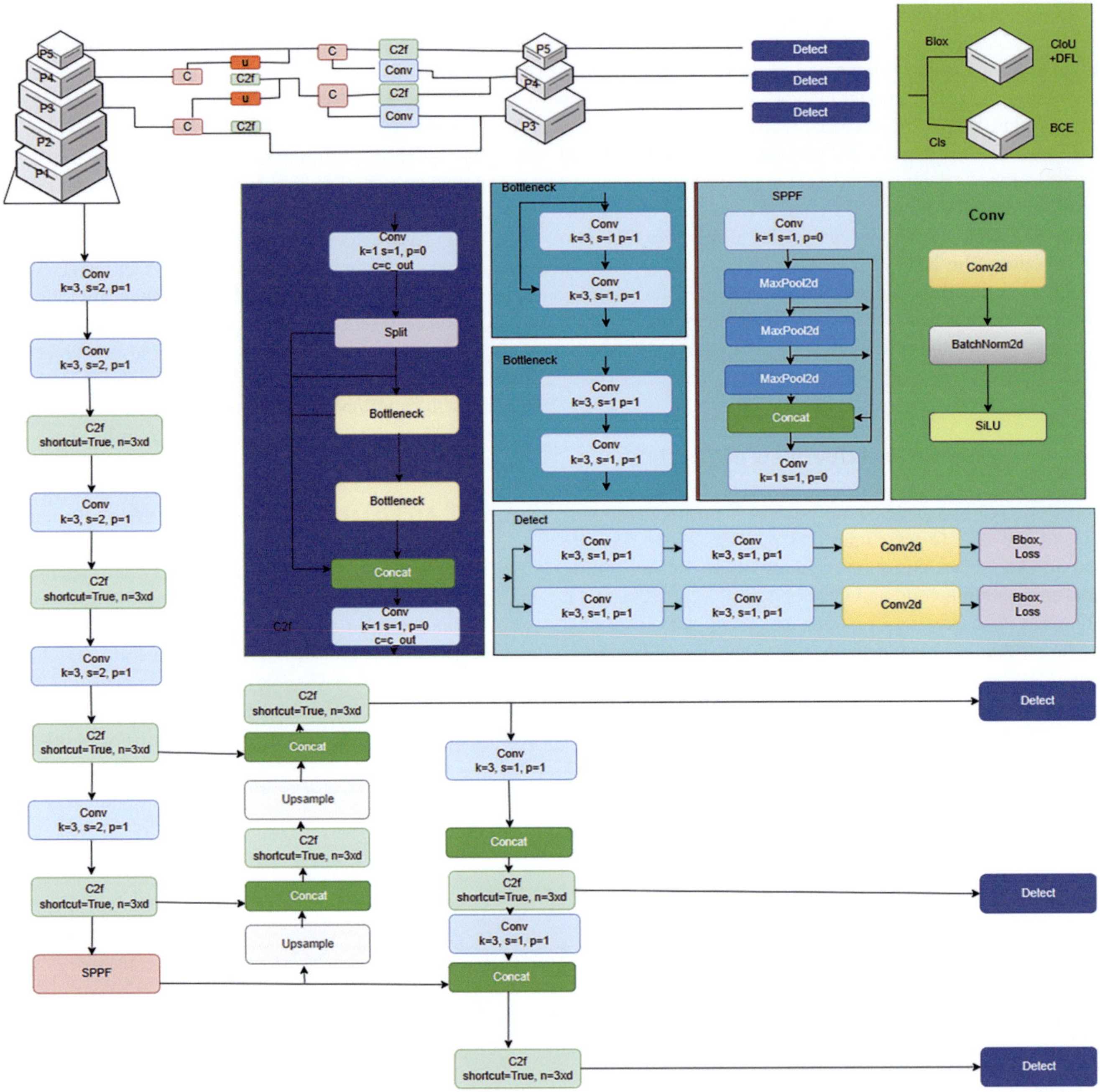

Fig. 3 Architecture of YOLOv8

reversible architecture of PGI is built into an auxiliary branch, adding no extra cost. The PGI's flexibility in selecting a suitable loss function for the target task also addresses issues faced by mask modeling. The PGI mechanism can be involved in deep neural networks of diverse lengths. Figure 4 indicates the architectural overview of YOLOv9. In their paper, Liu et al. [28] designed a generalized ELAN (GELAN) that balances the number of parameters, computational complexity, accuracy, and inference speed. This method permits users to select proper computational blocks for additional inference appliances. By incorporating PGI and GELAN, the authors developed YOLOv9. They conducted investigations using the MS COCO dataset, and the outcomes confirmed that YOLOv9 achieved top performance across all cases.

4 Result and Discussion

This section explores the performance of three YOLO versions (v5, v8, and v9) in detecting Banana Leaf Spot Disease (BLSD) based on a dedicated dataset. The results are presented, the F1 score, Precision, Recall, and mAP. The

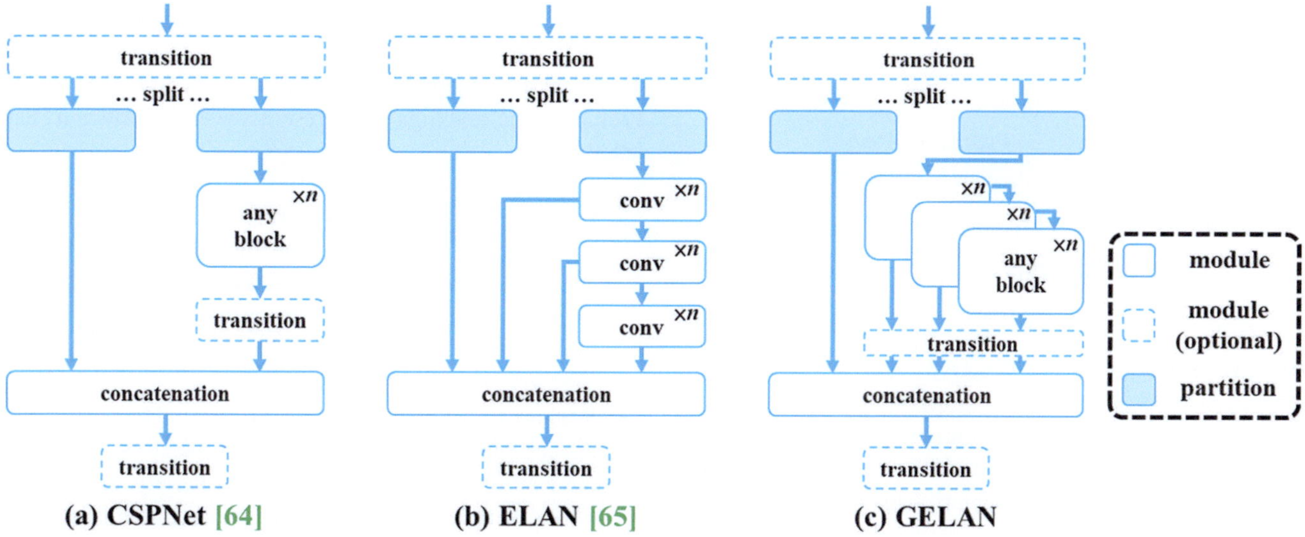

Fig. 4 Architecture of YOLOv9

visual expression of the models provides a clear view of their credibility. Evaluating the performance of any classification model, including those used for BLSD detection, requires robust metrics that capture various aspects of its effectiveness. This section describes three key metrics: F1 score, confusion matrix, and precision-recall (PR) curve.

The confusion matrix is a visual tool that delivers a thorough analysis of the classification results of a model. It presents the number of instances classified into each category, helping us to understand where the model functions well and where it might be producing errors.

True positives (TP): properly identified BLSD cases.

False positives (FP): Cases erroneously classified when healthy.

False negatives (FN): BLSD cases missed by the model.

True negatives (TN): Healthy leaves are correctly classified as such.

The PR curve visually depicts the trade-off between precision and recall across diverse thresholds. It allows us to understand how changes in the decision threshold, which determines whether a prediction is classified as BLSD or healthy, affect the model's performance.

Here's how to interpret the PR curve:

- High precision at low recall implies that the model prioritizes accuracy, but might miss some BLSD cases. As the threshold decreases, more cases are classified as BLSD, leading to increased recall but potentially sacrificing precision.
- An ideal curve would be closer to the top-left corner, displaying increased precision and recall across varying thresholds.

The Precision-Recall formula is described as follows:

$$\text{Precision} = \frac{\text{True Positives}}{\text{True Positives} + \text{False Positives}} \quad (1)$$

$$\text{Recall} = \frac{\text{True Positives}}{\text{True Positives} + \text{False Negatives}} \quad (2)$$

The F1 score, also understood as the harmonic mean of precision and recall, delivers a single numerical that value summarizes the overall model performance. This takes into account the following:

Precision: Measures the proportion of true positive predictions among all the positive predictions made by the model. An increased precision demonstrates that the model accurately identifies BLSD cases and minimizes false alarms.

Recall: This estimates the ratio of true positive forecasts among all confirmed samples in the dataset. An increased recall demonstrates that the model captures most of the actual BLSD cases and avoids missing important information. A higher F1 score (closer to 1) indicated a well-balanced model

Table 1 Performance comparison of different models on banana leaf spot disease on BananaLSD dataset

Dataset	Model	F1 score	Precision	Recall	mAP@0.5
BananaLSD dataset	YOLOv5s	0.65	0.899	0.85	0.638
	YOLOv8s	0.63	0.889	0.92	0.664
	YOLOv9	0.65	0.858	0.88	0.647

that performed excellently in terms of both precision and recall.

$$F_1 \text{ score} = \frac{2 \cdot \text{Precision} \cdot \text{Recall}}{\text{Precision} + \text{Recall}} \quad (3)$$

4.1 Result Analysis

The Table 1 compares the performance of three YOLO model variants-YOLOv5s, YOLOv8s, and YOLOv9-on the BananaLSD Dataset for detecting banana leaf spot diseases. YOLOv8s demonstrated the highest mAP@0.5 (0.664) and Recall (0.92), indicating superior overall detection and classification accuracy, particularly in minimizing false negatives. YOLOv5s achieved the highest Precision (0.899), making it the most reliable in producing accurate predictions without over-detection. YOLOv9 provided balanced performance across metrics, with an F1 Score equal to YOLOv5s (0.65) but slightly lower Precision and mAP@0.5. These results suggest that YOLOv8s is most suitable for tasks prioritizing comprehensive detection, while YOLOv5s is ideal for applications where prediction reliability is critical.

For YOLOv5, the confusion matrix Fig. 5a highlights a balanced performance in classifying true positives, false positives, and false negatives. The Precision-Recall curve Fig. 5b demonstrates YOLOv5's stability in maintaining high precision across varying recall levels, while the F1 curve Fig. 5c reflects its ability to achieve a balanced trade-off between precision and recall at different thresholds. For YOLOv8, the confusion matrix Fig. 6a reveals effective identification of diseased areas with minimal false negatives. Its Precision-Recall curve Fig. 6b emphasizes the model's ability to achieve high recall while maintaining competitive precision. The F1 curve Fig. 6c showcases Similarly, for YOLOv9 confusion matrix in Fig. 7a demonstrates balanced detection accuracy, albeit with a slight increase in false positives. The Precision-Recall curve in Fig. 7b highlights moderate precision variations across different recall levels. Finally, the F1 curve in Fig. 7c emphasizes consistent performance in maintaining a balance between precision and recall. These Figs. 5, 6, and 7 collectively illustrate the individual strengths and performance characteristics of each model.

The above figures show us the evaluation metrics of the YOLO models. These figures stand as a monument explaining the true validation of the YOLO models in detecting leaf diseases in Banana trees. From these figures, we can conclude how worthy these YOLO models are.

5 Challenges

The study of banana leaf spot diseases, including Cordana, Pestalotiopsis, and Sigatoka, presents several significant challenges:

1. **Limited Dataset Availability**: One of the immediate challenges is the restricted availability of high-quality labeled datasets for these diseases. The scarcity of comprehensive data hampers the development of robust detection models, which require diverse and representative samples to learn effectively.
2. **Variability in Disease Manifestation**: The appearance of leaf spot diseases can differ significantly relying on the environmental conditions, plant variety, and disease phase. This variability makes it challenging for detection models to generalize across different scenarios and to accurately identify all instances of the disease.
3. **Labor-Intensive and Time-Consuming Detection Methods**: Traditional methods for detecting banana leaf spot diseases are often labor-intensive and time-consuming. They typically require manual inspection and expert knowledge, which are not feasible for large-scale or real-time monitoring.
4. **Accuracy and Precision Limitations**: Current detection techniques are prone to inaccuracies and false positives/negatives. Ensuring high precision and recall rates in automated systems is challenging, particularly when distinguishing between similar symptoms and early-stage infections.
5. **Field Deployment Challenges**: Implementing advanced detection systems in real-world agricultural settings presents practical challenges. These include adapting the technology to diverse field conditions, integrating it with existing agricultural practices, and ensuring reliability and ease of use.
6. **Integration with Existing Management Practices**: Another challenge is to integrate automated disease detection with current disease management practices. Ensuring that the new techniques complement and enhance existing methods rather than disrupt them, is crucial for successful adoption.

Addressing these challenges is essential for advancing the field of disease diagnosis and ultimately enhancing the efficiency and effectiveness of crop protection strategies.

6 Future Work

The promising results obtained from our study underscore the potential of advanced detection techniques for improving banana leaf spot disease management. To further enhance

Fig. 5 Visual representation of YOLOv5 performance metrics in (**a**–**c**)

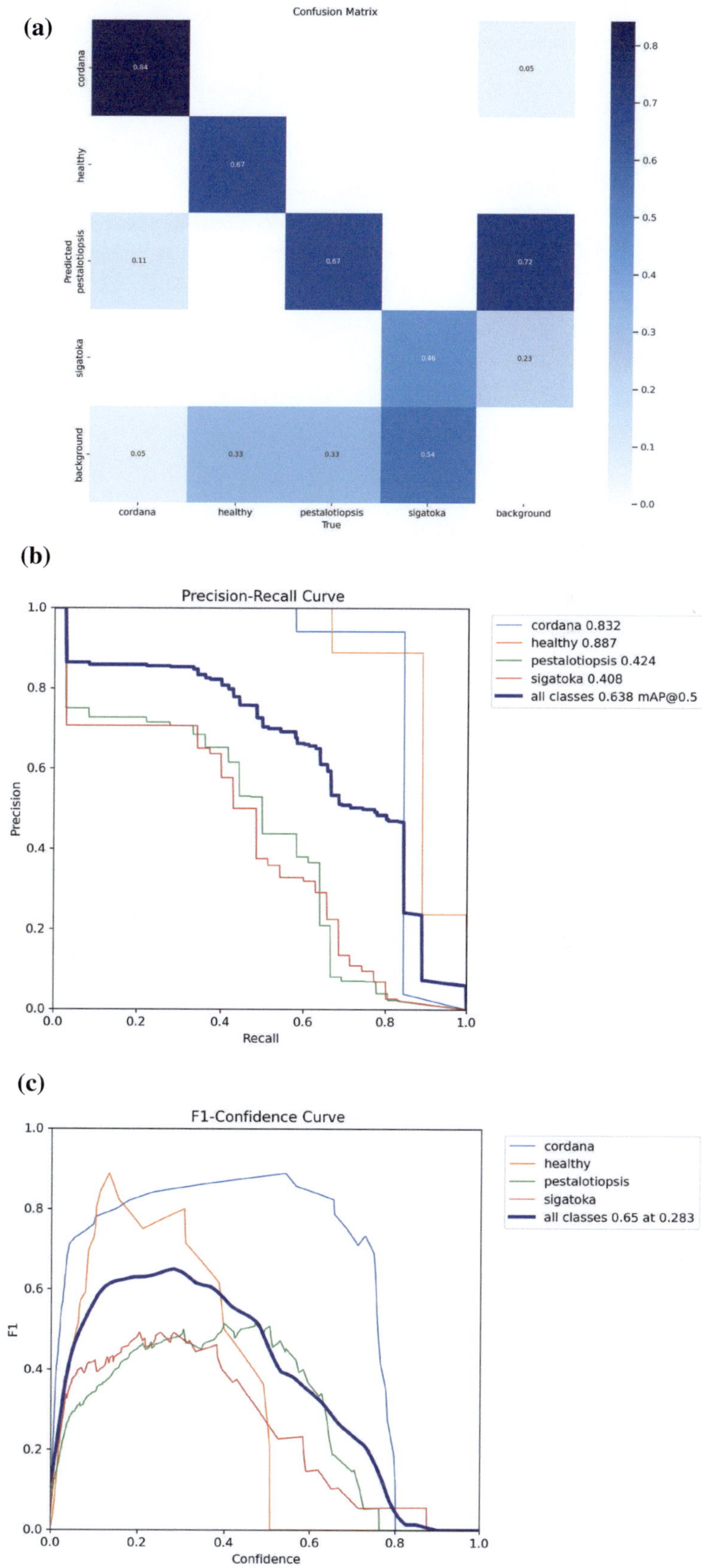

Fig. 6 Visual representation of YOLOv8 performance metrics

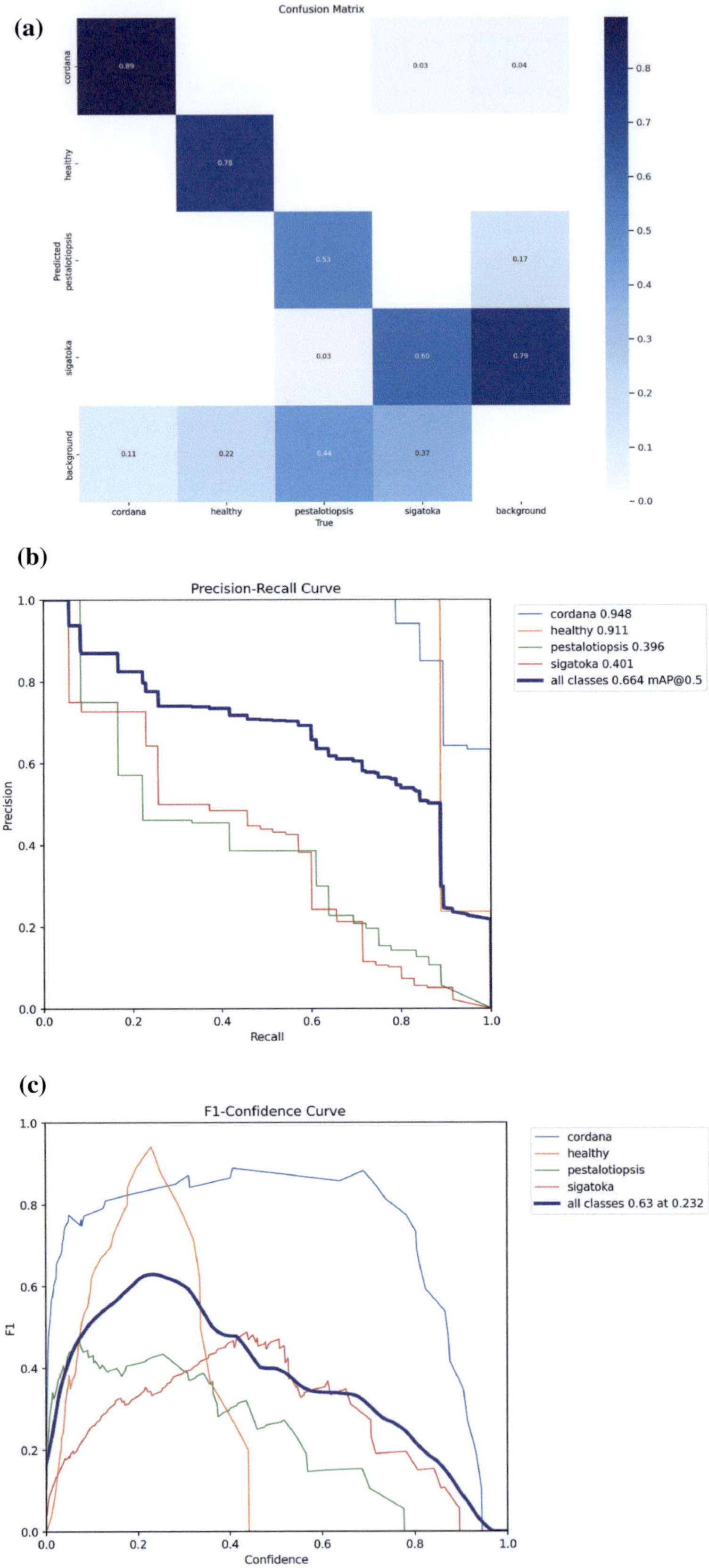

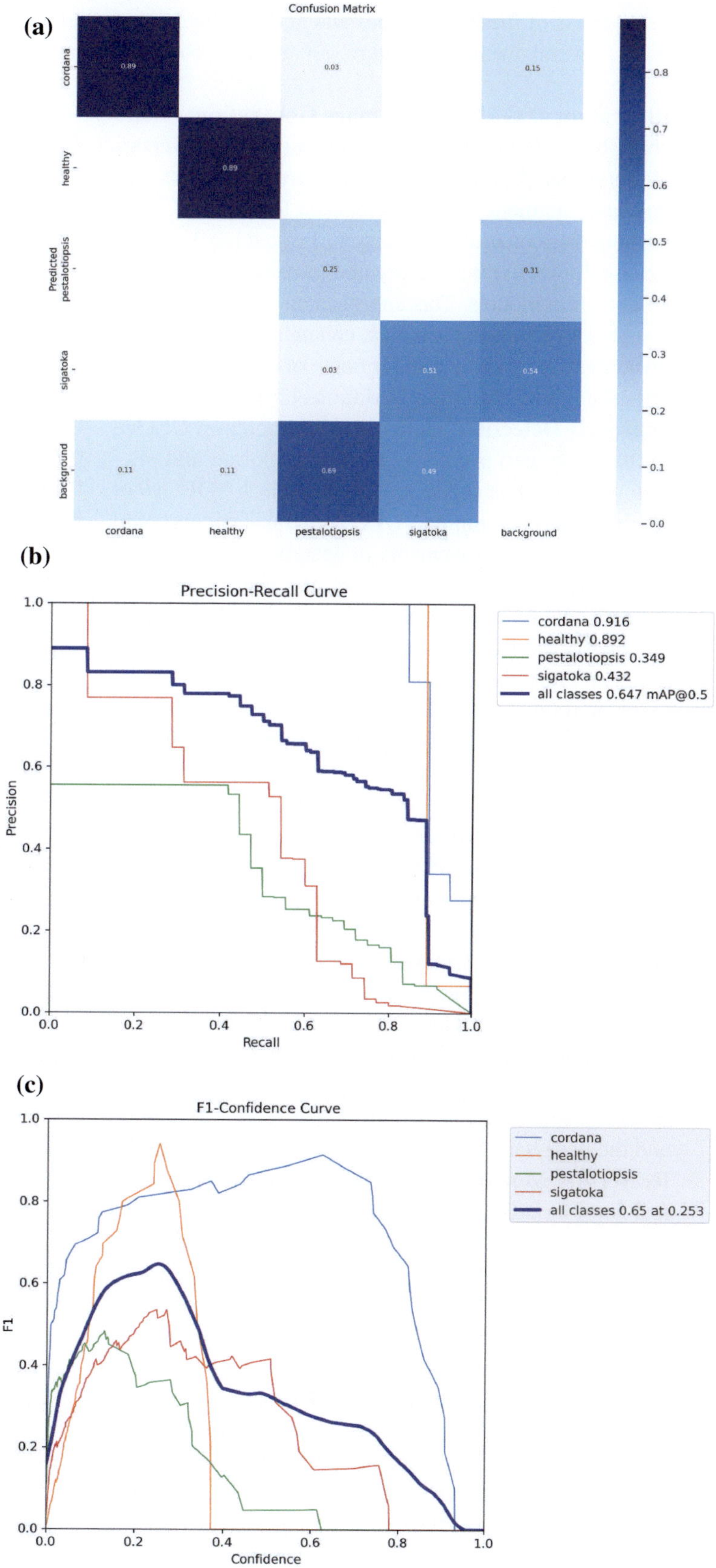

Fig. 7 Visual representation of YOLOv9 performance metrics

the effectiveness of these techniques and address the current limitations, several avenues for future research are proposed.

1. **Integration of GAN-Based Image Generation**: Future work should focus on leveraging Generative Adversarial Networks (GANs) to augment current datasets. By rendering synthetic images of banana leaf spots, GANs can help overcome the challenge of limited labeled data, thereby improving the validity and learning capability of the detection models. This approach can provide a more diverse set of training samples, enhancing the ability of the model to recognize a wider range of disease manifestations and potentially rare variations [29].
2. **Improved Detection Algorithms**: Although YOLOv8 has demonstrated promising results, exploring and integrating other advanced architectures such as YOLOv9 or hybrid models combining YOLO with transformers could lead to further improvements in detection accuracy and speed. Evaluating these models on the banana leaf spot dataset and comparing their performance with YOLOv8 provides valuable insights into achieving even higher precision and recall rates [30].
3. **Field Deployment and Real-World Testing**: To validate the practical applicability of the developed models, it is essential to conduct field trials. The Implementation of the detection system in real-world agricultural settings will provide insights into its performance under varying environmental conditions and operational challenges. Feedback from these trials can guide further refinement and adaptation of models.
4. **Enhanced Disease Classification and Severity Analysis**: Developing the scope of the analysis to possess not only just detection but also classification of disease severity can significantly aid targeted disease management. Incorporating severity assessment models could provide farmers with actionable insights into disease progression and the necessary interventions.
5. **Real-Time Monitoring and Automation**: Developing real-time monitoring systems integrated with detection models can facilitate automated disease management practices. Incorporating these systems into existing agricultural machinery or mobile platforms can streamline the process of disease identification and intervention, reduce manual labor, and increase efficiency.
6. **Cross-Crop and Multi-Disease Detection**: Exploring the applicability of the developed techniques to other crops and disease types could broaden the impact of research. Adapting these models for multi-disease detection and cross-crop applications could provide a versatile solution for various agricultural challenges.

By addressing these future research approaches, we aim to extend the area of agricultural disease control, ensuring more reliable and efficient methods for improving crop health and supporting food security worldwide.

7 Conclusion

Our research is a noteworthy advancement in the application of AI for the premature detection of BLSD in farming environments. By means of a comprehensive evaluation of the YOLOv5s, YOLOv8s, and YOLOv9 models, we demonstrated that deep learning methods are efficient in precisely detecting BLSD cases from a variety of datasets. Our approach offers a structured framework for handling difficult agricultural problems, starting with data preprocessing and ending with model selection. We unlock the door for more resilient and environmentally friendly agricultural research and practices by utilizing cutting-edge technologies and multidisciplinary teamwork, which will eventually improve food security and farmer livelihoods worldwide.

References

1. Almeyda, E., Ipanaqué, W.: Recent developments of artificial intelligence for banana: application areas, learning algorithms, and future challenges. Engenharia Agrícola **42**, 20210144 (2022)
2. Guo, J., Mao, K., Yuan, Z., Qin, Z., Xu, T., Bateni, S.M., Zhao, Y., Ye, C.: Global food security assessment during 1961–2019. Sustainability **13**(24), 14005 (2021)
3. Nobel, S.N., Imran, M.A., Bina, N.Z., Kabir, M.M., Safran, M., Alfarhood, S., Mridha, M.: Palm leaf health management: a hybrid approach for automated disease detection and therapy enhancement. IEEE Access (2024)
4. Sakurada, K., Shibuya, M., Wang, W.: Weakly supervised silhouette-based semantic scene change detection. In: 2020 IEEE International Conference on Robotics and Automation (ICRA), pp. 6861–6867. IEEE (2020)
5. Patel, A., Agravat, S.: Banana leaves diseases and techniques: a survey. In: Data Science and Intelligent Applications: proceedings of ICDSIA 2020, pp. 209–215. Springer (2021)
6. Ebimieowei, E., Wabiye, Y.-H.: Control of black sigatoka disease: challenges and prospects. African J. Agric. Res. **6**(3), 508–514 (2011)
7. Stevens, B., Diels, J., Brown, A., Bayo, S., Ndakidemi, P.A., Swennen, R.: Banana biomass estimation and yield forecasting from non-destructive measurements for two contrasting cultivars and water regimes. Agronomy **10**(9), 1435 (2020)
8. Kabir, M.M., Ohi, A.Q., Mridha, M.F.: A multi-plant disease diagnosis method using convolutional neural network. In: Computer Vision and Machine Learning in Agriculture, pp. 99–111 (2021)
9. Prodeep, A.R., Hoque, A.M., Kabir, M.M., Rahman, M.S., Mridha, M.: Plant disease identification from leaf images using deep CNN's efficientnet. In: 2022 International Conference on Decision Aid Sciences and Applications (DASA), pp. 523–527. IEEE (2022)
10. Islam, M.R., Oliullah, K., Kabir, M.M., Alom, M., Mridha, M.: Machine learning enabled iot system for soil nutrients monitoring and crop recommendation. J. Agric. Food Res. **14**, 100880 (2023)
11. Kabir, M.M., Hafiz, M.S., Bandyopadhyaa, S., Jim, J.R., Mridha, M.: Tea leaf age quality: age-stratified tea leaf quality classification dataset. Data Brief **54**, 110462 (2024)

12. Ramcharan, A., McCloskey, P., Baranowski, K., Mbilinyi, N., Mrisho, L., Ndalahwa, M., Legg, J., Hughes, D.P.: A mobile-based deep learning model for cassava disease diagnosis. Front. Plant Sci. **10**, 425916 (2019)
13. Kamdar, J., Jasani, M., Jasani, J., Praba, J., Georrge, J.J.: 11 artificial intelligence for plant disease detection: past, present, and future. In: Internet of Things and Machine Learning in Agriculture: Technological Impacts and Challenges, vol. 8, p. 223 (2021)
14. Camargo, A., Smith, J.: An image-processing based algorithm to automatically identify plant disease visual symptoms. Biosyst. Eng. **102**(1), 9–21 (2009)
15. Kamilaris, A., Prenafeta-Boldú, F.X.: Deep learning in agriculture: a survey. Comput. Electron. Agric. **147**, 70–90 (2018)
16. Mohanty, S.P., Hughes, D.P., Salathé, M.: Using deep learning for image-based plant disease detection. Front. Plant Sci. **7**, 1419 (2016)
17. Ramcharan, A., Baranowski, K., McCloskey, P., Ahmed, B., Legg, J., Hughes, D.P.: Deep learning for image-based cassava disease detection. Front. Plant Sci. **8**, 1852 (2017)
18. Devi, K.S., Srinivasan, P., Bandhopadhyay, S.: H2k-a robust and optimum approach for detection and classification of groundnut leaf diseases. Comput. Electron. Agric. **178**, 105749 (2020)
19. Hassanien, A.E., Gaber, T., Mokhtar, U., Hefny, H.: An improved moth flame optimization algorithm based on rough sets for tomato diseases detection. Comput. Electron. Agric. **136**, 86–96 (2017)
20. Sood, M., Singh, P.K., et al.: Hybrid system for detection and classification of plant disease using qualitative texture features analysis. Procedia Comput. Sci. **167**, 1056–1065 (2020)
21. Bedi, P., Gole, P.: Plant disease detection using hybrid model based on convolutional autoencoder and convolutional neural network. Artif. Intell. Agric. **5**, 90–101 (2021)
22. Ibarra, N.C., Rivera, M.P., Manlises, C.O.: Detection of panama disease on banana leaves using the YOLOv4 algorithm. In: 2023 15th International Conference on Computer and Automation Engineering (ICCAE), pp. 209–214. IEEE (2023)
23. Fu, L., Yang, Z., Wu, F., Zou, X., Lin, J., Cao, Y., Duan, J.: YOLO-Banana: a lightweight neural network for rapid detection of banana bunches and stalks in the natural environment. Agronomy **12**(2), 391 (2022)
24. Cai, L., Liang, J., Xu, X., Duan, J., Yang, Z.: Banana pseudostem visual detection method based on improved YOLOv7 detection algorithm. Agronomy **13**(4), 999 (2023)
25. Jocher, G., Chaurasia, A., Qiu, J.: Ultralytics YOLOv8. https://github.com/ultralytics/ultralytics
26. Han, T., Cao, T., Zheng, Y., Chen, L., Wang, Y., Fu, B.: Improving the detection and positioning of camouflaged objects in YOLOv8. Electronics **12**(20), 4213 (2023)
27. Ma, B., Hua, Z., Wen, Y., Deng, H., Zhao, Y., Pu, L., Song, H.: Using an improved lightweight YOLOv8 model for real-time detection of multi-stage apple fruit in complex orchard environments. Artif. Intell. Agric. (2024)
28. Liu, H., Sun, F., Gu, J., Deng, L.: Sf-YOLOv5: a lightweight small object detection algorithm based on improved feature fusion mode. Sensors **22**(15), 5817 (2022)
29. Showrov, A.A., Aziz, M.T., Nabil, H.R., Jim, J.R., Kabir, M.M., Mridha, M., Asai, N., Shin, J.: Generative adversarial networks (GANs) in medical imaging: Advancements, applications and challenges. IEEE Access (2024)
30. Jiang, P., Ergu, D., Liu, F., Cai, Y., Ma, B.: A review of yolo algorithm developments. Procedia Comput. Sci. **199**, 1066–1073 (2022)

MIX
Papier aus verantwortungsvollen Quellen
Paper from responsible sources
FSC® C105338

If you have any concerns about our products,
you can contact us on
ProductSafety@springernature.com

In case Publisher is established outside the EU,
the EU authorized representative is:
Springer Nature Customer Service Center GmbH
Europaplatz 3, 69115 Heidelberg, Germany

Printed by Libri Plureos GmbH
in Hamburg, Germany